Prussian Blue-Type Nanoparticles and Nanocomposites

Prussian Blue-Type Nanoparticles and Nanocomposites

Synthesis, Devices, and Applications

edited by

Yannick Guari

Joulia Larionova

Pan Stanford Publishing

Published by

Pan Stanford Publishing Pte. Ltd.
Penthouse Level, Suntec Tower 3
8 Temasek Boulevard
Singapore 038988

Email: editorial@panstanford.com
Web: www.panstanford.com

British Library Cataloguing-in-Publication Data
A catalogue record for this book is available from the British Library.

Prussian Blue-Type Nanoparticles and Nanocomposites: Synthesis, Devices, and Applications

ISBN 978-981-4800-05-1 (Hardcover)
ISBN 978-0-429-02474-0 (eBook)

Contents

Preface

Prussian blue is the oldest coordination polymer discovered accidentally at the beginning of the 18th century. This compound and its analogues have been investigated through centuries for their exciting physical and chemical properties and remain very contemporary objects of study especially when considered at the nanoscale or within nanocomposites.

Indeed, Prussian blue and its analogues at the nanoscale are intriguing molecule-based materials made of metal ion nodes and molecular building blocks and presenting controlled size and shape that attract a great deal of attention for the last two decades. This interest is mainly due to many of the specific features that originate from the intrinsic molecular nature of these materials, such as definable and flexible molecular structures, porosity, low density, tuneable physical and chemical properties, and soft-chemistry routes for their synthesis. Moreover, because of their high surface-to-volume ratio, they possess unique size- and shape-dependent electrical, optical, magnetic, and catalytic properties that are strikingly different from those of their bulk counterparts.

Addressing the study of Prussian blue and its analogues not in the bulk state but in the form of nanoparticles isolated or integrated within nanocomposites allows us to reinvestigate the properties of these materials and consider these with new variants. Moreover, the fundamental understanding of these properties at the nanoscale makes it possible to consider improving the properties of Prussian blue and its analogues as pigments, electrode material, radioactive caesium sorbent material, etc., and to consider new applications in the field of nanomedicine as agents for diagnosis and/or therapy.

This book reviews the synthetic aspects related to Prussian blue and its analogues at the nanoscale (nanoparticles and nanocomposites), their properties and their potential applications in various fields. The contributors are specialists in the field of research related to Prussian blue and its analogues, including its history, its magnetic and optical properties, and its potential involvement in the resolution of major societal challenges of the 21st century in the fields of health or energy. They discuss their research and provide an overview of the field.

Yannick Guari
Joulia Larionova
January 2019

Chapter 1

The History of Prussian Blue*

Alexander Kraft

Gesimat GmbH, Koepenicker Str. 325, 12555 Berlin, Germany

kraftalex@aol.com

Prussian blue is a nanoporous deep blue pigment invented in Berlin in 1706 by Johann Jacob von Diesbach and Johann Conrad Dippel. Prussian blue (iron(III) hexacyanoferrate(II) or ferric ferrocyanide) was one of the first purely synthetic pigments. It was less expensive and more readily available or more easily produced as compared to ultramarine or other blue pigments which were in use at the time as a blue color in paintings. More than 300 years after its invention, Prussian blue is still a very interesting material for cutting-edge research.

1.1 Introduction

Although Prussian blue was invented more than 300 years ago and has ever since been an object of scientific inquiry, it remains

*In memory of Dr. Karl-Heinz Heckner (1937–2016), cofounder of Gesimat GmbH in 1998.

Prussian Blue-Type Nanoparticles and Nanocomposites: Synthesis, Devices, and Applications
Edited by Yannick Guari and Joulia Larionova

ISBN 978-981-4800-05-1 (Hardcover), 978-0-429-02474-0 (eBook)
www.panstanford.com

an interesting and important research object today. This can, e.g., be seen in the increasing number of scientific articles on research with Prussian blue. According to Chemical Abstracts Services (CAS®), in the past 40 years the number of scientific articles with Prussian blue as a research subject has risen from 21 in 1976 to 316 articles per year in 2016 [1].

In 2008, we had already reported on the invention and history of Prussian blue [2]. Meanwhile, some new information on the early history of this material has been discovered. Therefore, it is worth to come back to this subject again.

1.2 Prelude

Before we discuss the invention of Prussian blue in 1706, it has to be remarked that Prussian blue may have existed on Earth more than 3.5 billion years earlier. Spanish scientist Marta Ruiz Bermejo and coworkers from the Centro de Astrobiologia in Madrid believe that Prussian blue was an abundant material at certain times in the prebiotic phase of Earth [3]. They have shown that Prussian blue is easily formed during spark-discharge experiments in a plausible artificial prebiotic atmosphere of methane and other gases in the presence of saline aerosols of iron salts. In the prebiotic era, Prussian blue may have covered a large portion of the Earth's surface, producing a real "blue planet" Earth. The prebiotic Prussian blue later decomposed due to changes in the Earth's pH and temperature releasing hydrogen cyanide gas and leaving behind thin layers of iron oxide on the Earth's surface. Cycles of Prussian blue formation and decomposition may have occurred several times. The so-called banded iron formations, which are still found in some regions of the Earth (such as in Western Australia, Minnesota and Michigan), could be the remains of the early presence of Prussian blue [4]. Furthermore, prebiotic Prussian blue can directly be linked with the origin of life because its decomposition product HCN can further react to amino acids under possible prebiotic terrestrial conditions.

1.3 The Invention of Prussian Blue

The invention of the new blue pigment Prussian blue by Johann Conrad Dippel (1673–1734) and Johann Jacob von Diesbach (ca. 1670–1748) took place in 1706 in Berlin [5]. Since 1701, Berlin was the capital of the newly founded Kingdom of Prussia. The new pigment was named after the country of its invention.

Johann Conrad Dippel had arrived in Berlin in autumn 1704. The 31-year-old was already well known throughout Germany because of his controversial theological writings. Born on castle Frankenstein in the southern part of the German region of Hassia as a son of a protestant preacher, Dippel later studied theology as generations of his ancestors had done. During his university time in Giessen and Strasbourg (1691–1696), Dippel passionately took part in the theological struggles between the orthodox Lutherans and the new pietistic protestant movement. After changing sides from the orthodox Lutherans to the pietists, he became a fanatical pietist for the rest of his life. From 1697 on, he published numerous polemic books under the pseudonym Christianus Democritus. These often aggressive and insulting books prevented him from becoming a Lutheran preacher in his home region of Hassia.

Figure 1.1 Johann Conrad Dippel (1673–1734), one of the inventors of Prussian Blue in 1706 (*source*: University Giessen, Germany).

Around 1700, after some years of theological dispute, Dippel became interested in alchemy. Like many alchemists of the time, he concentrated his attempts on transmuting base metals into gold or finding a cure-all universal remedy. In a relatively short time, Dippel earned a reputation as a promising alchemist. This led to his call to Berlin by Count August David zu Sayn-Wittgenstein (1663–1735), who was one of the leading figures in the court of King Friedrich I in Prussia. Dippel rented a palatial house for his alchemy studies. Other radical pietists and some laboratory assistants were hired to help him with his experimental work, but Dippel could not control his theological disputes. He also wrote several of his polemic books in his few years in Berlin.

Born around 1670, Johann Jacob von Diesbach was of Swiss origin. He had served as a mercenary soldier in the English army, which he left as a Lieutenant. Around 1701 he came to Berlin. Here he became active as a producer of pigments and dyes. The "von" in his name is a German nobility particle which indicates that the person belongs to a noble family. The old patrician Swiss house of Diesbach, originally from the city of Bern, is known since the 15th century. Since his business of producing dyes and pigments was not typical for a nobleman, in his daily life he was simply known as Johann Jacob Diesbach. Only in the parish registers of three different churches in Berlin we find him as Johann Jacob von Diesbach [6].

Today, we know of two historical sources which describe the coincidental first preparation of Prussian blue:

In 1731, the eminent physician and chemist Georg Ernst Stahl (1659–1734) published a book in support of his phlogiston theory of oxidation and reduction. In this book, he described the circumstances which led to the invention of Prussian blue [7]. Stahl, who was professor for medicine at the university of Halle from 1694 to 1715 came to Berlin in 1715 to serve as the personal physician to king Friedrich Wilhelm I.

The second historical source is the diary of Johann Christian Senckenberg which is currently being transcribed and published online. Senckenberg (1707–1772) was a German physician, naturalist and collector based in Frankfurt am Main. As a young man he was a friend and admirer of Dippel, who was more than 30 years his senior. In 1732 Senckenberg visited Dippel twice in Berleburg, Dippel's last place of residence. In his diary, which includes entries from 1730 until 1772, Senckenberg also reported on his travels to Berleburg, including accounts of many stories which were told to him by Dippel. On

August 22, 1732, Dippel reported to Senckenberg, among some other things, the story of the invention of Prussian blue [8].

From these two sources, the following story of the invention of Prussian blue can be told: In 1706, von Diesbach was working in the laboratory of Dippel in Berlin, perhaps in the Friedrichstadt neighborhood. One day von Diesbach was preparing a Florentine lake according to a well-established recipe. For this purpose, he had to extract the red dye carminic acid (as we call it today) from dried and crushed cochineal insects, which were imported from New Spain (present day Mexico). This extraction is performed in a heated aqueous solution of alum. After filtration, an aqueous solution of potash is added. The addition of this alkaline solution to the slightly acidic solution of red carminic acid and alum leads to the precipitation of a white alumina powder. Carminic acid readily absorbs on the surface of the alumina, resulting in a red powder. After filtration and drying, a red lake pigment, the Florentine lake, is the final product [9]. The name Florentine lake comes from the invention of this material by a monk in Florence in the 17th century. This Florentine lake was to be sold to painters. The hue of the red color of the Florentine lake shifts if certain metal salts are added to the extraction solution. The addition of iron sulfate (vitriol of iron in von Diesbach's time) shifts the color of Florentine lake from bright red to a more violet tone. Von Diesbach must have produced such a variant of Florentine lake, because only this explains the addition of a substantial amount of iron ions in the solution to which he added the potash solution.

According to Stahl, Dippel produced his later "infamous" animal oil ("oleum Empyreumaticum animale") in Berlin. With this animal oil, Dippel believed he had found a universal remedy. Dippel's animal oil had a penetrating and extremely disgusting smell. This mixture of organic compounds such as pyrroles, pyridines, and nitriles was prepared by dry distillation from animal substrates such as bones, hoofs, horns, or dried blood. Aynsley and Campbell wrote of Dippel's animal oil medicine [10]: "A glance at the list of principal constituents is enough to convince one of the heroic nature of the cure." For the rectification of his animal oil Dippel mixed it with potash ("Sale Tartari"). After repeated distillation, the remaining potash was stored away in a glass container. This potash was contaminated with cyanide and/or hexacyanoferrate, which led to the unexpected precipitation of Prussian blue when this source of potash was used by von Diesbach for the production his Florentine lake.

Figure 1.2 Georg Ernst Stahl (1659–1734, left) and Johann Christian Senckenberg (1707–1772, right) were both contemporaries of Dippel and von Diesbach. They both reported slightly different stories of the invention of Prussian Blue (*sources*: Leopoldina, Halle, Germany; Wikimedia Commons).

According to Senckenberg's diary, once during his time in Berlin (1704–1707) Dippel prepared a large amount of a "sal volatile" by dry distillation of calcined sal tartari (potash) and dried ox blood. The remaining mixture of about six pounds of mixed sal alcali tartari and sal sanguine was not thrown away, as Dippel would have done, but collected and stored away by Dippel's young assistant Roesser. However, on the label, Roesser had only written "Sal tartari" (i.e., potash) and not "ex Sale alcali tartari constans et sale sanguinis." In need of potash, von Diesbach used the wrongly labeled previously heated mixture of calcined sal tartari and dried ox blood for the preparation of his violet Florentine lake. However, much to his surprise, a blue color "caeruleum Berolinense" (Prussian blue) resulted. We know today, that this potash was contaminated with cyanide and/or hexacyanoferrate. This led to the precipitation of Prussian blue when this source of potash was used by von Diesbach for the precipitation of his Florentine lake from an Fe^{2+} ion-containing aqueous solution. Without informing Dippel, von Diesbach drew up a contract with some painters for the delivery of the new blue pigment. However, he could not produce this material anymore after the source of contaminated sal tartari was finished. Therefore, he came to Dippel and reported his problem. Dippel told him to use sal tartari and bovine blood and it did work out.

This is what Stahl and Senckenberg reported about the invention of Prussian blue. Both stories differ somewhat—most importantly, with respect to the cause for the accidental formation of hexacyanoferrate. According to Stahl it was the result of Dippel's animal oil production but according to Senckenberg it was due to the preparation of a sal volatile. Since Senckenberg's information came directly from Dippel, who was one of original inventors, we should consider his story to be more credible than Stahl's.

1.4 The Early History of Prussian Blue until 1725

1.4.1 The Secret Production of Prussian Blue by Frisch and von Diesbach between 1706 and 1724

In early 1707 Dippel was arrested and held for about a week in the Hausvogtey prison in Berlin at the request of the Swedish ambassador. Dippel had published a new theological book, which contained some harsh criticism of the Swedish Lutheran church and the Swedish king Karl XII. Dippel was released on bail provided by Count Wittgenstein and soon fled from Berlin to the Netherlands.

After Dippel had left Berlin, von Diesbach was searching for another partner with whom he could exploit the invention of Prussian blue. The new partner he found was the teacher Johann Leonhard Frisch (1666–1743), with whom von Diesbach was already well acquainted. Some years earlier, he had been hired by Frisch as his teacher in chemical laboratory techniques.

When Frisch came to Berlin in 1698, he had already lived an adventurous life. However, no one could foresee that he would become an eminent scientist. Frisch was born to a well-established family of clergyman and lawyers in the Upper Palatinate region of Germany. He was raised in Franconia, mainly in the city of Nuremberg. Later, Frisch studied theology in Altdorf (near Nuremberg), Jena, and Strasbourg. After 1688, he traveled throughout Europe, where he visited France, Italy, Switzerland, and Hungary. Between his travels, he was active as a translator in the Imperial army, as a private teacher for young noblemen and as a manager of several agricultural enterprises in Germany. His last long journey led him to the Netherlands. On his way back home to Nuremberg, he stopped in Berlin, intending to stay for only a short time. However, a friend offered him a job as a school teacher at the most prestigious school in Berlin, the Gymnasium to the Grey Monastery. Frisch accepted this job and eventually stayed in Berlin for the remaining 45 years of his life.

Figure 1.3 Johann Leonhard Frisch (1666–1743): Frisch and von Diesbach were the first commercial producers of Prussian Blue (*source*: Leopoldina, Halle, Germany).

Frisch had a steady career at the gymnasium. His job as a teacher gave him enough time to follow his scientific interests. In December 1706, he became a member of the Royal Prussian Society of Sciences. His scientific work included linguistic studies, which culminated in the publication of several dictionaries, and the study of insects and birds. These latter studies resulted in two encyclopedic books published in several volumes. His

13-volume illustrated work on insects was completed in 1738. His 12-volume work with illustrations of German birds was completed in 1763 by three of his sons and one of his grandsons 20 years after his death. In the *Miscellanea Berolinensia*, the Latin language scientific journal of the Prussian Society of Sciences, Frisch authored as many as 51 articles, in which he made among other things important early contributions to parasitology. Frisch was a true universal scientist, and last but not least, Frisch was also a gifted entrepreneur who earned a considerable amount of money from different business ventures.

We know of Frisch's and von Diesbach's collaboration in the production and marketing of Prussian blue from letters Frisch had sent from Berlin to the president of the Royal Prussian Society of Sciences Gottfried Wilhelm Leibniz (1646–1716) in Hannover between 1708 and 1716 [11]. Frisch and von Diesbach started their collaboration with chemical experiments designed to improve the new product. A treatment of the blue precipitate with a strong acid, followed by rinsing with rainwater and drying was developed. This process dissolved part of the precipitated aluminum oxide to produce a more pure Prussian blue powder with a deeper blue color for the same amount of material.

At least as early as March 1708, Prussian blue was produced by von Diesbach and Frisch. In their collaboration, von Diesbach was responsible for the chemical work to produce the Prussian blue and Frisch was the businessman, marketing and selling the new blue pigment. He sold Prussian blue for 30 thaler per pound. Von Diesbach and Frisch both earned a lot of money with this business. Because it was mostly Frisch who was publicly active in relation to Prussian blue, many people in Berlin at that time thought that Frisch was also the inventor of Prussian blue, which, of course, he was not.

Prussian blue was soon used by the local painters in Berlin, including important members of the Royal Academy of Arts. Several paintings painted in 1710 and in the following years in Berlin have been identified to contain the pigment Prussian blue [12]. To date, the oldest known picture to be painted with Prussian blue, however, was painted in the Netherlands in 1709 [13]. The painter Pieter van der Werff (1665–1722) may have bought Prussian blue for his painting "Entombment of Christ" from Dippel, who lived in the Netherlands since 1707.

Living in Amsterdam and Maarssen near Utrecht, Dippel worked as a physician and alchemist. The respectable Vredenhoef house that he owned in Maarssen on the river Vecht still exists today. From a letter of Heinrich Hasperg to Leibniz, written in 1714 [14], it can be concluded that Dippel was also producing Prussian blue during his stay in the Netherlands. Dippel left the Netherlands in 1714 and moved to the Danish town of Altona near Hamburg. Very probably, he also produced Prussian blue there. In 1719, he was imprisoned on the Danish island of Bornholm. He could not produce Prussian blue anymore and Frisch and von Diesbach in Berlin finally had a monopoly on the manufacture of Prussian blue.

Soon after starting the business, Frisch began to sell Prussian blue not only in Berlin and in the Kingdom of Prussia but also in other German regions. In 1710, Prussian blue was sold in England for the first time. In 1711, it was delivered to Switzerland and then in 1716 to Russia. In 1714, two ultramarine-producing factories in Paris closed because of the large amounts of Prussian blue that were delivered by Frisch to that city. Ultramarine, which is produced from lapis lazuli mined in Afghanistan, was favored by painters, but it was also extremely expensive. Prussian blue was sold for only 10% of the price of ultramarine. In 1709, the first fake Prussian blue, perhaps based on indigo blue, appeared on the market.

In 1710, the first publication on Prussian blue appeared in the first volume of the Miscellanea Berolinensia, the new scientific journal of the Royal Prussian Society of Science in Berlin [15]. The whole volume was published in Latin, according to the conventions of the time. The two-page article on Prussian blue was written by Frisch, but published anonymously under the title Notitia Coerulei Berolinensis nuper inventi as a Notice on the newly invented Berlin Blue. This was not a scientific article as one might expect for a scientific journal. It was more or less an advertisement. Frisch described, among other things, the excellent properties of Prussian blue for use by painters. In addition, he mentioned that Prussian blue is nontoxic and costs only a tenth of the price of ultramarine. He also hinted that it is possible to buy Prussian blue from the book dealer of the Society of Sciences in Berlin [16].

1.4.2 The Quest for the Prussian Blue Recipe

It is no wonder that many people were interested in the manufacturing procedure for Prussian blue. In several letters of the time, physicians, chemists, apothecaries and other natural scientists discussed making Prussian blue. Of course, people tried to develop their own Prussian blue recipes or buy the secret recipe from Frisch and von Diesbach. In 1712, a group of Italians wanted to buy the recipe from Frisch. In return, Frisch requested at least 12,000 thalers. The Italians were not willing to pay this sum. But how long could Frisch, von Diesbach, and Dippel protect their secret?

An early handwritten Prussian blue recipe [17] was recently found in Dresden in a convolute of papers from Johann Friedrich Böttger (the alchemist who invented European porcelain). Because Böttger died in 1719, this recipe must have come from 1719 or earlier, a time in which only von Diesbach, Frisch, and Dippel and perhaps some of their few coworkers might have been in possession of the recipe. How Böttger acquired the procedure is unknown.

According to the recipe, three aqueous solutions are prepared. For the first solution, saltpeter and cream of tartar are thoroughly mixed and deflagrated. The product of this reaction (mainly potash) is mixed with dried beef blood and calcinated. The resulting mass is washed with water to produce solution number one. The second solution is prepared from alum and "corcionelle." The third solution is made by dissolving iron sulfate in water. In the next step, the second and third solutions are mixed and poured into solution number one. The Prussian blue precipitates and is filtered and treated with hydrochloric acid. It is then washed again with water and finally dried. Because of the acid treatment, this procedure resembles the procedure used by von Diesbach and Frisch.

In Dresden, Prussian blue has been used for the decoration of black lacquered, basically red stoneware, a forerunner of the delicate white porcelain. Black lacquered stoneware was a product of the famous Meissen porcelain factory between 1710 and 1719. It is, however, not known whether this Prussian blue was produced locally in Dresden or bought from Frisch in Berlin.

Another early recipe for Prussian blue was sent in a letter from Detlef Klefeker (1675–1750) in Hamburg to the mineralogist and chemist Johann Friedrich Henckel (1678–1744) in Freiberg in Saxony in the year 1722 [18].

Klefeker was a wealthy merchant who dealt with exotic trees, such as orange trees, banana plants and pomegranate trees. At the time, there was a large demand to equip the orangeries of the wealthy Europeans. Klefeker was also a hobby-alchemist. In 1736, he even published an alchemical book. It is not entirely clear, but Klefeker or his friend the physician Wichert from Hamburg had bought the recipe for Prussian blue for the sum of 2,000 thaler before 1722. It was not conveyed from whom the recipe was purchased. However, it may have been Dippel, who lived in nearby Altona from 1714 to 1719 or one of the collaborators of Frisch and von Diesbach in Berlin. Klefeker, who held the title of a Prussian Garden Commissair, had many businesses in Berlin and often traveled to that city.

According to Klefeker's recipe, dried ox blood was mixed with potash. The mixture was then calcinated and washed out with water to produce a first solution. A second solution was prepared from alum and ground cochineal insects. The third solution was made from alum and iron sulfate. All three solutions were then combined. The resulting blue precipitate was washed with water and dried. Cochineal was still used in this preparation method. Because the acid treatment step is missing in this recipe, it may have been the recipe used by Dippel.

Obviously, it was not so easy to successfully repeat the experiment. In 1723, Henckel reported in an article in a scientific journal from Silesia [19] that he owned a Prussian blue recipe that used blood, potash, alum, vitriol of iron and water, but he was not able "to get the blue down from heaven." Certainly, the calcination step, in which hexacyanoferrate is produced, has to be done properly to successfully produce Prussian blue.

When the Prussian blue production recipe was finally published in a scientific journal for the first time in 1724, several recipes were already circulating throughout Europe.

1.4.3 Caspar Neumann's Letters to John Woodward

In 1706, the year of the invention of Prussian blue, Caspar Neumann (1683–1737), a German apothecary, was employed at the Royal Court Apothecary shop in Berlin. He stayed there until 1711. Therefore, he would have watched with interest the market introduction and success of Prussian blue. Perhaps he would have thought about the possibility of producing this material on his own. As a travelling apothecary and accomplished piano player, he had become a minor favorite of King Friedrich I. In 1711, as a special favor by the king, Caspar Neumann was sent on a long trip through Europe to improve his education as an apothecary and chemist. This expensive travel was planned for several years and was paid for entirely by the king. In 1713, after visiting many parts of Germany and the Netherlands, Neumann arrived in London. Shortly after his arrival, he heard the news that the Prussian King had died. The new King Friedrich Wilhelm I, was much more economical, perhaps even stingy, compared to his father. He immediately cut government spending. Without advance warning, Neumann's travel costs were no longer paid. He did not even receive money to return to Berlin.

As a result, Neumann stayed in England and attempted to find a means to make a living. Luckily, he met a Dutchman named Abraham Cyprianus (1656–1718). Cyprianus had been called to England in 1696 by the First Physician to the English King to have himself freed from a bladder stone. Cyprianus, who was also called "lithotomus expertissimus," was the principal expert in the removal of bladder stones in Europe. He lived as a wealthy surgeon in London. Cyprianus was also a dedicated hobby-chemist with a well-equipped laboratory. Among other things, he searched for bladder stone-dissolving chemicals. Cyprianus hired Neumann as his laboratory assistant. For the next seven years, Neumann worked day by day in the laboratory of Cyprianus. He performed and documented the experiments for Cyprianus and discussed them with his mentor. Neumann learned English very quickly and became acquainted with a many English chemists, physicians, and apothecaries. He was also invited to take part in the meetings of the apothecary guild and the Royal Society.

After some time, Neumann decided to stay in England. To finally settle his affairs in Berlin, he made a short trip to the Prussian capital in 1716. There he met the influential Georg Ernst Stahl, the First Physician to the King since 1715 and one of the leading chemists in Europe. In his discussion with Neumann, Stahl recognized what a first rate chemical expert and apothecary Neumann had become. Therefore, he tried to convince Neumann to return to Berlin. He was successful in doing so, offering Neumann the post as head of the Royal Court Apothecary shop in Berlin and the right to continue his educational travel through Europe, once again completely paid for by the king. Neumann could not decline this offer. From 1719 on, Caspar Neumann was Royal Court Apothecary in Berlin.

In 1723, Neumann wrote two letters [20] to John Woodward (1665–1728), a physician and one of the most renowned members of the Royal Society in London. In his first letter from June of that year, he expressed his gratitude to the Royal Society and mentioned that he wanted to present something curious or useful to the society. As such he offered a recipe for the manufacture of Prussian blue which would be produced by his friend Frisch. Neumann added that he had not learned the recipe from Frisch. Rather, after acquiring some knowledge about the principal ingredients, he had developed the preparation procedure by himself. If the Royal Society was interested, he would have sent the process with his next letter. However, the Royal Society was not to publish the recipe under Neumann's name, but rather under a different name.

Figure 1.4 In 1723 Caspar Neumann (1683–1737, left) sent a Prussian blue recipe to John Woodward (1665–1728, right) in London. Woodward published this recipe in 1724 (*sources*: Leopoldina, Halle, Germany, and Royal Society, London, UK).

In a second letter from November 1723, Neumann sent the Latin description of the preparation process, Praeparatio Caerulei Prussiaci, to Woodward.

1.4.4 John Woodward Publishes the Prussian Blue Recipe

The addressee of Caspar Neumann's letters, John Woodward, was famous for his studies in geology, but he was not a chemist in the least. Nevertheless, he presented the preparation procedure for Prussian blue on January 9, 1724, before the Royal Society. He asked John Brown (?–1735), the leading chemist of the Royal Society at that time, to put the Prussian blue recipe to the test.

In Neumann's recipe, saltpeter and cream of tartar are mixed and detonated. This mass is then mixed with finely powdered dried ox blood to be heated in a crucible. After calcination, the material is thrown into hot water for leaching out of the blood lye. After filtration, this solution is mixed with a solution of iron sulfate and a solution of alum. The blue precipitate is filtered, and after treatment with hydrochloric acid, it is washed and dried. Cochineal is not used in this process.

John Brown repeated the experiment described in the recipe from Woodward. He confirmed the production of a wonderful blue-colored powder. However, he did not stop there. Brown investigated if other animal parts besides blood could be used. He found that the replacement of blood by beef also produced Prussian blue. Further experiments showed that if iron sulfate is replaced by other metal salts Prussian blue is not produced. Therefore, iron sulfate was shown to be essential. However, by producing hexacyanoferrate compounds of silver, mercury, copper, bismuth, and lead, he performed the first documented research on the so-called Prussian blue analogs, which continue to represent a major research field today. John Brown presented his experimental results in two meetings of the Royal Society on April 2 and 16, 1724.

In the January/February 1724 issue of the Philosophical Transactions of the Royal Society, two articles on Prussian blue were published. The first was the Latin description of the preparation procedure, which was printed under the title

"Preparation of Prussian blue sent from Germany to John Woodward" (originally in Latin) [21]. Immediately following was the publication by John Brown titled "Observations and experiments upon the foregoing preparation" [22]. This piece was written in English.

This issue of the Philosophical Transactions was printed in mid-1724. From this moment on, the recipe for the preparation of Prussian blue was publicly available. The producers from Berlin had lost their production monopoly. In addition, the new recipe was the first in which no red dye was used for preparation. From this point on, the product was less expensive and of better quality thanks to this major improvement by Caspar Neumann.

Only one year later, in 1725, the French chemist Etienne Francois Geoffroy (1672–1731), also known as Geoffroy the Elder, communicated to the French chemists the recipe of the Prussian blue production and published some new information. He found that Prussian blue production could be achieved from other parts of animals such as horn, hair, skin, or hoof, in addition to dried blood and flesh. He also reported to the Academy of Science in Paris that Prussian blue (Bleu de Prusse in French) was produced in London in large amounts and that the Prussian blue from London was of much better quality than the product from Berlin [23, 24]. He speculated that the producers from Berlin were not working hard enough because until recently they could always sell all the Prussian blue they produced. However, he also considered it possible that the Prussian blue from London was prepared in a different way from the product made in Berlin. Today, we can suppose that both were true. In London, Prussian blue was produced mainly in the Borough of Southwark and it was reported to be an "odious stinking business" [25].

In the next few years, Prussian blue production was established over the entire continent of Europe. Often the pigment was sold under different names such as Paris Blue or Milori Blue, usually named after the production location, the owner of the production facility, or based on an advertising idea. The production technology for the pigment changed greatly over time. In 2004 Chiharu Asai analyzed about 100 methods published between 1724 and 1904 and documented the increasing improvements in product quality that resulted [26].

1.5 The Further History of Prussian Blue

In the 18th century, Prussian blue was not only used as a pigment for painters. Some decades after its invention it was also applied to the dyeing of textiles, following the work of the French chemist Pierre-Joseph Macquer (1718–1784) published in 1749 [27]. Normally, a pigment such as Prussian blue cannot be used for dying textiles. However, by immersing textiles first in a solution of iron sulfate and in a second step in a hexacyanoferrate solution (or vice versa) leads to the precipitation of Prussian blue inside the textile fibers. In 1752, Macquer was also the first who produced blood-lye salt, solid yellow potassium ferrocyanide $K_4Fe(CN)_6$, from Prussian blue and potassium hydroxide [28].

Later Prussian blue was also used for bluing textiles, a major application field until the mid 20th century. In 1927, copper phthalocyanine was developed as an alternative blue pigment [29] in some aspects superior to Prussian blue. Nonetheless, Prussian blue is still used as a low cost blue pigment today, sold under the commercial name iron blue.

In 1751, the chemist Andreas Sigismund Marggraf (1709–1782) from Berlin described for the first time the test for dissolved iron in acidified aqueous solutions by adding blood lye, i.e., aqueous potassium ferrocyanide solution. The formation of a Prussian blue precipitate indicates the presence of dissolved iron [30].

Starting in 1724 and continuing for about 250 years, chemists tried to define the composition, stoichiometry, and structure for Prussian blue. Eminent scientists such as Joseph Priestley (1733–1804), Carl Wilhelm Scheele (1742–1786), Claude-Louis Berthollet (1748–1822), Joseph Louis Gay-Lussac (1778–1850), and Jöns Jakob Berzelius (1779–1848) were among the researchers in the field [31]. In 1782 Scheele discovered hydrogen cyanide by heating Prussian blue with sulfuric acid [32], and in 1811 Gay-Lussac's determination of the composition of hydrogen cyanide HCN [33] led to the conclusion that Prussian blue contained cyanide. Because of the lack of modern analytical methods, the details of the crystal structure and even of the analytical composition of Prussian blue were for a long time only partially resolved.

The first structural hypothesis for Prussian blue was presented by British physicist James Fargher Keggin (1905–1993) and chemist Frank Douglas Miles (1885–1968) in 1936 with the help of X-ray powder patterns [34]. In 1962, US chemist Melvin Bernard Robin showed that the deep blue color of Prussian blue arises from the charge transfer between $Fe^{II}(CN)_6^{4-}$ and Fe^{III} ions [35]. Finally, in the 1970s Swiss chemist Andreas Ludi et al. [36] published a detailed structure and confirmed the composition as $Fe_4[Fe(CN)_6]_3\ xH_2O$ (x = 14–16), which is now widely accepted as correct.

Turnbull's Blue was an alternative name for Prussian blue. It emerged in the first half of the 19th century. For more than 130 years it was discussed to be ferrous ferricyanide, a compound different from Prussian blue (ferric ferrocyanide). The starting point for the later production of Turnbull's Blue was the availability of the red potassium ferricyanide compound. In 1822 the synthesis of this compound was for the first time described by German chemist Leopold Gmelin (1788–1853) who was professor at the university in Heidelberg. He obtained potassium ferricyanide by oxidizing yellow ferrocyanide which has been dissolved in water with chlorine gas [37]. In the same publication, he also described the formation of Prussian blue by the reaction of dissolved ferricyanide with dissolved Fe(II)-salts.

In 1828, Turnbull's Blue was introduced as a product by Scottish chemical manufacturing company Turnbull & Ramsay [38]. The chemical company of the Turnbull family was started in the Glasgow area in Scotland by John Turnbull (1740–1817) in the 1790s. His son John Turnbull Jr. (1765–1835) was leading the company in the time of the introduction of the Turnbull's Blue product. The company which had changing partners over time (e.g., Robert Arthur or William Ramsay) was flourishing the whole 19th century. It came in decline in the 20th century and in 1968 the company was wound up voluntarily.

Turnbull's Blue became the common name for the blue product which is obtained after mixing ferricyanide with Fe^{2+}-containing solutions. In 1963, with the help of Mössbauer spectroscopy it was finally determined that Prussian blue and Turnbull's Blue are the very same compound, i.e., ferric ferrocyanide [39, 40].

Figure 1.5 John Frederick William Herschel (1792–1871), inventor of the cyanotype process (*source*: University of Cambridge, UK).

The use of Prussian blue as a painter's pigment since the early 18th century and as a material for coloring textiles was followed by other uses. An important example is the cyanotype or blueprint process invented by English astronomer and chemist John Frederick William Herschel (1792–1871) in 1842 [41]. This was a commercially successful photocopying process in use about 100 years from 1843 until the early 1940s [42].

In 1867 the German pathologist Max Perls (1843–1881) developed a method to detect iron in tissue called Perls' Prussian blue stain. In the presence of iron, the addition of separate solutions of potassium ferrocyanide and acid to the corresponding tissue lead to the formation of Prussian blue, which is clearly visible in the microscope [43]. This method is still much in use today.

Radioactive cesium ^{137}Cs, a component of atomic fallout, is a hazardous radionuclide. In 1963, the Croatian physician and pharmacologist Vladimir Nigrovic (1934–2008) discovered that by administering Prussian blue the excretion of Cs from the body can be enhanced substantially [44]. After successful experiments with rats, German physician Horst Heydlauf proposed the use of orally administered Prussian blue as an effective antidote against thallium poisoning in 1969 [45]. Today, Prussian blue is a routinely used antidote against poisoning with thallium and radioactive cesium.

Figure 1.6 Vernon D. Neff (1932–2013), wiring a Prussian Blue display cell to a potentiostat (*source*: Kent State University Libraries. Special Collections and Archives).

In 1978, Vernon Duane Neff (1932–2013) of Kent State University in Ohio, USA published a short notice in which he described the spontaneous deposition of thin films of Prussian blue from aqueous precursor solutions containing mixtures of ferric and ferricyanide ions onto conducting substrates [46]. These films can be switched reversibly by electrochemical means between different colored oxidation states in a process called electrochromism. Especially interesting for practical use is the alternation between colorless and blue oxidation states. One possible application is the construction of so-called smart windows which can reversibly change their light transmission between very high and low values. Electrochromic devices with thin Prussian blue films with sizes up to 2.4 m^2 have meanwhile been developed [47]. The seminal work of V. D. Neff [48] was the starting point for the extensive research on the electrochemistry of Prussian blue.

In 1982, Japanese chemist Kingo Itaya and coworkers invented a more reliable, reproducible method for the deposition of thin films of Prussian blue by electrochemical deposition from aqueous solutions of ferric and ferricyanide ions [49].

Because of its redox activity and the highly reversible intercalation/deintercalation of certain cations, Prussian blue is also a promising material for energy storage devices. V. D. Neff was also the first who presented research on a Prussian blue battery in 1985 with both cathode and anode made of Prussian blue [50]. In 2004, a design for a potassium battery employing a potassium anode and a Prussian blue–based cathode was presented by the Iranian chemist Ali Eftekhari [51]. Research on the use of Prussian blue as an active material in energy storage and conversion devices such as batteries [52], fuel cells [53], microbial fuel cells [54], thermoelectric generators [55] and supercapacitors [56] remains a field of cutting-edge research today.

A technically interesting property of Prussian blue is its ability to catalyze the reduction of hydrogen peroxide [57]. This property can be used in glucose biosensors, as shown in 1994 by Russian chemist Arkady A. Karyakin and coworkers. They presented research on electrodes modified with glucose oxidase and Prussian blue as a selective electrocatalyst for hydrogen peroxide reduction [58]. Glucose oxidase produces hydrogen peroxide from glucose. Hydrogen peroxide can be detected electrochemically at Prussian blue with high sensitivity and selectivity. Therefore, they denoted Prussian blue as an "artificial peroxidase" [59]. Later, many oxidase enzymes other than glucose oxidase have been coupled to Prussian blue modified electrodes to expand the possible use of Prussian blue sensors to many other analytes, such as cholesterol, oxalate, galactose, or glutamate to name but a few [60]. Prussian blue is also an excellent material as a sensor for some electroinactive cations such as potassium [61] or ammonium [62] because of its selective intercalation property. The use of Prussian blue as an active material for sensors is still a hot research topic [63].

In the past decade, research on Prussian blue has expanded to such diverse fields such as nanomedicine [64] or environmental remediation [65]. Therefore, more than 300 years after its invention, Prussian blue is more than ever in the focus of contemporary research and development.

References

1. SciFinder® of Chemical Abstracts Services (CAS®) of the American Chemical Society in February 2017.
2. Kraft, A. (2008). On the discovery and history of Prussian blue, *Bull. Hist. Chem.*, **33**, pp. 61–67.
3. Ruiz-Bermejo, M., Osuna-Esteban, S., and Zorzano, M.-P. (2013). Role of ferrocyanides in the prebiotic synthesis of α-amino acids, *Orig. Life Evol. Biosph.*, **43**, pp. 191–206.
4. Ruiz-Bermejo, M., Rogero, C., Menor-Salvan, C., Osuna-Esteban, S., Martin-Gago, J. A., and Veintemillas-Verdaguer, S. (2009). Thermal wet decomposition of Prussian Blue: Implications for prebiotic chemistry, *Chem. Biodiversity*, **6**, pp. 1309–1322.
5. Berger, J. E. (ca. 1730). *Kernn Aller Fridrichs-Städtischen Begebenheiten*, (handwritten manuscript, Berliner Staatsbibliothek Preussischer Kulturbesitz, Ms. Bor. Quart. 124), p. 26.
6. Kraft, A. (2016). Addendum to "On the discovery and history of Prussian blue," *Bull. Hist. Chem.*, **41**, pp. 1–2.
7. Stahl, G. E. (1731). Experimenta, Observationes, Animadversiones, CCC Numero Chymicae et Physicae, (Ambrosius Haude, Berlin), pp. 280–283.
8. Senckenberg, J. C. (1732). *Observationes in me ipaso factae* (diary) volume 2, (handwritten manuscript, Universitätsbibliothek Johann Christian Senckenberg, Goethe Universität Frankfurt am Main, Na 31,2, transcribed and edited by Vera Faßhauer), p. 406.
9. Eastaugh, N., Walsh, V., Chaplin, T., and Siddall, R. (2004) *Pigment Compendium: A Dictionary of Historical Pigments* (Elsevier, Amsterdam), pp. 118–119.
10. Aynsley, E. E., and Campbell, W. A. (1962). Johann Konrad Dippel, 1673–1734, *Med. Hist.*, **6**, pp. 281–286.
11. Fischer, H. L. (ed.) (1896) *Joh. Leonh. Frisch's Briefwechsel mit G. W. Leibniz* (P. Stankiewicz, Berlin).
12. Bartoll, J. (2006/2010). Frühe Spuren des Berliner Blaus auf den Gemälden in den preußischen Königsschlössern, *Jahrbuch Stiftung Preussische Schlösser und Gärten*, **8**, pp. 219–227.
13. Bartoll, J., Jackisch, B., Most, M., Wenders de Calisse, E., and Vogtherr, C. M. (2007). Early Prussian blue. Blue and green pigments in the paintings by Watteau, Lancret and Pater in the collection of Frederick II of Prussia, *Techne*, **25**, pp. 39–46.

14. Hasperg, H. (1714). Letter from Paris from August 17, 1714 to G. W. Leibniz in Hannover, *Leibniz, G. W.: Sämtliche Schriften und Briefe*, Transkriptionen des Briefwechsels 1714, n° 39, pp. 87–90 (Internet version from October 24, 2016: http://www.gwlb.de/Leibniz/Leibnizarchiv/Veroeffentlichungen/Transkriptionen1714bearb.pdf).

15. Frisch, J. L. (1710). Notitia coerulei berolinensis nuper inventi, *Miscellanea Berolinensia*, **1**, pp. 377–378.

16. Kraft, A. (2011). Notitia Coerulei Berolinensis Nuper Inventi—On the 300th anniversary of the first publication on Prussian blue, *Bull. Hist. Chem.*, **36**, pp. 3–9.

17. Loesch, A. (ed.) (2013). *Sächsisch schwartz lacquirtes Porcelain* (Sandstein-Verlag, Dresden) pp. 179–180.

18. Anon (1794). *Mineralogische, Chemische und Alchymistische Briefe von reisenden und andern Gelehrten an den ehemaligen Chursächsischen Bergrath J.F. Henkel. Zweyter Theil* (Walther, Dresden) pp. 185–195.

19. Henckel, J. F. (1723). Von Herrn, D. Joh. Fridr. Henckels, [...], aus Saltz-Kraut und Sode neu-erfundenen blauen Farbe, *Sammlung von Natur- und Medicin- wie auch hierzu gehörigen Kunst- und Literatur-Geschichten*, **18**, pp. 421–427.

20. Kraft, A. (2009). On two letters from Caspar Neumann to John Woodward revealing the secret method for preparation of Prussian blue, *Bull. Hist. Chem.*, **34**, pp. 134–140.

21. Woodward, J. (1724). Praeparatio Caerulei Prussiaci ex Germania missa ad Johannem Woodward, *Philos. Trans. R. Soc.*, **33**, pp. 15–17.

22. Brown, J. (1724). Observations and experiments upon the foregoing preparation, *Philos. Trans. R. Soc.*, **33**, pp. 17–24.

23. Geoffroy, E.-F. (1725). Observations sur la Preparation de Bleu de Prusse ou Bleu de Berlin, *Mem. Acad. R. Sci.*, pp. 153–172.

24. Geoffroy, E.-F. (1725). Nouvelle Observations sur la Preparation de Bleu de Prusse, *Mem. Acad. R. Sci.*, pp. 220–237.

25. Campbell, R. (1747). *The London Tradesman: Being a Compendious View of All the Trades, Profession, Arts, Both Liberal and Mechanic [...]* (T. Gardner, London), p. 106.

26. Asai, C. (2004). Handmade Berlinerblau, *ZKK Zeitschrift für Kunsttechnologie und Konservierung*, **18**, pp. 268–292.

27. Macquer, P. J. (1749). Memoire sur une nouvelle espece de teinture bleue, dans laquelle il n'entre ni pastel ni indigo, *Mem. Acad. R. Sci.*, pp. 255–264.

28. Macquer, P. J. (1752). Examen chymique de bleu de Prusse, *Mem. Acad. R. Sci.*, pp. 60–77.

29. de Diesbach, H., and von der Weid, E. (1927). Quelques sels complexes des o-dinitriles avec le cuivre et la pyridine, *Helv. Chim. Acta*, **10**, pp. 886–888.

30. Marggraf, A. S. (1751). Examen Chymique de l'Eau, *Histoire de l'Academie Royale des Sciences et Belles-Lettres de Berlin*, **7**, pp. 131–157.

31. Coleby, L. J. M. (1939). A history of Prussian blue, *Ann. Sci.*, **4**, pp. 206–211.

32. Scheele, C. W. (1782). Über die färbende Materie in Berlinerblau, *Neue Schwed. Akadem. Abhandl.*, **3**, pp. 256–266.

33. Gay-Lussac, J. L. (1815). Recherches sur l'acide prussique, *Ann. Chim.*, **95**, pp. 136–231.

34. Keggin, J. F., and Miles, F. D. (1936). Structures and formulae of the Prussian blues and related compounds, *Nature*, **137**, pp. 577–578.

35. Robin, M. B. (1962). The color and electronic configuration of Prussian blue, *Inorg. Chem.*, **1**, pp. 337–342.

36. Buser, H. J., Schwarzenbach, D., Petter, W., and Ludi, A. (1977). The crystal structure of Prussian blue: $Fe_4[Fe(CN)_6]_3$ $x H_2O$, *Inorg. Chem.*, **16**, pp. 2704–2710.

37. Gmelin, L. (1822). Über ein besonderes Cyaneisenkalium, und über eine neue Reihe von blausauren Eisensalzen, *J. für Chem. und Physik*, **34**, pp. 325–346.

38. Webster, J. W. (1839). *A Manual of Chemistry, Containing the Principal Facts of the Science* (Marsh, Capen, Lyon, Webb, Boston), p. 418.

39. Duncan, J. F., and Wigley, P. W. R. (1963). The electronic structure of the iron atoms in complex iron cyanides, *J. Chem. Soc.*, pp. 1120–1125.

40. Fluck, E., Kerler, W., and Neuwirth, W. (1963). The Mössbauer effect and its significance in chemistry, *Angew. Chem.*, **2**, pp. 277–278.

41. Herschel, J. F. W. (1842). On the action of the rays of the solar spectrum on vegetable colours, and on some new photographic processes, *Philos. Trans.*, **132**, pp. 181–214.

42. Ware, M. (1999). *Cyanotype: The History, Science and Art of Photographic Printing in Prussian Blue* (National Museum of Photography, Film and Television, Bradford, UK).

43. Perls, M. (1867). Nachweis von Eisenoxyd in gewissen Pigmenten, *Virchows Arch.*, **39**, pp. 42–48.

44. Nigrovic, V. (1963). Enhancement of the excretion of radiocaesium in rats by ferric cyanoferrate(II), *Int. J. Radiat. Biol.*, **7**, pp. 307–309.

45. Heydlauf, H. (1969). Ferric-cyanoferrate (II): An effective antidote in thallium poisoning, *Eur. J. Pharmacol.*, **6**, pp. 340–344.

46. Neff, V. D. (1978). Electrochemical oxidation and reduction of thin films of Prussian blue, *J. Electrochem. Soc.*, **125**, pp. 886–887.

47. Kraft, A., Rottmann, M., and Heckner, K.-H. (2006). Large-area electrochromic glazing with ion-conducting PVB interlayer and two complementary electrodeposited electrochromic layers, *Sol. Energy Mater. Sol. Cells*, **90**, pp. 469–476.

48. Ellis, D., Eckhoff, M., and Neff, V. D. (1981). Electrochromism in the mixed-valence hexacyanides. 1. Voltammetric and spectral studies of the oxidation and reduction of thin films of Prussian blue, *J. Phys. Chem.*, **85**, pp. 1225–1231.

49. Itaya, K., Akahoshi, H., and Toshima, S. (1982). Electrochemistry of Prussian blue modified electrodes: An electrochemical preparation method, *J. Electrochem. Soc.*, **129**, pp. 1498–1500.

50. Neff, V. D. (1985). Some performance characteristics of a Prussian blue battery, *J. Electrochem. Soc.*, **132**, pp. 1382–1384.

51. Eftekhari, A. (2004). Potassium secondary cell based on Prussian blue cathode, *J. Power Sources*, **126**, pp. 221–228.

52. Kuperman, N., Padigi, P., Goncher, G., and Solanki, R. (2017). High performance Prussian blue cathode for nonaqueous Ca-ion intercalation battery, *J. Power Sources*, **342**, pp. 414–418.

53. Selvarani, G., Prashant, S. K., Sahu, A. K., Sridhar, P., Pitchumani, S., and Shukla, A. K. (2008). A direct borohydride fuel cell employing Prussian blue as mediated electron-transfer hydrogen peroxide reduction catalyst, *J. Power Sources*, **178**, pp. 86–91.

54. Fu, L., You, S.-J., Zhang, G.-Q., Yang, F.-L., Fang, X.-H., and Gong, Z. (2011). PB/PANI-modified electrode used as a novel oxygen reduction cathode in microbial fuel cell, *Biosens. Bioelectron.*, **26**, pp. 1975–1979.

55. Nonoguchi, Y., Murayama, T., Ishizaki, M., Kanaizuka, K., Kurihara, M., Hata, K., and Kawai, T. (2014). SWNT composites with compositionally tunable Prussian blue nanoparticles for thermoelectric coordination programming materials, *Chem. Lett.*, **43**, pp. 1254–1256.

56. Lisowka-Oleksiak, A., and Nowak, A. P. (2007). Metal hexacyanoferrate network synthesized inside polymer matrix for electrochemical capacitors, *J. Power Sources*, **173**, pp. 829–836.

57. Itaya, K., Shoji, N., and Uchida, I. (1984). Catalysis of the reduction of molecular oxygen to water at Prussian blue modified electrodes, *J. Am. Chem. Soc.*, **106**, pp. 3423–3429.

58. Karyakin, A. A., Gitelmacher, O. V., and Karyakina, E. E. (1994). A high-sensitive glucose amperometric biosensor based on Prussian blue modified electrodes, *Anal. Lett.*, **27**, pp. 2861–2869.

59. Karyakin, A. A., Karyakina, E. E., and Gorton, L. (2000). Amperometric biosensor for glutamate using Prussian blue-based artificial peroxidase as a transducer for hydrogen peroxide, *Anal. Chem.*, **72**, pp. 1720–1723.

60. Ricci, F., and Palleschi, G. (2005). Sensor and biosensor preparation, optimisation and applications of Prussian blue modified electrodes, *Biosens. Bioelectron.*, **21**, pp. 389–407.

61. Krishnan, V., Xidis, A. L., and Neff, V. D. (1990). Prussian blue solid-state films and membranes as potassium ion-selective electrodes, *Anal. Chim. Acta*, **239**, pp. 7–12.

62. Hermes, M., and Scholz, F. (1997). The electrochemical determination of ammonium based on the selective inhibition of the low-spin iron(II)/(III) system of Prussian blue, *J. Solid State Electrochem.*, **1**, pp. 215–220.

63. Kong, B., Selomulya, C., Zheng, G., and Zhao, D. (2015). New faces of porous Prussian blue: Interfacial assembly of integrated hetero-structures for sensing applications, *Chem. Soc. Rev.*, **44**, pp. 7997–8018.

64. Long, J., Guari, Y., Guerin, C., and Larionova, J. (2016). Prussian blue type nanoparticles for biomedical applications, *Dalton Trans.*, **45**, pp. 17581–17587.

65. Parajuli, D., Kitajima, A., Takahashi, A., Tanaka, H., Ogawa, H., Hakuta, Y., Yoshino, K., Funahashi, T., Yamaguchi, M., Osada, M., and Kawamoto, T. (2016). Application of Prussian blue nanoparticles for the radioactive Cs decontamination in Fukushima region, *J. Environ. Radioact.*, **151**, pp. 233–237.

Chapter 2

Structure and Magnetism of Prussian Blues and Analogues: A Historical Perspective

Michel Verdaguer

Institut Parisien de Chimie Moléculaire, UMR CNRS 8232, Université Pierre et Marie Curie-Sorbonne Université, Case 42, 4 place Jussieu, Paris, F75232, France

michel.verdaguer@upmc.fr

2.1 Introduction: Miscellanies

The goal of the editors of this volume is to present to the reader a historical, artistic, and scientific vision of Prussian blue (PB). The present section is devoted to the history of magnetism of Prussian blue and its analogues (PBA). Writing about magnetism's history of Prussian blue demands that (i) Prussian exists and there is some agreement on its crystal and electronic structure; (ii) reliable magnetic measurements have been made on well characterized compounds and ... published; (iii) and there is some understanding of the origin and mechanism of magnetic properties if one wants to improve them to go forwards for applications.

Prussian Blue-Type Nanoparticles and Nanocomposites: Synthesis, Devices, and Applications
Edited by Yannick Guari and Joulia Larionova

ISBN 978-981-4800-05-1 (Hardcover), 978-0-429-02474-0 (eBook)
www.panstanford.com

(i) Prussian blue exists and I met it. The history of its discovery, often considered as a typical example of a serendipitous finding in science, is remarkably presented in this book by Alexander Kraft (Kraft, 2008a; Kraft, 2008b; Kraft, 2011). One can consider that three main papers announced to the world the availability of Prussian blue: an anonymous information in Berlin in 1710, (Anonymous, 1710), two communications at the Royal Society in 1724, (Woodward, 1724; Brown, 1724) followed soon after by two communications by Geoffroy at the Royal Academy of Science in Paris (Geoffroy, 1725). From 1709, Prussian blue was produced on a large scale for painters, first in Berlin and the Netherlands and then in whole Europe. Immediately, it was used by contemporary artists (Watteau, Lancret, Pater, van der Werft, ...). It is amazing that as soon as the pigment became available in a given country, its *Gentianella* colour (and its prize compared to Ultramarine) (Kraft, 2008b) attracted attention of the most talented painters who used it at once and produced masterpieces. The great wave of Kanagawa of Hokusai (1831) (Fig. 2.1) is a marvelous example of this worldwide recognition: Imported in Japan from 1829, "berorin" (Berlin) blue was used in the woodblock series of 36 views of Fuji-san which dates from 1830–1833 and became extremely fashionable in Japan as monochrome *aizuri-e* (Bayou, 2008).

The chemical history of Prussian blue (the synthesis, the composition, the stoichiometry, and the structure, ...) follows the history of chemistry itself, from Stahl and the phlogistic, to the structure by single-crystal X-ray diffraction by Ludi et al. (Buser et al., 1977) through Macquer, Priestley, Scheele, Hassenfratz and Lavoisier, Proust, Berthollet, Berzelius, Gay-Lussac, ... (Kraft, 2008a). Section 2.2 of this chapter is devoted to the discovery of the nature and of the structure of Prussian blue.

(ii) Magnetic measurements came much later and in some cases the compounds were not clearly and cleanly characterized, their structures unknown or ignored and the useful instrumentation unavailable. Section 2.3 is devoted to them.

Figure 2.1 Hokusai, Under the great wave of Kanagawa (Woodblock Print by Uchida Art Co, Kyoto Handicraft Center. On my wall, a photograph by M. Verdaguer).

(iii) Unravelling the origin of the magnetism needed even more time. A remarkable feature is that the poor ferromagnetic properties of Prussian blue itself (its long range (Curie) temperature is a modest $T_C = 5.6$ K) and the brilliant hue of the pigment, so fascinating for many since the beginning of the 18th century, are related to the same electronic origin, that is, mixed valency in an *a priori* simple, face cubic centred structure (Section 2.4). The magnetic properties of bimetallic Prussian blue analogues (same basic structure and different transition metal ions) are in Section 2.5. The amazing photomagnetic properties are evoked in Section 2.6.

2.2 Composition and Structure of Prussian Blue

It is useful to begin by a synthetic warning: Nothing is simple in the chemistry of Prussian blue. The blue precipitate can be obtained from the reaction of solutions of different ions and it is naïve to believe that the composition of the precipitate (or of the colloidal solid, or the distribution of ill-defined nanosized objects) corresponds to the stoichiometry expected on the paper from the electroneutrality of the solid. Water, adsorbed ions (cations and anions) and other species at the surface, inserted species in vacancies, redox process, aging of the products interfere widely in the elemental composition and therefore in the physical properties, including magnetic. Such problems hindered for a long time determination of the nature, the structure and obviously the magnetism of Prussian blues. A recent review by Samain et al., (Grandjean et al., 2016) dedicated to the understanding of the fading of Prussian blue pigment in paintings, carefully analyses the importance of the "recipes" (18th century recipes and modern ones) on the composition and properties of the corresponding compounds, which vary amazingly widely.

Prussian blues (and their analogues) must therefore be prepared very carefully to obtain pure samples having well-defined and reproducible properties, including magnetic.

Among the important questions to decipher along the years were: Are Turnbull and Prussian blues the same blue? Which is the difference between soluble and insoluble Prussian blues? Which are the exact composition, formulation, and structure of Prussian blue? Perfect face-centred cubic or presence of vacancies? Which is the origin of the bright blue colour? And later, what are the oxidation and spin states of the iron ions? Where are the Fe(II) and where are the Fe(III)? Which role do the water molecules and the vacancies play? The chemical, physical and structural answers to these entangled problems came slowly. They were in many cases contradictory or even ignored. In the description of the different works, we chose to provide the reader with substantial quotations of the original papers so that (s)he is closest of the steps of an investigation in progress.

2.2.1 Composition and Formulation

We give here the present accepted understanding based on the analyses and crystallographic studies of Ludi, Buser, Güdel and coworkers (Buser et al., 1977; Lüdi and Güdel, 1971). The recognized composition and formulation of Prussian blue (also known as "insoluble" Prussian blue) is today $\{Fe_4[Fe(CN)_6]_3\}\cdot 14\text{-}16H_2O$, that is ferric hexacyanidoferrate(II) with a variable amount of water molecules. The oxidation and spin states of iron and the charge of the solid have been explicited, $\{Fe_4^{III}[Fe^{II}(CN)_6]_3\}^0 \cdot 14\text{-}16H_2O$, high spin Fe^{III} and low spin Fe^{II}. This represents the formula per conventional unit cell. The stoichiometry is Fe^{III}: Fe^{II} = 4/3. The amount of water molecules is variable. Based on one Fe^{III} ion, the formulation becomes $\{Fe^{III}[Fe^{II}(CN)_6]_{3/4}\}^0 \cdot 3.5\text{-}4H_2O$. Finally, the presence of the intrinsic vacancies present in the network (an important parameter, as discussed below) can be explicited as $\{Fe_4^{III}[Fe^{II}(CN)_6]_3\square_1\}^0 \cdot 14\text{-}16H_2O$.

The usual way to represent nowadays the synthesis of Prussian blue in water, from $[Fe(CN)_6]^{4-}(aq)$ and $Fe^{3+}(aq)$ is

$$3\,[Fe^{II}(CN)_6]^{4-}(aq) + 4Fe^{3+}(aq) \rightarrow \{Fe_4^{III}[Fe^{II}(CN)_6]_3\}^0 \cdot 14\text{-}16H_2O. \quad (2.1)$$

The reaction of the hexacyanidoferrate(III) on Fe^{2+} previously thought to give Turnbull blue (ferrous hexacyanidoferrate(III), $\{Fe_4^{II}[Fe^{III}(CN)_6]_{8/3}\square_{4/3}\}^0$ leads also to Prussian blue. In such a way, Turnbull blue does not exist. However, we shall discover thanks to Robin and Day that the situation is more subtle. In the history of the understanding of the nature and of the electronic structure of Prussian blue, this was one of the paradox to clear up (Davidson, 1937) (see Section 2.2.3).

The early literature on Prussian blue distinguishes "insoluble" and "soluble" Prussian blue. Everybody agrees on the fact that such a distinction based on solubility has no real meaning and that "soluble" is a misnomer. Both kinds of compounds are very insoluble. "Soluble" Prussian blue is obtained when the source of $[Fe(CN)_6]^{4-}(aq)$ is $K_4[Fe(CN)_6]$ and more generally when potassium or alkali ions are present during the synthesis. The solid then presents some potassium, or other alkalis in its analysis. The approximate composition of the compound formed is then

$$[Fe^{II}(CN)_6]^{4-}(aq) + 4Fe^{3+}(aq) + K^+(aq) \rightarrow \{K^IFe^{III}[Fe^{II}(CN)_6]\}^0 \cdot xH_2O \quad (2.2)$$

with $x < 10$. The K^I:Fe^{III}:Fe^{II} stoichiometry in (2.2) is 1:1:1. Nevertheless, Lüdi and Güdel warned us that "potassium can be considered as an impurity of these often colloidal precipitates.

Thus, the polynuclear cyanides containing potassium or other alkali cations are non-stoichiometric compounds rather than compounds showing a definitive formula as far as the alkali ions are concerned" (Lüdi and Güdel, 1971). Indeed, the preparation of a Prussian blue either potassium-free or with a controlled potassium stoichiometry needs special synthetic care. My experience is that it is nevertheless possible to get alkali derivatives of Prussian blue analogues with a definite formula. In the early literature, the study of the "soluble" version of Prussian blue ($\{K^{I}Fe^{III}[Fe^{II}(CN)_6]\}$) is generally accompanying the one of the "insoluble" counterpart ($\{Fe_4^{III}[Fe^{II}(CN)_6]_3\}^0$).

Finally, the general formulation of analogues of Prussian blues, made of a transition metal A^{q+}, an hexacyanidometalate $[B(CN)_6]^{p-}$ and an alkali cation C^+, should be $\{C_pA_q[B(CN)_6]_r\}^0 \cdot xH_2O$, with simple relationships between p, q and r depending on the valence of the metallic ions A, B, C. The general formulation of Ludi and Güdel, forgetting the cation considered as an impurity, leads to the simplified formulation $A_k[B(CN)_6]_l \cdot xH_2O$, with $l < k$.

2.2.2 Structure

One of the difficulties encountered to determine the crystallographic structure of Prussian blues is that reactions (2.1) and (2.2) produce neutral solids, which precipitates rapidly as amorphous, colloidal or polycrystalline powders when reactants present usual concentrations (≥0.01 mol l^{-1} for example); dilute concentrations can lead to very slow precipitation (Davidson, 1937). The solubility products are very small. Obtaining single crystals suitable for X-ray diffraction is not an easy task (Buser et al., 1977) and the first structural results were proposed on the basis of powder analyses. Furthermore, the presence of potassium (or alkali) ions in the synthetic solution provides a mixture of other species, having the same colour, but with variable stoichiometries.

"Although Prussian blue has been known for a long time as an important pigment, its chemistry has never been satisfactorily explained. The whole subject is surprisingly confused in spite of the large volume of work which has been published" wrote Keggin and Miles in the introduction of their 1936 *Nature* paper entitled "Structure and formulae of Prussian blues and related compounds" (Keggin and Miles, 1936). It can be of interest, in the present historical perspective, to note that the two researchers belonged to the Imperial Chemical Industries and that their research laboratories were located in Ardeer (Ayrshire, Scotland), in a factory created by Alfred Nobel and wisely surrounded by sand hills for the large-scale production of dynamite. The analyses were performed in most part in the Dyestuff Laboratories in Blackley, near Manchester.

On the basis of powder X-Ray diffraction and chemical analysis, the authors proposed three classes of definite compounds having the structures given in Fig. 2.2.

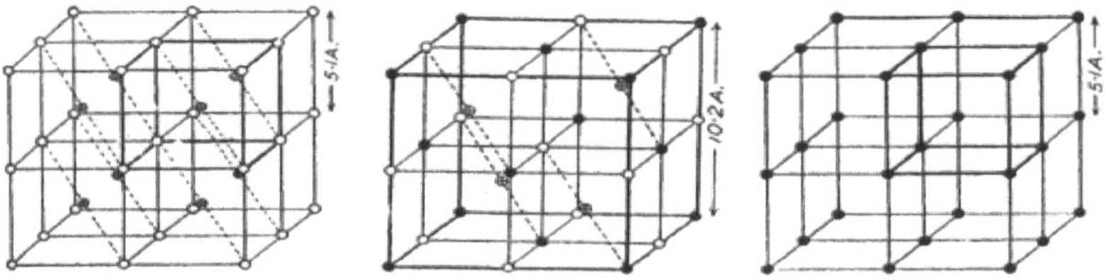

(a) (b) (a)

Figure 2.2 Keggin and Miles model: (a) ferrous ferrocyanido, $\{C^{I}_{2}Fe^{II}[Fe^{II}(CN)_6]\}$; (b) alkali ferric ferrocyanido $\{C^{I}Fe^{III}[Fe^{II}(CN)_6]\}$, (soluble Prussian blue); (c) ferric ferricyanido, $\{Fe^{III}[Fe^{III}(CN)_6]\}$ (Berlin green). [● = Fe^{II}; ⊗ = C^{I}; ○ = Fe^{III}]. Reproduced with permission from Keggin and Miles (1936).

In Keggin's paper, cation C^{I} is shown by R and the valence by ′ (Fe^{II} = Fe″, Fe^{III} = Fe‴): (a) $C^{I}_{2}Fe^{II}[Fe^{II}(CN)_6]\}$: "Fe″$R_2$Fe″$(CN)_6$ where R represents Na, K, NH_4, Rb, or Cs." The authors remark that these "white ferrous-alkali ferrocyanides" are "intermediate in the preparation of commercial blues"; (b) $\{C^{I}Fe^{III}[Fe^{II}(CN)_6]\}$: "Fe‴RFe″$(CN)_6$", where R represents an alkali metal or ammonia. Commercial pigments all tend to be of this type"; (c) $\{Fe^{III}[Fe^{III}(CN)_6]\}$: Fe‴Fe‴$(CN)_6$, the structure of Berlin green, the oxidation product of Prussian blue. No alkali atoms are present.

The structures are simply described as follows. For $C^{I}Fe^{III}[Fe^{II}(CN)_6]$ (b), "the structure is cubic, a = 10.2 Å. The iron atoms are arranged, ferrous and ferri alternately, at the corners of a cubic lattice of 5.1 Å edge. The alkali atoms occur at the centres of alternate small cubes. It is possible that the small cubes not containing alkali may contain water molecules." The "alternate small cubes" are known today as "octants."

"In this class are compounds of all the alkali metals and ammonium; with the exception of lithium and caesium." The chemists accustomed to the chemistry of Prussian blues know that lithium ion, too small, is easily washed out during the purification process. Nevertheless, electrochemistry allows insertion of lithium (Flambard et al., 2017). As for caesium, the ion fits quite well the size of the "small cubes" and provides valuable derivatives. For $C^{I}_{2}Fe^{II}[Fe^{II}(CN)_6]$ (a), "the structure is again cubic, a = 5.1 Å and is closely related to that of the prussian blues ... all cubes contain an alkali atom." For $Fe^{II}[Fe^{III}(CN)_6]$ (c), the "structure also is very similar to that of prussian blue; ... there are no alkali atoms present... This unit cell is again 5.1 Å and it is possible that water occurs in the small cubes." In the three cases, the Fe^{III}:Fe^{II} stoichiometry is 1:1.

At the end of the paper, a fourth family is announced, proposing a model: "In addition, blues containing no alkali have been prepared, and X-ray photographs show that the Fe-CN skeleton structure is cubic and similar to that already described (a = 10.2 Å approx.). Here *it seems that additional iron, or iron in combination with a negative ion, can replace the alkali metal.*" One can recognize here the "insoluble" Prussian blue, $A_k[B(CN)_6]_l x \cdot H_2O$ in the formulation of Ludi and Güdel.

The Keggin and Miles model remained the reference for the so-called "Prussian blue structure" for a long time. Examples can be found in van Bever (van Bever, 1938), Weiser (Weiser et al., 1937) (Weiser et al., 1941; Weiser et al., 1942) or Schriver and coworkers (Shriver et al., 1965; Brown et al., 1968; Shriver and Brown, 1969). The later are dedicated to bond isomerization in peculiar PB analogues. Schriver even drew conclusions on the existence and the position of the additional metallic ion in the octant, which indeed is not there. New data appeared progressively showing the limit of the Keggin's model and the necessity to move to a "modified model." Among the new data were the measurements of the density of the material. It should be clear that for a given stoichiometry $A_k[B(CN)_6]_l$, ($l < k$), the density is different (i) if all octahedral sites are occupied by heavy metals, alternately by A and $[B(CN)_6]$ (that is $A_k[B(CN)_6]_k$) with extra metallic ions added in the octants; (ii) or if the stoichiometry is achieved through $(k - l)$ vacancies in place of $[B(CN)_6]$. This observation is due to Güdel, then student in Bern (Güdel, private communication). The experimental and calculated densities are always present as an important piece of information in the articles of Ludi and coworkers. The structural and electronic characterization of Prussian blue and analogues was indeed a long-term research project in the laboratory of Ludi at the University of Bern. The "modified model" appeared first based on crystallographic data of $Co_3[Co(CN)_6]_2$ and $Cd_3[Co(CN)_6]_2$ analogues (Ludi and Güdel, 1968) (Lüdi and Güdel, 1971) and then on X-ray diffraction on single crystals of Prussian blue itself (Buser et al., 1977). We summarize briefly the main features of the "modified model". "Whereas the simple Keggin-Miles model correctly reflects the polymeric nature of Prussian Blue, the actual structure is more complicated. The next best approximation is still using a cubic face-centred unit cell... A number of positions in this unit cell are only partially occupied, the vacant sites being randomly distributed. Any non-random distribution necessarily leads to a cubic primitive or lower symmetry lattice. The X-ray data clearly demonstrate that in general the lattice is cubic primitive." (Buser et al., 1977). Such a description implies the presence of vacancies *intrinsic* to PBs, whenever the stoichiometry A_k:B_l corresponds to $l/k < 1$. It is *not* a "defect" structure as too often claimed in the literature. Many important properties - including magnetic are determined by these vacancies.

In Buser's paper, different single crystals were obtained by slow diffusion of water vapour on a solution of Prussian blue in concentrated hydrochloric acid (Buser et al., 1977) and carefully washed. Skipping the conditions of intensities' collection and the crystallographic analysis we concentrate on the results, given in Fig. 2.3 and in Table 2.1. The main conclusions are as follows. According to the density $d_{exptl} = 1.75–1.81$ g m^{-3} one unit of $Fe_4[Fe(CN)_6]_3 \cdot xH_2O$ is contained in the cubic primitive elementary cell with $a = 10.166$ Å, $d_{calcd} = 1.78$ g cm^{-3} for $x = 15$. Non-face-centred reflections indicate a deviation from the well-known model in the space group *Fm3m* (O_h^5). The $Fe(CN)_6$ positions are only partly occupied. Partial ordering of the corresponding vacancies causes the deviation from *Fm3m* symmetry. The least-squares refinement in the space group O_h^1

Pm3m is around 0.04. The average distances are Fe^{II}-C = 1.92 Å, C-N = 1.13 Å, and Fe^{III}-N = 2.03 Å.

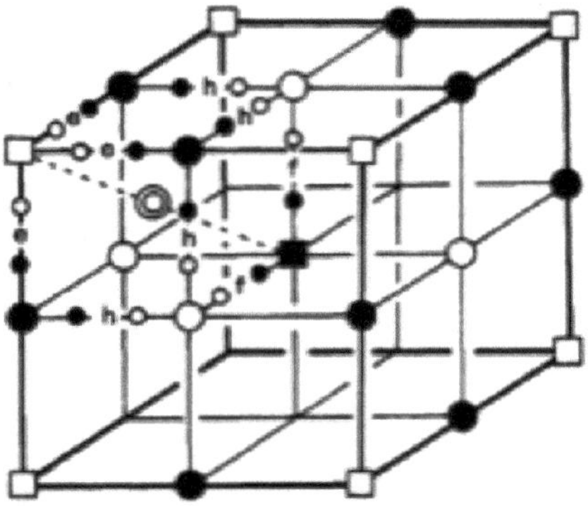

Figure 2.3 The unit cell of Prussian Blue in space group *Pm3m*. The atom types and crystallographic positions are partially indicated by the following symbols: (□) Fe(III), la (origin); (○) Fe(III), 3c; (□) Fe(II), lb; (●) Fe(II), 3d; (◎) O, 8g; (o) C, 6e, 6f, 12h; (•) N or O, 6e, 6f, 12h (cf. Table 2.1). Reproduced with permission from Buser et al. (1977).

Table 2.1 Atom distribution in the unit cell of Prussian blue and average coordination of the Iron in the Space groups *Fm3m* and *Pm3m*

	Fm3m			*Pm3m*		
Atom	Pos.	Occ.	Coord.	Pos.	Occ.	Coord.
4 Fe^{III}	4a	1	$FeN_{4.5}$–$O_{1.5}$	1a	1	$FeN_{6-2p}O_{2p}$
				3c	1	$FeN_{4+(2p/3)}O_{2-(2p/3)}$
3 Fe^{II}	4b	¾	FeC_6	1b	*P*	FeC_6
				3d	1–*p*/3	FeC_6
18 CN	24e	¾	—	6e	1–*p*/3	
				6f	*p*	
				12h	1–*p*/3	
6 O	24e	¼	—	6e	*p*/3	
				6f	1–*p*	
				12h	*p*/3	
8 O	8	1 (zeolitic)	—	1	—	

Notes: Pos. = Position; Occ. Occupation; Coord. = Coordination; *p* is the occupancy factor of site 1b. The notations are the ones in International Tables for X-ray Crystallography, The Kynoch Press, Birmingham, 1952 [adapted from (Buser et al., 1977)].

Besides the clarification of the presence of the $[Fe(CN)_6]$ vacancies and their possible ordering, this paper is also important since it allows to evidence two kinds of water molecules, the ones coordinated to the A atoms close to a vacancy (8) and zeolithic ones, present at the centres of the octants. The nature of the water molecules was an important concern in previous structural studies by van Bever, Weiser, Schriver, limited by the resolution of the available techniques at the time of measurements.

Later, the mean positions of water molecules and the hydrogen bond network were refined by neutron diffraction in a $[Mn^{II}_3[Co^{III}(CN)_6]_2$ analogue (Beall et al., 1977) and in deuterated samples of Prussian blue itself (Herren et al., 1980), as shown in Fig. 2.4.

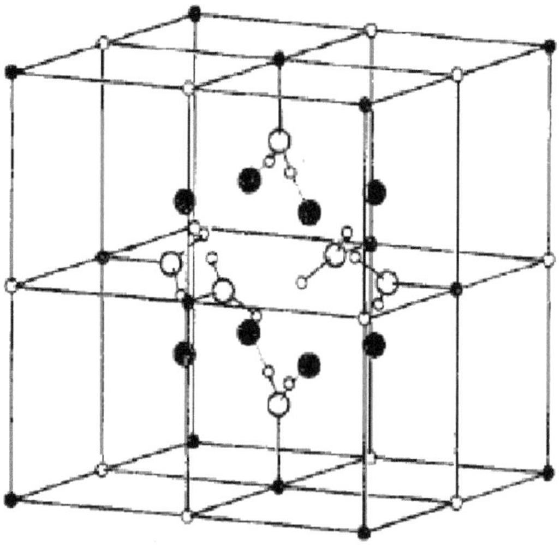

Figure 2.4 Ortep plot of the unit cell of Prussian blue displaying part of the hydrogen-bond network. Cyanide ions are omitted for clarity. [(●) Fe^{III}, (o)Fe^{II}, (O) coordinated O, (●) D_2O. Reproduced with permission from Herren et al. (1980).

This important result of neutron diffraction, more sensitive to light atoms (C, N, O) than X-rays is still a mean image. The exact structure of the different possible vacancies was determined later by paramagnetic NMR of PB analogues (Flambard et al., 2009; Flambard et al., 2017).

2.2.3 The Prussian Blue's Paradox (Prussian or Turnbull?)

"The Prussian blue paradox" is the title of a paper by Davidson in the *Journal of Chemical Education* in 1937 (Davidson, 1937). It points out that when ferric ion is mixed with hexacyanidoferrate(II), the following redox equilibrium occurs:

$$Fe^{3+} + [Fe^{II}(CN)_6]^{4-} \underset{2}{\overset{1}{\rightleftarrows}} Fe^{2+} + [Fe^{III}(CN)_6]^{3-} \quad (2.3)$$

The equilibrium constant $K = [Fe^{III}(CN)_6]\ [Fe^{2+}]/[Fe^{II}(CN)_6]\ [Fe^{3+}]$. With thermodynamic data provided by Müller (Müller and Stanisch, 1909), Davidson used $K \approx 10^5$. The present values of standard potential for the two redox couples Fe^{3+}/Fe^{2+} and $[Fe^{III}(CN)_6]^{3-}/[Fe^{II}(CN)_6]^{4-}$ (and therefore of K) are presently slightly different but the demonstration remains valid as far K remains large: The equilibrium is strongly displaced in direction 1.

The reactions of the four species are able to give different precipitates:

$$\text{"Prussian"} \rightleftarrows Fe^{3+} + [Fe^{II}(CN)_6]^{4-} \rightleftarrows Fe^{2+} + [Fe^{III}(CN)_6]^{3-} \rightleftarrows \text{"Turnbull"} \quad (2.4)$$

Initial state	1	1	0	0
Equilibrium	$1-x$	$1-x$	x	x
or	0.0032	0.0032	0.9968	0.9968

since $K = x^2/(1-x)^2 \approx 10^5$, then $x \approx 0.9968$.

Following this thermodynamical reasoning, the authors working with ferricyanides were firmly in favour of the existence of Turnbull blue (ferrous ferricyanide) obtained through

$$3Fe^{2+}_{aq} + 2[Fe^{III}(CN)_6]^{3-}_{aq} \rightleftarrows \{Fe^{II}_3[Fe^{III}(CN)_6]_{2(solid)}\}^0 \quad (2.5)$$

rather than Prussian blue (ferric ferrocyanide) obtained by

$$4Fe^{3+} + 3[Fe^{II}(CN)_6]^{4-} \rightleftarrows \{Fe^{III}_4[Fe^{II}(CN)_6]_{3\,(solid)}\}^0. \quad (2.6)$$

The slow kinetics of precipitation of solid PB displaces continuously equilibria (2.4) towards "Prussian."

For sake of simplicity, we forget water in the above formulae. We do not discuss the existence of the third white precipitate, the most insoluble $\{Fe_2^{II}[Fe^{II}(CN)_6]_{2(solid)}\}^0$, which is the usual industrial intermediate giving rise to Prussian blue by partial oxidation in air, since it needs special experimental conditions to appear.

Müller and coworkers (Müller and Stanisch, 1909) established the formula of PB as $\{Fe_4^{III}[Fe^{II}(CN)_6]_3$ with a Fe^{II}: Fe^{III} ratio = ¾ and said that Prussian and Turnbull blues are different compounds. Later, Weiser et al. (Weiser et al., 1942) indicated that "the results of X-ray diffraction investigations, density determinations, and measurements of oxidation-reduction potential indicate that *Prussian blue and Turnbull's blue are identical chemical individuals*." Nevertheless, "the results obtained do not enable one to decide whether the composition of Prussian blue and Turnbull's blue should be represented by the formula $Fe_4[Fe(CN)_6]_3$ or by the formula $Fe_3[Fe(CN)_6]_2$." And the authors finally arrived at the conclusion that "chemical evidence suggests that both Prussian blue and Turnbull's blue may be represented by the formula $Fe_3[Fe(CN)_6]_2$".

An infra-red study of Emschwiller (Emschwiller, 1954) lead to an opposite conclusion: The frequency of the band related to the cyanido stretching of Prussian and Turnbull blues is the same (2075 cm^{-1}) and corresponds clearly to a ferrocyanide (in the range 2068–2090 cm^{-1} measured in Mn, Co, Ni, Zn PB analogues) rather that to a ferricyanide (in the range 2158–2176 cm^{-1} for the same analogues). The identity of Prussian and Turnbull blues was further confirmed by the Mössbauer spectra which were published after 1962 (see Section 2.3.2).

2.3 Magnetism of Prussian Blue and Analogues

We focus here on the first few articles dealing with magnetism but several reviews are available. For example, in a chapter on modern perspectives of the chemistry of metal cyanide compounds, Dunbar et al. summarized in 1997 the magnetic properties of Prussian Blue analogues (Dunbar and Heintz, 1997). On the other hand, in a series dedicated to new molecular magnetic materials, Girolami and me summed up our independent and converging steps on high Curie temperature magnetic Prussian blues analogues (Verdaguer and Girolami, 2005).

Along the years, one can distinguish several approaches to the magnetism of Prussian blues: (i) Are Prussian blues magnetic and how? (Davidson and Welo, 1927); (ii) can magnetism help to discover and to understand the nature of the compounds (Davidson and Welo, 1927), the structure of the network, the valences and the spin states of the metals (that is *magnetochemistry*)? (Davidson and Welo, 1927), (iii) the structure being known, is it possible to tune the electronic structure to prepare new PBAs with new magnetic properties (that is *molecular magnetism*) (Michel Verdaguer et al., 1999a). We examine in turn the different phases. We focus on the magnetic results of PBs. In many cases magnetism was studied on PB analogues and we cannot ignore them since these studies were important in the understanding of the PBs themselves.

2.3.1 First Magnetic Measurements (1927–1956)

The first magnetic investigations on Prussian blue were published in 1927 in Davidson and Welo (1927) under the title "The nature of Prussian blue." The introduction presents the main conclusions of Müller and coworkers on the "simple" structures; "Soluble Prussian Blue is potassium ferric ferrocyanide, $KFe^{III}[Fe^{II}(CN)_6]$" ... "Insoluble Prussian Blue is ferric ferrocyanide, $Fe_4^{III}[Fe^{II}(CN)_6]_3$ in which the potassium of the soluble blue is replaced by ferric ion" ... "Turnbull's Blue is made up of three molecules of the soluble blue in which two of the potassium atoms have been replaced by ferrous iron, hence $KFe^{II}\{Fe^{III}[Fe^{II}(CN)_6]\}_3$ or $KFe^{II}\{Fe_3^{III}[Fe^{III}(CN)_6]\}_3$." The authors quote also the hypotheses by Reihlen (Reihlen and Zimmermann, 1927) and Cambi (Cambi and Clerici, 1927) of a "constitutional formula for Soluble Prussian Blue in which the two iron atoms are combined in the form of a complex polynuclear ion:"

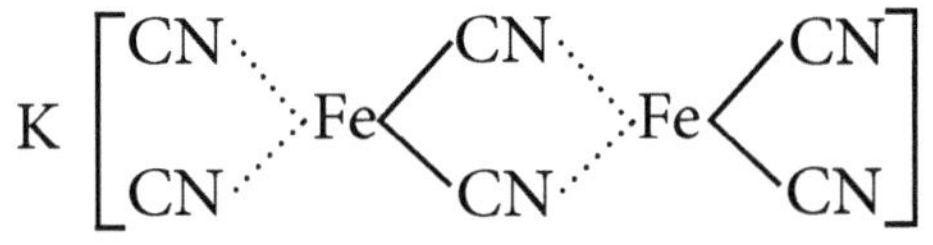

Scheme 2.1

The authors advertised clearly: "The present paper deals with the results obtained in the attempt to distinguish between the simple and the more complex structures". Two techniques are used: conductivity in solution and magnetism. The measurement methods are not described. The conclusions are also very clear: "Neither the conductivity measurements in solution nor the magnetic measurements in the solid state have produced any evidence of the existence of the postulated polynuclear ion".

As for magnetism, the paper presents (in two tables) and discusses carefully previous solid state magnetic data of the components of Prussian blues, salts of Fe^{III} and $Fe^{II}(CN)_6$, so that the authors are able to compute the magnetic susceptibilities of the PBs in different structural hypotheses. They assume that the ferrocyanido ions are diamagnetic, that the paramagnetism originate only from the ferric ions and that the Pascal constants for diamagnetism corrections are applicable. The magnetic susceptibility are measured at three temperatures (ambient, melting ice (+2.1°C) and solid CO_2 (−70°C), in the range 200–300 K. The results are shown in Fig. 2.5 and Table 2.2.

Table 2.2 Comparison of experimental and computed values of magnetic parameters for soluble and insoluble Prussian blues

	C_a(exp)	C_a(calc)	P(exp)	P(calc)	Θ(exp)
Soluble	4.05	3.8–5.0	8.3	27.4–32.4	+22
Insoluble	3.92	3.8–5.0	27.8	27.4–32.4	+14.7

Source: Adapted from Davidson and Welo (1927).
Note: C_a is the Curie constant and P the number of Bohr magnetons ($P = 14{,}07\sqrt{C_a}$).

The authors conclude that (i) the two compounds follow a Curie–Weiss law $\chi_M = C_a/(T - \Theta)$ in the temperature range explored, with close values for soluble ($C_a = 4.05$; $\Theta = +22$ K) and insoluble ($C_a = 3.92$; $\Theta = +14.7$ K) species; (ii) the computed values are consistent with the formulae proposed by Müller: Soluble PB as a double potassium ferric salt of hydroferrocyanic acid in which all of the paramagnetism is due to the ion Fe^{III}; and in insoluble PB, only four of the seven iron atoms per molecule play a part in the paramagnetism with an equal contribution; (iii) the formulation as polynuclear compounds is inconsistent with the known magnetic properties of iron salts.

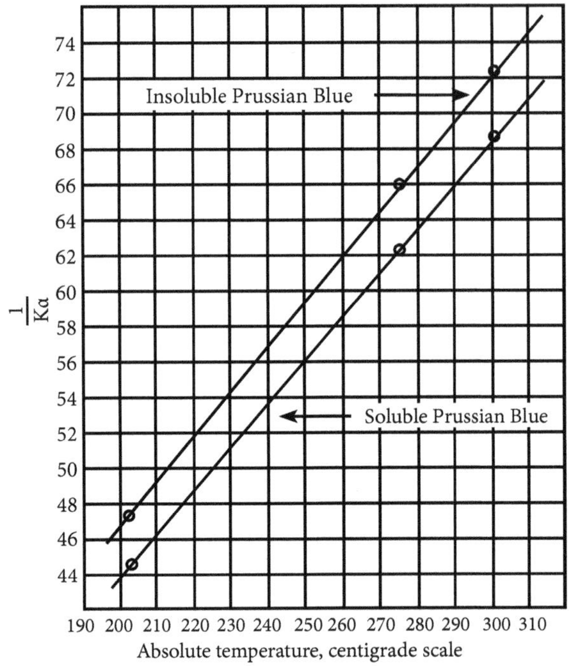

Figure 2.5 Thermal variation of the molar magnetic susceptibilities $\chi_M(K_a)$ of soluble and insoluble Prussian blues. Reproduced with permission from Davidson and Welo (1927).

With our present knowledge, we realize that this is the first demonstration that *Prussian blue is magnetic*: (i) The magnetic properties confirm the chemical analyses and formulae of Müller, that is the correct stoichiometries of the compounds and the existence of two sites of iron, one diamagnetic $Fe^{II}(CN)_6$, the other paramagnetic, $Fe^{III}(NC)_6$; the two sites were called some time later "carbon hole" or "nitrogen hole"; it is unfortunate that these conclusions were ignored by many of the following workers; (ii) the range of temperature explored is too narrow and at too high values to evidence the weak exchange interactions and the polynuclear nature of PB; nevertheless the value of the Curie–Weiss constant $\Theta = +22$ K could have announced some ferromagnetic interaction; (iii) to exclude the polynuclear nature from these measurements was therefore risky (see Section 2.2.2); (iv) the role of water is not tackled, which can be understood in the absence of elemental analysis.

This early work was followed by more extensive studies of Prussian Blue and some analogues in 1940 (Richardson and Elliott, 1940), but it was not until much later, in 1968, that Prussian Blue was shown to be a ferromagnet with T_C = 5.6 K (Ito et al., 1968).

In 1940, Richardson and Elliott started from the structures of Keggin, Miles and van Bever to analyse the magnetic properties of PB analogues of ferri and cobalticyanides (Richardson and Elliott, 1940). They propose to answer the question "whether the chemical bonds in these compounds are ionic or covalent, or a combination of the two forms." They report chemical analyses in line with the proposed formulae "within a few tenths of a per cent." They use a Gouy balance at the unique temperature of 20°C. The magnetic moments obtained from the formula $\mu = \sqrt{(\chi_m + 100)T}$ are given in Table 2.3, in Bohr units (magnetons). In the formula, χ_m is the molar susceptibility, T the absolute temperature, and 100 the diamagnetic correction).

Table 2.3 Experimental (column 2) and calculated magnetic moments in Bohr magnetons: spin only moment $[\{4s(s + 1)]\}^{1/2}$, column 3]; spin and orbital moment $[\{4s\,(s + 1) + l(l + 1)]\}^{1/2}$, column 4].

Compound	μ exp.	μ spin only	μ spin+ orb.	μ obsd. range
$K_2MnFe(CN)_6$	6.05	5.92	5.92	5.9
$K_2CoFe(CN)_6$	4.90	3.87	5.20	5.0–5.2
$K_2CuFe(CN)_6$	1.85	1.73	3.00	1.9–2.0
$KCoCo(CN)_6$	5.30	3.87	5.20	5.0–5.2
$KCoCo(CN)_6.2H_2O$	5.30	3.87	5.20	5.0–5.2
$KNiCo(CN)_6$	3.10	2.83	4.47	3.2

One can realize that the observed moments agree with those calculated on the assumption that each substance contains a ferro- or cobalticyanide ion in which the transition element is forming d^2sp^3 bonds to the cyanides; the other heavy metals exist as "simple ions" as proposed by Pauling and that "the experimental values are in accord with previous observations of the magnetic moments of these ions; manganese, nickel, and copper have very little orbital moment, while that of cobalt is nearly unquenched." Other hypotheses by the authors failed to reproduce the data.

The conclusion is about structure and bonding: The results are in agreement with the assumption that covalently bonded ferrocyanide or cobalticyanide ions are present in the compounds containing ferrous or cobaltic ions, respectively. All other heavy metals are present as "simple ions."

A new significant step occurred much later, in 1956, with two companion articles in *Physical Review*. The first one is only half a page and announces the second article with more detailed results. The goal of the authors is to explore new ferromagnetic materials.

The paper of Holden et al. (Holden et al., 1956) reports that the authors "have observed large positive susceptibilities and ferromagnetic remanence at liquid helium temperatures in a number of complex cyanides of iron transition group metals." "The field used in the measurement of susceptibility was about 10 oersteds. The maximum moment was about 0.2 to 2/cc. For all the substances in Table, the observed remanence was about $B_r \simeq 5$ gauss." Their data are reported in Table 2.4.

Table 2.4 Absolute temperature of maximum susceptibility of cyanides.

	K^+	Zn^{++}	Cr^{+++}	Mn^{++}	Fe^{+++}	Fe^{++}	Co^{++}	Ni^{++}	Cu^{++}
Ferricyanide $(Fe(CN)_6)^{-3}$	0			2°	15°	3.5°	12°	19°	11°
Ferrocyanide $(Fe(CN)_6)^{-4}$			2°		3.5°				
Cobalticyanide				0	0				
Manganicyanide	0	0	0	3 ± 5°		0	16°	17°	0
Chromicyanide		0		0	0	0	0	0	0

The authors are respected scientists of the famous Bell Telephone Laboratories at Murray Hill. One of them, Anderson, received later the Nobel Prize in Physics. Their paper is rather surprising: (i) The authors write that their apparatus, previously described, is "incapable of precision measurements on ferromagnetic substances, both because of the difficulty of calibration"; (ii) the table reports a parameter (the temperature of maximum susceptibility) which is not fully pertinent to characterize 3D order in magnetic materials (the 3.5 K was considered as T_C by (Mayoh and Day,

1974)); (iii) the authors describe clearly the synthesis of the compounds but not the elemental analysis and they write: "The existence, as chemical individuals, of some of the substances ostensibly produced in these reactions has been questioned. It is also well known that such precipitates often contain potassium, and that the valence states of the heavy metal ions are ambiguous. This is particularly so for Prussian and Turnbull's blues: $(Fe^{+3})_4[(Fe(CN)_6)^{-4}]_3$ and $(Fe^{+2})_3[(Fe(CN)_6)^{-3}]_2$." The preceding discussions on the Prussian/Turnbull problem are ignored. The paper brought nevertheless two good news: Physicists of magnetism begin to be interested by Prussian blue analogues and they are able to measure magnetism down to 1 K.

The companion paper by Bozorth et al. (Bozorth et al., 1956), from the same laboratory, describes the magnetization of "rare-earths orthoferrites" and "iron-group cyanides." It suffers the same problems for the characterization of the materials but reports useful data, not only for ferrites, which became later widely used magnetic materials, but also for compounds of interest for us, Prussian blue analogues.

The apparatus used is described: "A null method of measurement is used. Movement of a pendulum in a field gradient is detected with strain gauges and balanced with a coil. In the space containing the specimen and the balancing coil the gradient varies linearly with distance" (Fig. 2.6).

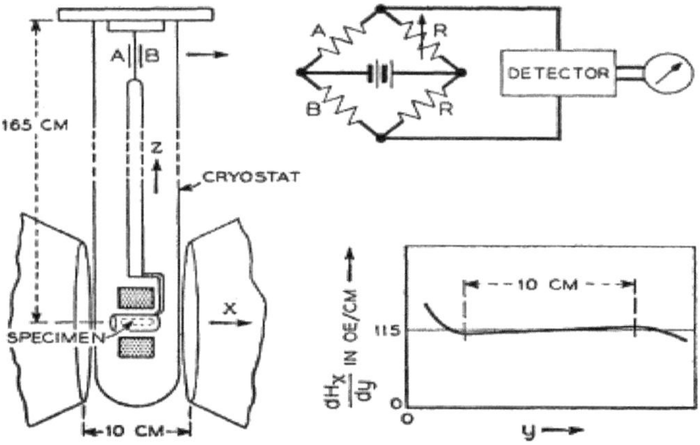

Figure 2.6 Schematic diagram of apparatus. Tapered pole faces cause movement of pendulum into plane of paper. Strain gauges (A,B) are on strip lying in plane of paper, 90° to position shown. Reproduced with permission from Bozorth et al. (1956).

The analyses for PB analogues (but not for Prussian blue), are honestly presented in a table (see Table 2.5). At least for chemists, they are frightening. The so-called remainder "inert material" (besides the proposed formula) is varying from a minimum of 15.3% to 44.9%, which means that the measured samples contain a non-negligible amount of inert impurities.

Table 2.5 Analysis of ferricyanides, expressed as $M_aFe_bC_6N_6$ + remainder (inert material). Ferromagnetic moments and Curie points

M ion	*a*, atoms of M	*b*, atoms of Fe	Percent remainder	n_B per C_6N_6	θ, K
Mn	2.27	0.99	24.5	3.03	11
Co	1.52	0.91	30.7	0.93	21
Ni	1.72	1.03	44.9	2.78	27
Cu	1.86	0.96	28.8	0.69	16
Zn	1.64	1.05	15.3	0.08	...

The magnetic results are gathered in Table 2.6.

With our present understanding of the systems, it is possible to recognize in the tabulated "$Fe^{III}C_6N_6Fe^{III}$" the ferri-ferricyanide (Berlin green), known to be unstable and in "$Fe^{III}C_6N_6Fe^{II}$", Prussian blue. Furthermore, the article gives the corresponding figures of thermal variation of the magnetization. We reproduce here two of them (Fig. 2.7).

Table 2.6 Ferromagnetic moments, n_B in Bohr magnetons per C_6N_6 group, and Curie points in K

	Fe^{III}		Mn^{II}		Fe^{III}		Co^{II}		Ni^{II}		Cu^{II}		Zn^{II}	
	n_B	θ	n_B	θ	n_B	θ	n_B	θ	n_B	θ	n_B	θ	n_B	θ
$Cr^{III}C_6N_6$							0.3	13						
$Mn^{III}C_6N_6$			0.8	50	0.85	10	0.4	21	0.7	37	0?	.....	0	....
$Fe^{III}C_6N_6$	3.85	22	1.9	11	3.5	8	0.7	21	1.15	27	0.45	16	0	....
$Co^{III}C_6N_6$			2.1	5	0.8	5	0.3	3	....		0?	5?	....	....

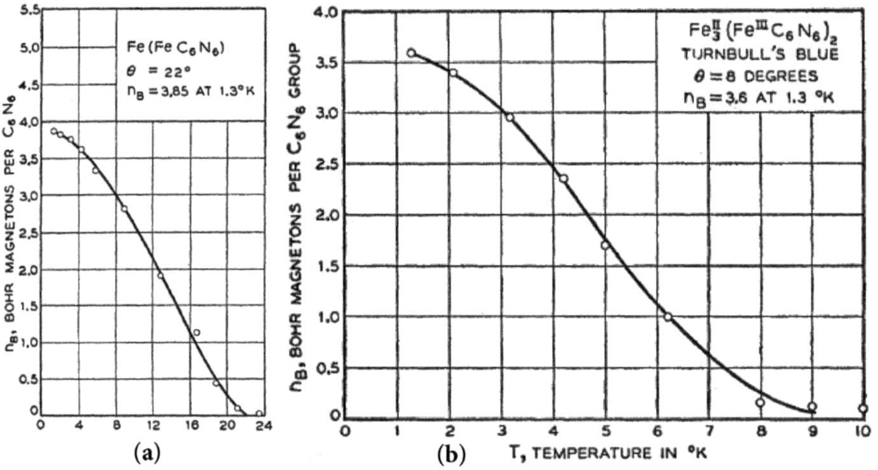

Figure 2.7 Ferromagnetism of several cyanides: (a) Berlin green; (b) Turnbull blue. Reproduced with permission from Bozorth et al. (1956).

These results are much more informative than the preceding ones. The ferromagnetic behaviour is well established. Approximate Curie temperatures and Bohr magnetons are given, even if, with the proposed stoichiometry, they are still far from the expected exact values. Of special interest for us is the indication that "The same result (as Turnbull blue) is obtained when a ferric salt is added to a potassium ferrocyanide; thus Turnbull's blue and Prussian blue seem to have identical ferromagnetic behaviour. Previous work (Weiser et al., 1942) indicated that they have the same composition and structure. However, as noted in a table (Table 2.6 above), the Curie constants of these materials are not the same." Once more, the fact that the compounds are not really pure impedes to reach a firm conclusion. We have made progress in the knowledge of magnetic PBs but several questions still remain unsolved.

2.3.2 First Mössbauer Measurements (1962–1968)

"The most compelling evidence for the formulation of Prussian blue as ferric ferrocyanide rests on the results of the many

Mössbauer spectral studies of this material." Indeed, Mössbauer spectroscopy observed for the first time in 1957 by R. Mössbauer (who received the Nobel Prize in Physics in 1961) (Mössbauer, 1958) allows to characterize readily the surrounding of iron in a material (oxidation state and spin state). The first most conclusive article is attributed to Ito et al., who recorded the Mössbauer spectra in the temperature range 1.6–300 K (Ito et al., 1968). The first spectra of Prussian blues were published some years earlier by Epstein in 1962 (Epstein, 1962) together with the spectra of the possible components of Prussian blues Fe^{2+}, Fe^{3+}, $\{Fe^{II}(CN)_6\}^{4-}$, and $\{Fe^{III}(CN)_6\}^{3-}$—compulsory to interpret the PB spectra.

The data obtain by Epstein are given in Fig. 2.8.

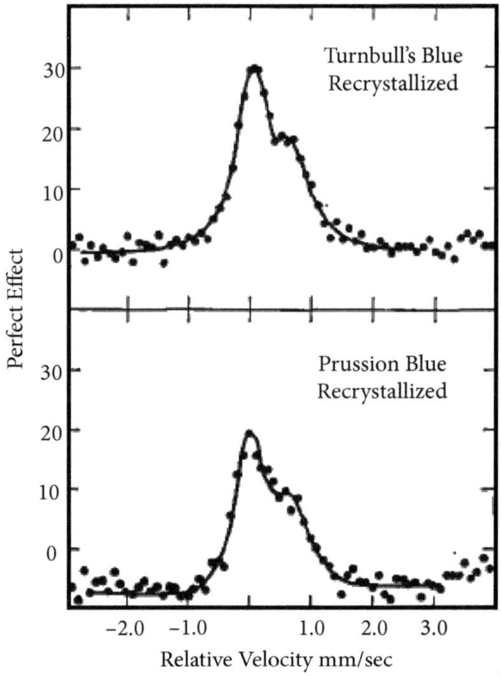

Figure 2.8 Comparative Mössbauer spectra at room temperature, Prussian blue, Turnbull's blue. Reproduced from Epstein (1962).

They are recorded on recrystallized samples and a home-made spectrometer at room temperature. The spectra are not enough resolved to bring important information. For the ferric ferrocyanide, "although the presence of two close peaks makes their positions somewhat uncertain, the results of several runs confirmed the anomalous δ of the iron cyanide ion so that it cannot be positively identified as either iron II or III." ... "The same complete spectrum was obtained for recrystallized Turnbull's blue (our own preparation), which is nominally ferrous ferricyanide, but which shows no ferrous peaks. In both cases, also, the ferric peaks are anomalously low as shown in Figure." A table compares the quadrupole splitting splitting ΔE and isomer shift δ for ferro- and ferricyanides, ferrous and ferric salts. For $Fe_4[Fe(CN)_6]_3$, the authors report no ΔE and δ = −0.02 for the $[Fe^{II}(CN)_6]$ peak and +0.55 for Fe^{III} peak. One year later, Duncan and Wigley published and compared a series of Mössbauer spectra of samples prepared by themselves from fully reduced $Fe^{II}Fe^{II}$ species to fully oxidized $Fe^{III}Fe^{III}$ ones (Duncan and Wigley, 1963). With spectra considered as excellent or good, their main conclusions are that "there is an observable difference between the two types of iron atom in any particular compound. The cation is invariably spin-free (*that is high spin*), and the anion spin-paired (*that is low spin*). In the oxidation of K,Fe[Fe^{II}(CN)], only the cation changes its valency state." In the same year, Fluck et al. published a review on the Mössbauer effect and its significance in chemistry with a substantial account on Prussian blues (Fluck et al. 1963): "The positions of the lines in these spectra make it possible to draw conclusions about the nature of the atom's bonds. The efficiency of this method is demonstrated by illustration with numerous iron compounds." The article gives an impressive list of Mössbauer parameters of quadrupole splittings ΔE and isomer shifts δ. As for Prussian blue, the authors display the spectra of three blues (Prussian, soluble and insoluble, Turnbull (Fig. 2.9). They decompose the spectrum in three curves: 1 corresponds to the ferrocyanide ions, present in the three compounds, and the doublet 2 + 3 to the Fe^{3+} ions. The areas below the curves 1 and (2 + 3) allows to determine the ratio Fe^{II}:Fe^{III} found to be close to 3:4, for insoluble Prussian and Turnbull blues and 1:1 for soluble Prussian blue, which corresponds to the stoichiometries of the compounds. Hence the conclusions: The iron atoms in "insoluble Prussian Blue have well-defined oxidation levels. "Turnbull's Blue" shows the identical spectrum. "The deep color of Prussian Blue cannot therefore be attributed to oscillating valencies" reaching the same conclusions as Robin (see Section 2.4).

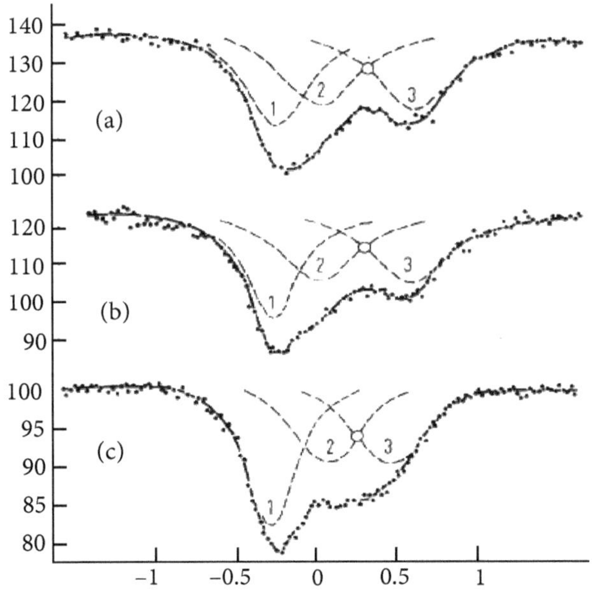

Figure 2.9 Mössbauer spectra of (a) insoluble Prussian blue, (b) Turnbull's blue, (c) soluble Prussian blue. Ordinate: transmission (u.a.), Abscissa: velocity of the source relative to the absorber (mm·sec^{-1}). Reproduced from Fluck et al. (1963), with permission.

Five years later, Ito and coworkers published Mössbauer studies (temperature range 1.6–300 K) and magnetization data (1.6–25 K) on the same three blues, either industrial or carefully home-prepared. Figure 2.10 displays the beautiful Mössbauer spectra at 1.6 K. "The spectra are well resolved and show a superposition of a hyperfine split level and a single line for all the compounds", that is what is expected when the compounds are in the ordered ferromagnetic state.

Ito's main conclusions are as follows: On the one hand, "the results below their Curie temperatures of 5.5 ± 0.5 K show that all of them are ferric ferrocyanide: One kind of irons is Fe^{3+} with high spin and the other is Fe(II) with low spin. It is concluded that Prussian blue and Turnbull's blue have the same electronic structure." On the other hand the results "seem to be consistent with the chemical formulas $KFe[Fe(CN)_6]$ for soluble Prussian blue and $Fe_4[Fe(CN)_6]_3$ for both insoluble Prussian blue and Turnbull's blue".

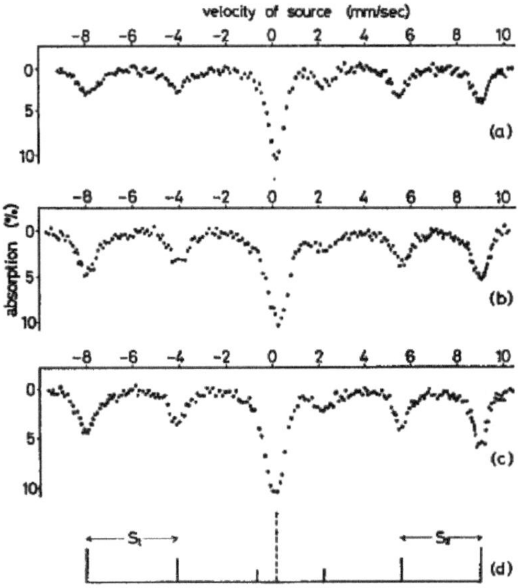

Figure 2.10 Mössbauer spectra at 1.6 K for (a) soluble Prussian blue; (b) insoluble Prussian blue; (c) Turnbull's blue. In (d) are shown the line positions for Fe^{II} (dotted line) and Fe^{III} (solid line). Reproduced with permission from Ito et al. (1968).

Most of the observations necessary to establish the nature of Prussian blues were therefore gathered after Mössbauer Ito's study. More generally, such a study points out the importance of the use of Mössbauer spectroscopy in molecular magnetism as often underlined by Gütlich (2012).

In the following, we reach another step, the *understanding* of two important related properties of PBs: colour and magnetism.

2.4 The Key for Understanding Colour and Magnetism in Prussian Blue: Electron Transfer

Prussian blue can be described as a class II mixed valency compound in the Robin and Day classification with "valences firmly trapped" (Robin and Day, 1967). Three articles can be considered to have contributed historically to an explanation of the colour and of the magnetism of Prussian Blue: a pioneering paper by Robin (Robin, 1962), who tackled the electronic structure of Prussian blue and electron transfer through a molecular orbital (MO) approach, followed by a series of two articles by Day and Mayoh about the calculation of the coupling integrals between donor and acceptor ions to determine (i) the metal to metal charge transfer band intensities (Mayoh and Day, 1974); and (ii) the ferromagnetic coupling energy due to mixed valence interaction (Mayoh and Day, 1976).

2.4.1 Electronic Structure of Prussian Blue

Robin starts his analysis by summarizing the previous results (face cubic centred structure of PB, existence of two different surroundings for the Fe ions: $Fe^{II}(CN)_6$ and $Fe^{III}(NC)_6$ with five unpaired electrons, and paramagnetism in the range 200–300 K. So that it is possible to "draw an orbital diagram using the orbitals of $Fe^{II}(CN)_6$ and $Fe^{III}(NC)_6$ ions taken separately" (Robin and Day, 1967). The proposed orbital diagram is shown in Fig. 2.11.

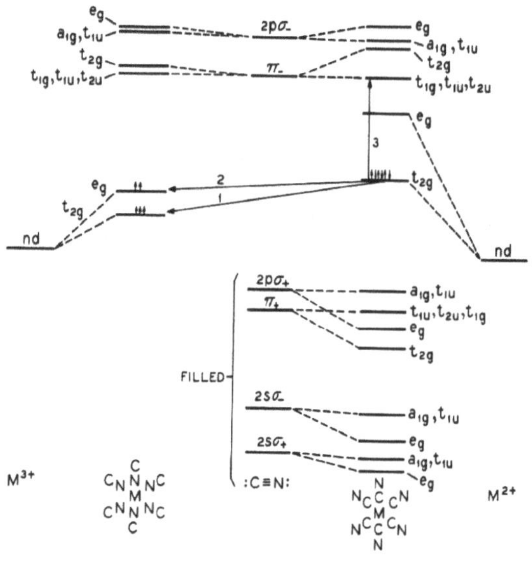

Figure 2.11 "A molecular orbital scheme for class II Prussian blue. The symmetry labels on any column of levels are appropriate only for the molecule listed at the foot of that column and the transitions labelled 1, 2 and 3 correspond to the bands labelled in the same way in Fig. 2.12. Reproduced from Robin and Day (1967).

In the figure, from left to right are shown the energy levels of d orbitals of Fe^{3+}, the MOs of high spin $Fe^{III}(NC)_6$ weakly splitted by the $Fe(NC)_6$ octahedral crystal field, t_{2g}^3, e_g^2 configuration, at the centre, the MOs of the ligand cyanide, then the MOs of low spin $Fe^{II}(CN)_6$ strongly splitted by the $Fe(CN)_6$ octahedral crystal field, t_{2g}^6, e_g^0 configuration and finally on the right, the d orbitals of Fe^{2+}. Relying on previous spectroscopic data of hexacyanidoferrates (II) and (III), their interpretation by Tanabe and Sugano (Tanabe and Sugano, 1954), Robin justifies the electronic configuration sketched in Fig. 2.11 by electron repulsion effects.

2.4.2 Mixed Valency, Electron Transfer and Electronic Spectrum

It is necessary at this point to recall the basics of the theory of mixed valence effects, even if their actual formulation came after the paper by Robin. We summarize the detailed demonstrations of reference (Robin and Day, 1967). Another approach can be found in chapter 3 of reference (Launay & Verdaguer 2018). They rely first of a simple one-electron description of electronic structure. In the case of PB, where the central Fe^{II} ion A (ψ_A^{II}) is surrounded by 6 Fe^{III} ions B, numbered from $i = 1$ to 6 (ψ_{Bi}^{III}), located at the same distance d, it is possible to write the zeroth-order wavefunction of the ground state as

$$\Psi_0' = \psi_A^{II} \prod\nolimits_{i=1,6} \psi_{Bi}^{III}. \tag{2.7}$$

When one electron is transferred from the A site, becoming $Fe^{III}(\psi_A^{II})$ to one of the B sites (ψ_{Bj}^{III}), the excited wavefunctions become

$$\psi_k = \psi_A^{III} \sum\nolimits_j c_{kj}\, \psi_{B1}^{III} \psi_{B1}^{III} \psi_{B2}^{III} \cdots \psi_{Bj}^{III} \cdots \psi_{B6}^{III}. \tag{2.8}$$

They are six such functions. The c_{kj} coefficients are chosen to normalize Ψ_k and ensured that Ψ_k belongs to a representation of the symmetry point group. If one of the Ψ_k (let us say Ψ_l) has the same symmetry as Ψ_0', the ground state at first order is obtained by configuration interaction, mixing Ψ_0' and Ψ_l with the mixing matrix element $\langle \Psi_0' | V | \Psi_l \rangle$:

$$\Psi_0 = (1-\alpha^2)^{1/2}\,\Psi_0 + \alpha\Psi_l \quad (2.9)$$

When it is possible to describe ψ_A^{III} as an electron core (ϕ_A^C) and one electron outside the core (ϕ_A^*), that is $\psi_A^{III} = \phi_A^C\phi_A^*$, and the same for $\psi_{Bj}^{II} = \phi_{Bj}^C\phi_{Bj}^*$, (the cores supposed to be closed-shell), Ψ_0 can be written:

$$\Psi_0 = (\kappa/N)\,[(1-\alpha^2)^{1/2}\phi_A^* + \alpha\,\textstyle\sum_j c_{lj}\,\phi_{Bj}^*] \quad (2.10)$$

where κ is the product of all the core-shell wave function ϕ_A^C and ϕ_{Bj}^C; N is a normalization factor; Ψ_0 appears as the wave function of the excited (optical) electron; α is derived from the following expression:

$$\alpha^2 = N^2\,[1 - \{1 + [2E_l^2 - 2E_l\,(E_l^2 + 4V^2)^{1/2} + 4V^2]/4V^2\}^{-1}] \quad (2.11)$$

The values of E_l and α allow an easy distinction between the three Robin and Day classes of mixed valency systems: If E_l tends to zero, α and $(1-\alpha)^{1/2}$ tends to $N\sqrt{2}/2$ (class III). When E_l is very large, α tends to 0; $(1-\alpha)^{1/2}$ tends to 1; Ψ_0 tends to Ψ_0', the zeroth-order wavefunction, there are no interaction and localized valences (class I). In between one finds the class II situation with small E_l and finite value of α.

The mixed valence spectra corresponds to transition between the (first order) ground state to the excited state, after electron transfer:

$$(\kappa/N)\,[(1-\alpha^2)^{1/2}\,\phi_A^* + \alpha\textstyle\sum_j c_{lj}\,\phi_{Bj}^*] \rightarrow (\kappa/N)\,[\beta\phi_A^* + (1-\beta^2)^{1/2}\textstyle\sum_j c_{mj}\,\phi_{Bj}^*], \quad (2.12)$$

where β is the excited state equivalent of α for the ground state. When α and β are small, the electronic spectrum comprises the transitions specific to isolated A and B and the ones of mixed valence spectrum. Their intensity is of peculiar interest. It is proportional to the square of the transition moment integral μ. For allowed transitions, $\sum_j c_{mj}\phi_{Bj}^*$ is of different symmetry of $\sum_j c_{lj}\phi_{Bj}^*$ and is zero:

$$\mu = \text{Cte} \int [(1-\alpha^2)^{1/2} \phi_A^* + \alpha \sum_j c_{lj} \phi_{Bj}^*] \mathbf{M} \sum_j c_{mj} \phi_{Bj}^* d\tau, \tag{2.13}$$

where **M** is electric moment operator and the constant takes into account the orbital degeneracies. The x component of μ can be written as

$$\mu_x = e\alpha \, \text{Cte} \int \sum_j c_{lj} \, \phi_{Bj}^* \, ex \sum_j c_{mj} \phi_{Bj}^* \, d\tau \tag{2.14}$$

$$\mu_x = e\alpha \, \text{Cte} \sum_j c_{lj} c_{mj} x_j, \tag{2.15}$$

where x_j is the x coordinate of the j site. Similar expressions run for y and z. With a system of regular geometry, the coefficients are determined by symmetry and x_j by the crystal structure, α is reached from the usual relation:

$$\mu^2 = 9.17 \; 10^{-4} \int \varepsilon_\lambda \; d \log\lambda,$$

where ε_λ is the molar absorption coefficient at wavelength λ and d is the distance between A and B centres where the electron transfer occurs.

The electronic spectrum or Prussian blue is reported in Fig. 2.12.

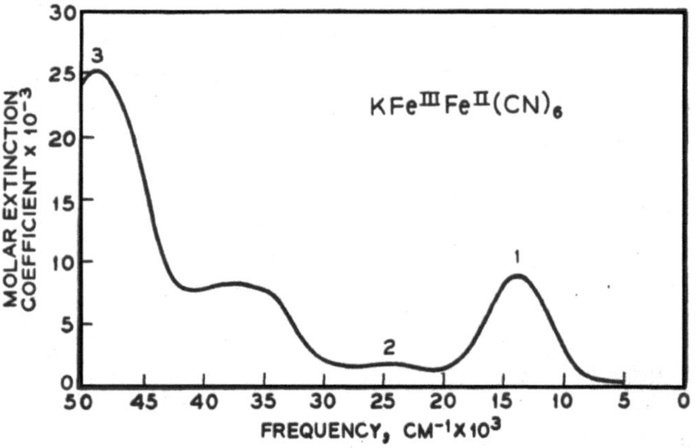

Figure 2.12 The electronic spectrum of soluble Prussian blue $KFe^{III}[Fe^{II}(CN)_6]$ as a colloidal dispersion in water. Reproduced with permission from Robin (1962).

After careful analyses and discussion, Robin proposes to assign the three labelled electronic transitions in the visible spectrum (Fig. 2.12) as shown in Fig. 2.11 and Table 2.7.

Table 2.7 Robin's analysis of the electronic spectrum of soluble Prussian blue

	Energy/cm^{-1}	ε	Transition
1	14,100	10,000	$Fe^{II}(t_{2g}) \rightarrow Fe^{III}(t_{2g})$
2	24,500	1,000	$Fe^{II}(t_{2g}) \rightarrow Fe^{III}(e_g)$
-	33–43,000	8,000	CN (π) $\rightarrow Fe^{III}(t_{2g})$
3	48,000	25,000	$Fe^{II}(t_{2g}) \rightarrow$ CN (π*, t_{1u}, t_{2u})

Transition 1 is the allowed mixed valence transition at the origin of the blue colour of PB. Transition 2 corresponds to a forbidden transition to $Fe^{III}(e_g)$ orbitals [see also (Mayoh & Day, 1974)]. Transition 3 is the allowed metal to ligand charge transfer transition, also present in $Fe^{II}(CN)_6^{4-}$.

For transition 1, at the maximum of absorption the energy is 14,700 cm^{-1}, the molar absorption coefficient ε_λ is 9,800, and the transition moment μ is 1.23 Å. The computed value of α^2 is then 0.01. The physical meaning is that the optical electron resides 99% of its time on the $Fe(CN)_6$ site (that is $KFe^{III}[Fe^{II}(CN)_6]$) and 1% in the $Fe(NC)_6$ site (that is $KFe^{II}[Fe^{III}(CN)_6]$). In other words, the ground state of Prussian blue is made from 99% of Prussian blue and 1% from Turnbull blue. Amazingly, Turnbull blue is reappearing as an excited state of Prussian blue! Robin points out also that this finding invalidates the earlier attempts to explain the colour of Prussian blue by a "resonance" or "oscillation of valence" between the two structures: $Fe^{II}(CN)_6Fe^{III} \leftrightarrow Fe^{III}(CN)_6Fe^{II}$.

The contribution by Robin shows that the ambiguous formulation of Prussian blue as either the ferric salt of ferrocyanic acid or as the ferrous salt of ferricyanic acid can be resolved by applying simple orbital theory to the explanation of the Prussian blue electronic spectrum.

Robin's analysis was followed later by Day and Mayoh, who proposed a perturbative method to provide a quantitative "estimate of the coupling integrals between donor and acceptor ions which are of fundamental importance in determining metal $\leftrightarrow$ metal charge transfer band intensities, mobilities of carriers in transport processes, and the ferromagnetic coupling energy due to mixed valence "interaction".

In their computation of the electronic spectrum, the level of agreement between observed and calculated transition moments for the metal → metal charge transfer excitations is most encouraging. They confirm in particular the 1% residence time of the optical electron in the $Fe(NC)_6$ site. They open the way to the explanation of the ferromagnetic interaction in PB, tackled in Mayoh and Day (1976).

2.4.3 Mixed Valency and Ferromagnetism

The need for a new theoretical model to understand the ferromagnetism of PB becomes obvious when observing that (i) PB presents the electronic configuration shown in Fig. 2.13: The two paramagnetic ferric ions are located at 10.18 Å apart and interact through a long diamagnetic NC-Fe^{II}-CN bridge. With strictly localized spins, the interaction is expected to be very weak and antiferromagnetic, thanks to the overlap of magnetic orbitals of the same symmetry on the two ferric ions.

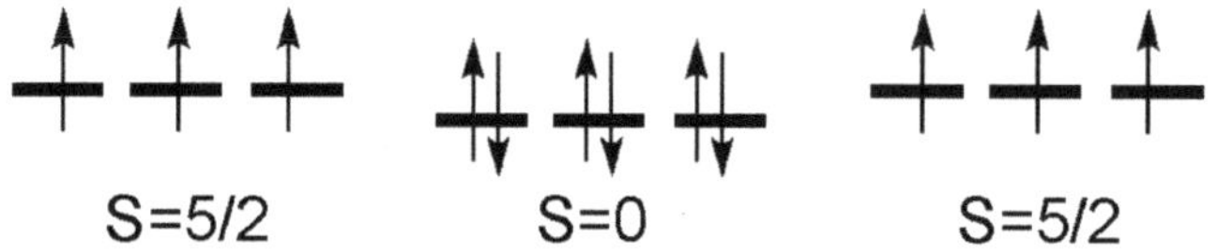

Figure 2.13 Geometry, bonding and spin structure along a Prussian blue edge.

Prussian blue is an intrinsic semiconductor having a room temperature resistivity around 10^7 ohm·cm. Then, the double exchange Zener model explaining the ferromagnetism of *metallic* compounds by the conducting electrons moving with unchanged spin in a network of fixed atoms having parallel spins is inapplicable. The model by Robin and Mayoh follows the mixed valency scheme described in Section 2.4.3: configuration interaction between the zeroth-order ground state (Eq. 2.7) and the first electron transfer excited state (Eq. 2.8). Figure 2.14 displays the ground and the excited states in the case of two neighbouring Fe ions, Fe^{II} and Fe^{III}.

The easiest electron transfer occurs through an optical electron not changing its spin in the transfer. Its spin orientation ($\beta, \downarrow$) is opposite to the five $\alpha, \uparrow$ electrons on Fe^{III} since no more $\alpha, \uparrow$ electron can be accommodated on the high spin Fe^{III}, to please Wolfgang Pauli. The excited spin state is $S_T = 5/2$.

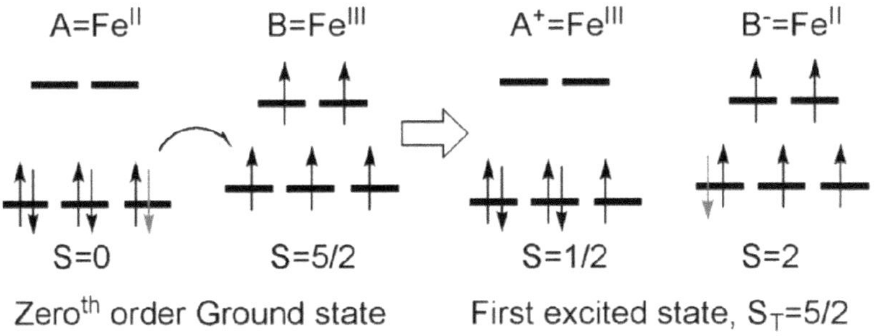

Figure 2.14 Scheme of electron transfer (β spin, grey) from a t_{2g}^6 low spin Fe^{II} ion A to a neighbouring $t_{2g}^3 e_g^2$ high spin Fe^{III} B, leading to a ferromagnetic interaction between Fe^{II} and Fe^{III}.

In the case of soluble Prussian blue, six peripheral Fe^{III} ions B_i are around the central Fe^{II} ion A as shown in Fig. 2.15.

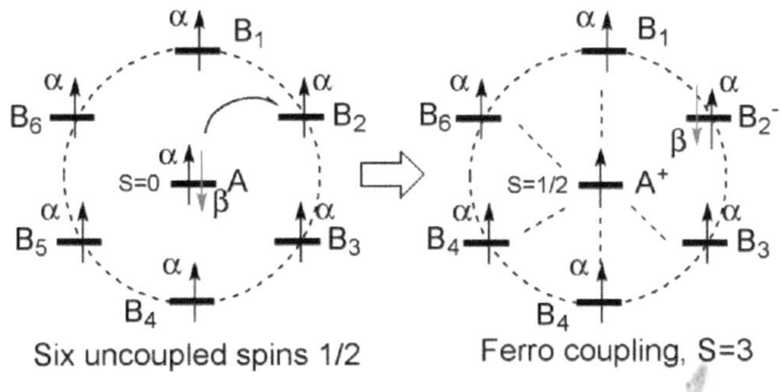

Figure 2.15 Schematic model of an electron transfer in a soluble Prussian blue where six paramagnetic ions B_i (i = 1–6) surround the diamagnetic central ion A at equal distances (dotted circle). Right: ground state. Left: excited state after electron transfer. The dotted lines emphasize the ferromagnetic interaction between the central and the peripheral spins.

The β spin is transferred to one of the B_i sites (here B_2^-). After electron transfer, in the excited state the central A^+ ion ($S = 1/2$) interacts with the $S = 1/2$ of the other B_i leading to a ferromagnetic interaction. The wave function Ψ_1 of the excited state is a combination of six such wave functions of appropriate symmetry (Eq. 2.8) and provides the improved ground state by configuration interaction (Eq. 2.9).

Day and Mayoh calculate the energy E_F of stabilization of the ferromagnetic state (the one of 6α parallel spins, ferromagnetically coupled) compared to the paramagnetic situation of six uncoupled spins. They define different spins configurations (a) 6α; (b) $5\alpha, 1\beta$; (c) $4\alpha, 2\beta$; (d) $3\alpha, 3\beta$; their equivalents when inverting α and β; their degeneracies ω_i, their wavefunctions Ψ_{Gi} (i = a–d) and their energies E_{Gi}

$$E_{Gi} = \langle \Psi_{Gi} | \mathbf{H} | \Psi_{Gi} \rangle$$

They compute the E_{Gi} as a function of c, the valence delocalization coefficient and the usual two-centres repulsion integrals J (Coulomb) and K (exchange) between central and peripheral orbitals for the configurations i = a – d. They arrive at the following expression of the energy E_F:

$$E_F = \sum_{i = b,c,d} (E_{Ga} - E_{Gi})\omega_i = (1/8)(5J_{11} + 10K_{12})c^4 \tag{2.16}$$

The remaining of the calculation is a numerical evaluation of J, K and c; which depends on the geometry and of the theoretical description of the orbitals. The resulting estimates of J_{11} and K_{12} are 11.2 eV and 0.54 eV. The ferromagnetic coupling energy E_F, from (Eq. 2.16) is therefore equal to 7.675 c^4 eV. The application of the simple relation between E_F and T_C:

$$E_F = kT_C, \tag{2.17}$$

where k is the Boltzmann constant, provides $T_C = 8.9 \times 10^4\ c^4$ K. For Prussian blue, c was "estimated from the intensity of the mixed valence band as 0.106 while the estimate based on the perturbation calculation was 0.083. The former value leads to a calculated Curie temperature of 11.2 K and the latter to 6.6 K". The agreement with the experimental value is very remarkable.

"We have demonstrated not only that the mixed-valence delocalization mechanism indeed predicts ferromagnetic coupling but, further, that the extent of ground-state valence delocalization as estimated from the optical spectrum or a perturbation calculation leads to a calculated Curie temperature very close to that observed". The same model was applied on remarkable Prussian blue molecular analogues $\{Fe^{II}(CN\text{-}Fe^{III}L)_6\}$ by Rogez, Mallah et al. (Rogez et al., 2000).

2.5 Magnetism of Bimetallic Prussian Blue Analogues

We have already quoted the studies by Richardson, Holden and others related to bimetallic magnetic Prussian blue analogues where the PB Fe^{3+} ion is replaced by divalent transition metals A^{2+} and $[Fe^{II}(CN)_6]$ is substituted by other hexacyanidometalates $[B^{III}(CN)_6]$, for example B = Co^{III}, Cr^{III}. The synthesis of such compounds and the understanding of the exchange interaction between A and B components was part of the work of the research groups of Babel, Girolami and mine and has been reviewed (Babel, 1986; M. Verdaguer et al., 1999b; Verdaguer & Girolami, 2005). The reader is invited to consult these references. I would like here to point out only two points:

(i) In most of these compounds, magnetism can be understood thanks to a simple Heitler and London model proposed by Kahn and Briat for *localized* electrons. In this way, the exchange interaction can be interpreted in first approximation by the concepts of magnetic orbitals (singly occupied molecular orbitals) and two electrons exchange pathways, which can be ferromagnetic (f) if the orbitals are orthogonal or antiferromagnetic (af) if the orbitals are overlapping. The coupling J between spins S_A and S_B ($\mathbf{H} = -J\, \mathbf{S}_A \cdot \mathbf{S}_B$ is then the sum of af and f terms $J = J_{AF} + J_F$. The high T_C PB analogues were obtained in such a way. For details, the reader is referred to Section 2.7.3 of the recent reference (Launay & Verdaguer, 2018).

(ii) The T_Cs depends not only on the electronic structure of A and B (Fig. 2.14) determining the J value but also on the stoichiometry and on the number z of the magnetic neighbours of A (the number of neighbours of $[B^{III}(CN)_6]$ is always 6).

$$T_C \propto z\,|J| \tag{2.18}$$

Here we find the importance of the intrinsic vacancies of the Ludi and Güdel "modified model" of PB and of the coordinated and zeolithic water molecules. Equation 2.18 holds quite well for $\{Ni_4[Cr(CN)_6]_{8/3}\square_{4/3}\}^0$ (z = 4, T_{C1} = 60 K) and $\{K_4Ni_4[Cr(CN)_6]_4\square_0\}^0$ (z = 6, T_{C2} = 90 K, T_{C2}/T_{C1} = 1.5). In the case of our high T_C Prussian blue analogue $\{V_4[Cr(CN)_6]_{8/3}\square_{4/3}\}^0\cdot$14-16$H_2O$, there are z = 4 neighbours of V, T_{C1} = 315 K. Instead in Holmes' and Girolami's $\{K_4V_4[Cr(CN)_6]_4\square_0\}^0\cdot 8H_2O$, there is less water, z = 6 and T_{C2} = 376 K. T_{C2}/T_{C1} = 1.2 instead of the expected 1.5 but playing on stoichiometry was a winning strategy to get the record T_C for PBs, above the boiling point of water.

2.6 Light and Magnetism: Photomagnetism of Prussian Blue Analogues

Another interesting and more recent piece of work related to magnetic Prussian blue analogues and pointing out the crucial role of vacancies in PBs came to light in 1996 when Hashimoto and coworkers published in *Science* their observations on the photoinduced magnetization of a Prussian blue analogue, $K_{0.2}Co_{1.4}[Fe(CN)_6]\cdot 6.9H_2O$, that is $K_{0.6}Co_4[Fe(CN)_6]_{2.9}\cdot 19H_2O$, in our formulation (Sato et al., 1996).

I commented this important result in a news and views (Molecular electronics emerges in molecular magnetism) (Verdaguer, 1996). Many researchers engaged later in this new chapter of photomagnetism to prepare new compounds and to understand the phenomenon. Basically, the photon induces an electron transfer from the diamagnetic Fe^{II} to diamagnetic Co^{III} present in the compound and transform them in a paramagnetic pair $Fe^{III}CNCo^{II}$. The increase of the number of magnetic neighbours enhances the T_C value.

$$Co^{III}\text{-NC-}Fe^{II}(CN)_5 \xrightarrow{\text{light}} Co^{II}\text{-NC-}Fe^{III}(CN)_5 \qquad (2.19)$$

An example of the spectacular photo-induced transformation of an almost diamagnetic Co-Fe cubic nanoparticle to an ordered ferrimagnet, the reader can consult Section 4.5.3.2 in (Launay & Verdaguer 2018). Two remarks can be made on the flexibility of the Co-Fe systems. In the one hand, the amount of the alkali ions allow to control the stoichiometry of the compound, the number of vacancies, the ligand field around the cobalt and therefore the number of diamagnetic pairs $Fe^{II}CNCo^{III}$ which can be phototransformed. As Mössbauer spectroscopy for mixed valency PBs, X-ray absorption (EXAFS, XANES, and XMCD) played a crucial role to identify the oxidation and spin states of the metallic ions in bimetallic systems and the photomagnetic mechanisms as shown in this volume by Bleuzen and coworkers. On the other hand in the photomagnetic process, low spin Co^{III} is transformed to high spin Co^{II}, the Co-ligand distances increase strongly (around 0.1 Å). This is allowed only if the cobalt ion is closed from a vacancy and can expand its coordination sphere with less constraints in the network. When there is no vacancy, in $K_4Co_4^{III}[Fe^{II}(CN)_6]_4 \cdot nH_2O$, the photomagnetic process in blocked.

The field was reviewed several times (Sato, 2003; Sato et al., 2007; Bleuzen et al., 2009), including a recent review on Co-Fe molecular systems, as pieces formally extracted from the cubic PB structure (molecular cubes and squares) (Aguila et al., 2016). See also the chapter by Bleuzen et al. in this book.

2.7 Conclusion

Prussian blues analogues and derivatives build an extensive family of exciting compounds. They are complex systems whose nature, structure and properties were difficult to historically establish since their chemistry, apparently simple, is indeed full of traps. We tried to show in this chapter (i) why the progresses were slow, due to the complexity of the systems, the ignorance of previous works and the non-availability of ad hoc physical techniques; (ii) how the difficulties were overcome by careful and reproducible preparations and analyses together with the appearance of new instrumentation. The present understanding of the PBs systems allow to use Prussian blues and analogues in a more and more rational way in an increase number of scientific fields (besides the initial artistic one): magnetism, electronics (sensors), electrochemistry and electrode materials, reactivity and catalysis, ...

References

Aguila, D., et al., 2016. Switchable Fe/Co Prussian blue networks and molecular analogues. *Chemical Society Reviews*, 45(1), pp. 203–224.

Anonymous, 1710. Notitia Caerulei Berolinensis Nuper Inventi. *Miscellanea Berolinensia ad incrementum Scientiarum*, 1, pp. 377–378.

Babel, D., 1986. Magnetism and Structure: Model Studies on Transition Metal Fluorides and Cyanides. *Comments on Inorganic Chemistry*, 5, pp. 285–320.

Bayou, H., 2008. *Hokusai, l'affolé de son art*, Paris: Musée Guimet, Réunion des musées nationaux.

Beall, G. W., et al., 1977. Crystal Structure of Trimanganese Bis(Hexakis (Cyano)Cobaltate) Dodecahydrate and Tricadmium Bis(Hexakis (Cyano)Cobaltate) Dodecahydrate by Neutron and X-Ray Diffraction. *Inorganic Chemistry*, 16(11), pp. 2715–2718. Available at: http://dx.doi.org/10.1021/ic50177a010.

van Bever, A. K., 1938. The Crystal Structure of Some Ferricyanides with Bivalent Kations. *Recueil des Travaux Chimiques des Pays-Bas*, 57(11), pp. 1259–1268. Available at: http://dx.doi.org/10.1002/recl.19380571108.

Bleuzen, A., et al., 2009. Photomagnetism in Clusters and Extended Molecule-Based Magnets. *Inorganic Chemistry*, 48(8), pp. 3453–3466.

Bozorth, R. M., Williams, H., J., and Walsh, D. E., 1956. Magnetic Properties of Some Orthoferrites and Cyanides at Low Temperatures. *Physical Review*, 103(3), pp. 572–578.

Brown, D. B., Shriver, D. F., and Schwartz, L. H., 1968. Solid-State Reactions of Iron(ii) Hexacyanochromate(iii). *Inorganic Chemistry*, 7(1), pp. 77–83. Available at: http://dx.doi.org/10.1021/ic50059a016.

Brown, J., 1724. Observations and Experiments Upon the Foregoing Preparation. *Philosophical Transactions of the Royal Society of London*, 33, pp. 17–24.

Buser, H. J., et al., 1977. The Crystal Structure of Prussian Blue: $Fe_4[Fe(CN)_6]_3$ xH_2O. *Inorganic Chemistry*, 16, pp. 2704–2710.

Cambi, L., and Clerici, A., 1927. Ferrous-Ferric Cyanides. *Atti accad. Lincei [6]*, 5, pp. 79–84.

Davidson, D., 1937. The Prussian Blue Paradox. *Journal of Chemical Education*, 14(5), p. 238. Available at: http://dx.doi.org/10.1021/ed014p238.

Davidson, D., and Welo, L. A., 1927. The Nature of Prussian Blue. *The Journal of Physical Chemistry*, 32(8), pp. 1191–1196. Available at: http://dx.doi.org/10.1021/j150290a007.

Dunbar, K. R., and Heintz, R. A., 1997. Chemistry of Transition Metal Cyanide Compounds: Modern Perspectives. *Progress in Inorganic Chemistry*, 45, pp. 283–391.

Duncan, J. F., and Wigley, P. W. R., 1963. The Electronic Structure of the Iron Atoms in Complex Iron Cyanides. *Journal of the Chemical Society*, pp. 1120–1125. Available at: http://dx.doi.org/10.1039/JR9630001120.

Emschwiller, G., 1954. Spectres infrarouges de ferrocyanures et de ferricyanures et constitution des bleus de Prusse. *Comptes Rendus Hebdomadaires des Séances de l'Académie des Sciences*, pp. 1414–1416.

Epstein, L. M., 1962. Mössbauer Spectra of Some Iron Complexes. *The Journal of Chemical Physics*, 36(10), pp. 2731–2737. Available at: http://dx.doi.org/10.1063/1.1732360.

Flambard, A., et al., 2017. Probing the Local Structure of Prussian Blue Electrodes by ^{113}Cd NMR Spectroscopy. *Dalton Transactions*, 46(19), pp. 6159–6162.

Flambard, A., Köhler, F. H., and Lescouëzec, R., 2009. Revisiting Prussian Blue Analogues with Solid-State MAS NMR Spectroscopy: Spin Density and Local Structure in $[Cd_3\{Fe(CN)_6\}_2]\cdot 15H_2O$. *Angewandte Chemie, International Edition*, 48, pp. 1673–1676.

Geoffroy, É.-F., 1725. Observations sur la Preparation de Bleu de Prusse ou Bleu de Berlin. *Mémoires de l'Académie Royale des Sciences*, pp. 153–172.

Grandjean, F., Long, G. J., and Samain, L., 2016. Characterization and Utilization of Prussian Blue and Its Pigments. *Dalton Transactions (Cambridge, England: 2003)*, 45(45), pp. 18018–18044.

Gütlich, P., 2012. Fifty years of Mössbauer spectroscopy in solid state research: Remarkable achievements, future perspectives. *Zeitschrift für anorganische und allgemeine Chemie*, 638(1), pp. 15–43. Available at: http://dx.doi.org/10.1002/zaac.201100416.

Herren, F., et al., 1980. Neutron Diffraction Study of Prussian Blue, $Fe_4[Fe(CN)_6]_3 \cdot xH_2O$. Location of water molecules and Long-Range Magnetic Order. *Inorganic Chemistry*, 19(4), pp. 956–959.

Holden, A. N., et al., 1956. New Low-Temperature Ferromagnets. *Physical Review*, 102(6), pp. 1453.

Ito, A., Suenaga, M., and Ono, K., 1968. Moessbauer Study of Soluble Prussian Blue, Insoluble Prussian Blue, and Turnbull's Blue. *Journal of Chemical Physics*, 48(8), pp. 3597–3599.

Keggin, J. F., and Miles, F. D., 1936. Structures and Formulae of the Prussian Blues and Related Compounds. *Nature*, 137, pp. 2704–2710.

Kraft, A., 2008a. On the Discovery and History of Prussian Blue. *Bulletin for the History of Chemistry*, 33(2), pp. 61–67.

Kraft, A., 2008b. On Two Letters from Caspar Neumann to John Woodward Revailing the Secret Method for Preparation of Prussian Blue. *Bulletin for the History of Chemistry*, 34(2), pp. 134–140.

Kraft, A., 2011. "Notitia Coerulei Berolinensis Nuper Inventi" on the 300th Anniversary of the First Publication on Prussian Blue. *Bulletin for the History of Chemistry*, 36(1), pp. 3–9.

Launay, J.-P., and Verdaguer, M., 2018. *Electrons in Molecules*: From *Basic Principles to Molecular Electronics*, Revised soft cover edition, Oxford University Press, Oxford. available at: https://global.oup.com/academic/product/electrons-in-molecules-9780198814597.

Ludi, A., and Güdel, H. U., 1968. Die Struktur der Hydrate von $Co_3[Co(CN)_6]_2$ und $Cd_3[Co(CN)_6]_2$. *Helvetica Chimica Acta*, 51(8), pp. 2006–2016.

Lüdi, A., and Güdel, H. U., 1971. Structural Chemistry of Polynuclear Transition Metal Cyanides. *Structure and Bonding*, 14, pp. 1–21.

Mayoh, B., and Day, P., 1976. Charge Transfer in Mixed-Valence Solids. Part VIII. Contribution of Valence Delocalization to the Ferromagnetism of Prussian Blue. *Journal of the Chemical Society, Dalton Transactions: Inorganic Chemistry (1972–1999)*, (15), pp. 1483–1486.

Mayoh, B., and Day, P., 1974. Charge Transfer in Mixed Valence Solids. Part VIII Perturbation Calculations of Valence Delocalization in Iron(II,III) Cyanides and Silicates. *Journal of the Chemical Society, Dalton Transactions*, pp. 845–852.

Mössbauer, R. L., 1958. Kernresonanzfluoreszenz von Gammastrahlung in Ir191. *Zeitschrift für Physik*, 151, pp. 124–143.

Müller, E., and Stanisch, T., 1909. Berlinerblau und Turnbullsblau. I. *Journal für Praktische Chemie*, 79(1), pp. 81–102. Available at: http://dx.doi.org/10.1002/prac.19090790106.

Reihlen, H., and Zimmermann, W., 1927. Über das Berlinerblau und die komplexen Metallcyanide. *Justus Liebigs Annalen der Chemie*, 451(1), pp. 75–108. Available at: http://dx.doi.org/10.1002/jlac.19274510104.

Richardson, J., and Elliott, N., 1940. The Magnetic Susceptibilities of Some Complex Cyanides with Transition Element Cations. *Journal of the American Chemical Society*, 62(11), pp. 3182–3183. Available at: http://dx.doi.org/10.1021/ja01868a080.

Robin, M. B., 1962. The Color and Electronic Configurations of Prussian Blue. *Inorganic Chemistry*, 1(2), pp. 337–342.

Robin, M. B., and Day, P., 1967. Mixed Valence Chemistry—A Survey and Classification. In Emeleus, H. J., and Sharpe, A. G., eds. *Advances in Inorganic Chemistry and Radiochemistry*. New York and London: Academic Press, pp. 247–422.

Rogez, G., et al., 2000. A Mixed-Valence Mixed-Spin Prussian-Blue-Like Heptanuclear Complex. *Angewandte Chemie, International Edition*, 39(16), pp. 2885–2887.

Sato, O., 2003. Optically Switchable Molecular Solids: Photoinduced Spin-Crossover, Photochromism, and Photoinduced Magnetization. *Accounts of Chemical Research*, 36(9), pp. 692–700.

Sato, O. et al., 1996. Photoinduced Magnetization of a Cobalt-Iron Cyanide. *Science (Washington, D. C.)*, 272(5262), pp. 704–705.

Sato, O., Maruyama, H., and Tao, J., 2007. Photoinduced Valence Tautomerism in Co Complexes. *Magune*, 2(9), pp. 456–461.

Shriver, D. F., and Brown, D. B., 1969. Environment of Interstitial Ions in a Prussian Blue Analog, $Co_3[Co(CN)_6]_2$. *Inorganic Chemistry*, 8(1), pp. 42–46. Available at: http://dx.doi.org/10.1021/ic50071a010.

Shriver, D. F., Shriver, S. A., and Anderson, S. E., 1965. Ligand Field Strength of the Nitrogen End of Cyanide and Structures of Cubic Cyanide Polymers. *Inorganic Chemistry*, 4(5), pp. 725–730. Available at: http://dx.doi.org/10.1021/ic50027a028.

Tanabe, Y., and Sugano, S., 1954. On the Absorption Spectra of Complex Ions. I. *Journal of the Physical Society of Japan*, 9(5), pp. 753–766. Available at: https://doi.org/10.1143/JPSJ.9.753.

Verdaguer, M., 1996. Molecular Electronics Emerges from Molecular Magnetism. *Science*, 272(5262), pp. 698–699.

Verdaguer, M. et al., 1999a. Molecules to Build Solids: High T_C Molecule-Based Magnets by Design and Recent Revival of Cyano Complexes Chemistry. *Coordination Chemistry Reviews*, 190–192, pp. 1023–1047.

Verdaguer, M., et al., 1999b. Room-Temperature Molecule-Based Magnets. *Philosophical Transactions of the Royal Society of London, Series A: Mathematical, Physical and Engineering Sciences*, 357(1762), pp. 2959–2976.

Verdaguer, M., and Girolami, G., 2005. Magnetic Prussian Blue Analogs. In Miller, J. S., and Drillon, M., eds. *Magnetism: Molecules to Materials V*. Weinheim: Wiley-VCH, pp. 283–346.

Weiser, H. B., Milligan, W. O., and Bates, J. B., 1941. Isothermal Dehydration of Heavy-Metal Iron Cyanides. *Journal of Physical Chemistry*, 45, pp. 701–706.

Weiser, H. B., Milligan, W. O., and Bates, J. B., 1942. X-Ray Diffraction Studies on Heavy-Metal Iron-Cyanides. *Journal of Physical Chemistry*, 46, pp. 99–111.

Weiser, H. B., Milligan, W. O., and Bates, J. B., 1938. X-Ray Diffraction Studies on Heavy Metal Ferrocyanide Gels. *The Journal of Physical Chemistry*, 42(7), pp. 945–954. Available at: http://dx.doi.org/10.1021/j100902a006.

Woodward, J., 1724. Praeparatio Caerulei Prussiaci ex Germania missa ad Johannem Woodward. *Philosophical Transactions of the Royal Society of London*, 33, pp. 15–17.

Chapter 3

Magnetic Properties of Prussian Blue Nanoparticles

Laure Catala and Talal Mallah

Institut de Chimie Moléculaire et des Matériaux d'Orsay, CNRS, Université Paris Sud, Université Paris-Saclay, 91405 Orsay Cedex, France

3.1 Introduction

Prussian blue (PB) is one of the most famous coordination compounds mainly because of its intense blue color [1, 2]. Analogues to Prussian blue were subsequently discovered and their structure determined, which helped understanding the electronic and geometrical structure of Prussian blue [3–8]. Since then, a tremendous development of these coordination systems took place both for the fundamental aspects and for applications in different domains.

Here, we focus on the magnetic properties of nanoparticles of PBA and particularly on the magnetic anisotropy that may become dominant when surface/volume ratio increases. In order to understand the magnetic behavior of PBA at the nanoscale, we

Prussian Blue-Type Nanoparticles and Nanocomposites: Synthesis, Devices, and Applications
Edited by Yannick Guari and Joulia Larionova

ISBN 978-981-4800-05-1 (Hardcover), 978-0-429-02474-0 (eBook)
www.panstanford.com

will first briefly present the general formulation of these systems because it has a large incidence on their physical behavior and particularly on their magnetic properties; then we will present the nature of the exchange coupling between metal ions and the main factors that govern magnetic anisotropy in mononuclear Ni^{II} and Co^{II} complexes. A short section is devoted to the preparation of nanoparticles made of one type of PBA network and of core–shell nanoparticles. Then, we focus on the magnetic behavior of pure and core–shell nanoparticles studying the effect of size and surface anisotropy. Then, the behavior of objects assembled as single layers on substrates isolated, organized as stripes that allow ultimately accessing those of single objects are presented.

It is not the objective here to present all the approaches that were developed to prepare PBA nanoparticles, neither to present the magnetic studies on all type of systems. We focus, mainly on nanoparticles with sizes small enough to present particular magnetic behavior such as magnetic anisotropy and we present the chemical approaches used to prepare such objects; the different synthetic approaches of nanoparticles of cyanide-bridged networks and their potential applications were recently reviewed [9, 10].

3.2 Structure, General Formulation and Impact on Some Physical Properties

The structure of PB was elucidated in 1974 [6, 8]. It is a three-dimensional network with coordination bonds spreading in the three dimensions of space; the two types of metal ions, differing here by their oxidation state, bridged by the cyanide ligand form a face centered cubic (fcc) structure like rock salt NaCl. At the difference of NaCl that holds by electrostatic interactions; the negative species, $[Fe^{II}(CN)_6]^{4-}$, are linked to the positive ones, Fe^{III}, by their nitrogen ends forming coordination bonds.

The neutral form of general formula $[Fe^{III}\{Fe^{II}(CN)_6\}_{0.75}]\bullet xH_2O$ contains $[Fe^{II}(CN)_6]^{4-}$ vacancies and is usually written in the following form $[Fe^{III}(H_2O)_{1.5}\{Fe^{II}(CN)_6\}_{0.75}\square_{0.25}]\bullet xH_2O$ highlighting (1) the presence of vacancies within the fcc network and (2) the average number of water molecules linked to Fe^{III} that is related to the density of vacancies. The $[Fe^{II}(CN)_6]^{4-}$ molecules are surrounded by six Fe^{III} ions each, while, because of the presence of vacancies, the Fe^{III} ions are on average surrounded by 4.5 $[Fe^{II}(CN)_6]^{4-}$ molecules and 1.5 water molecules that complete their coordination sphere to 6. The distance between the iron ions is around 5 Å, which corresponds to a unit cell parameter of 10 Å. The unit cell has 8 tetrahedral sites that can accommodate alkali ions (K^I, Rb^I or Cs^I) [11]. If half of the tetrahedral sites (or more) are occupied by an alkali ions A^I, no vacancies are present and Fe^{III} is linked to 6 nitrogen atoms leading to the following formula $[A^IFe^{III}\{Fe^{II}(CN)_6\}]\cdot xH_2O$. It is possible to have intermediate cases with less than half of the tetrahedral sites occupied by alkali ions that can be formulated as follows: $[A^I_yFe^{III}(H_2O)_{3(1-y)/2}\{Fe^{II}(CN)_6\}_{(3+y)/4}\square_{(1-y)/4}]\cdot xH_2O$ (Fig. 3.1). Finally, because of the porosity of the network, water molecules are present.

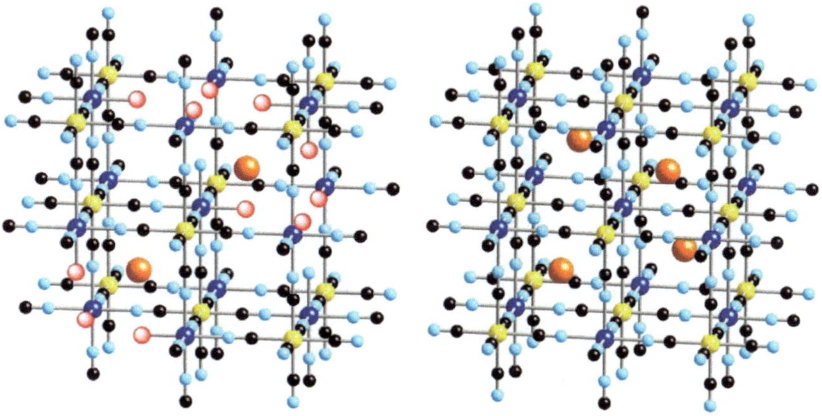

Figure 3.1 Schematic view of the cubic structure of Prussian blue with (left) and without (right) vacancies (Fe^{II}, yellow; Fe^{III}, dark blue; C, black; N, light blue; alkali ion, brown).

Prussian blue analogs (PBAs) are (in the majority of cases) bimetallic materials that possess the same structure as PB but where the metal present in the carbon site is in most cases at oxidation state +III, while in the nitrogen site a metal with an oxidation degree of +II is present. The general formula for the neutral compounds is thus $[M^{II}_N(H_2O)_2\{M^{III}_C(CN)_6\}_{2/3}\square_{1/3}]\cdot xH_2O$, while for those containing alkali ions is $[A^I_yM^{II}_N(H_2O)_{2(1-y)}$

$\{M_C^{III}(CN)_6\}_{(2+y)/3}\square_{(1-y)/3}]\bullet xH_2O$. It is worth noting that when the oxidation state of the two metal ions is +II, all the tetrahedral sites can be occupied by an alkali ion. Furthermore, dications such as Mg^{II} and Ca^{II} can also be inserted.

The above discussion on the diverse possibilities for a PBA networks to contain metallic ions with different oxidation states and to potentially insert different alkali or non-alkali cations in different proportion is crucial for the physical and chemical behavior of these systems.

PB and PBAs can be used for diverse applications, including gas and ions sensors, batteries, electrochromic devices, and, more important, in biology. One example that illustrates the flexibility of their oxidation states and their ability to insert tunable proportion of alkali ions is the case of electrochromic devices [12–16]. It has already been demonstrated that a device made by two electrodes namely $[K^IFe^{III}\{Fe^{II}(CN)_6\}]$ and $[K_2^IZn^{II}\{Fe^{II}(CN)_6\}]$ separated by an electrolyte can change color from blue to colorless by applying a potential between the two electrodes. The mechanism involves the transfer of an electron and a K^I ion from $[K_2^IZn^{II}\{Fe^{II}(CN)_6\}]$ (colorless) to $[K^IFe^{III}\{Fe^{II}(CN)_6\}]$ (blue) to give $[K^IZn^{II}\{Fe^{III}(CN)_6\}]$ and $[K_2^IFe^{II}\{Fe^{II}(CN)_6\}]$ that are colorless [17]. The device is optimized when the materials are in the form of nanoparticles with size lower than a few tens of nanometers, which highlights the role of size reduction and the necessity to be able to prepare nanoparticles with controlled size. Thus, not only the intrinsic chemical properties of PB and PBAs are important but also the mastering of their chemistry at the nanoscale. This aspect is, obviously, crucial for biological applications.

The role of the cations is also illustrated in photomagnetic systems, particularly those based on the cobalt-iron metallic combination [18–25]. Briefly, $[Co^{II}(H_2O)_2\{Fe^{III}(CN)_6\}_{2/3}]\square_{1/3}\bullet xH_2O$ is a paramagnetic material that becomes a ferrimagnet below 21 K (see below for magnetic properties). The insertion of Rb^I allows preparing the diamagnetic material $[Rb_y^ICo^{III}(H_2O)_{3(1-y)/2}\{Fe^{II}(CN)_6\}_{(3+y)/4}\square_{(1-y)/4}]\bullet xH_2O$, ($y \neq 0$ and smaller than 1) where M_N is at the oxidation state +III and M_C at +II. If the latter compound (diamagnetic) is cooled down below 21 K and illuminated by red light, an electron transfer from the diamagnetic Fe^{II} ion to the diamagnetic Co^{III} ion occurs leading to paramagnetic Fe^{III} and Co^{II} that are antiferromagnetically coupled. The result is a switch from a non-magnetic to a magnetic state upon light illumination, thanks to the presence of Rb^{I} in the tetrahedral sites that allowed the stabilization of the switchable diamagnetic material.

The effect of the alkali ions was also observed on the magnetic behavior of the bulk compounds. The ordering temperature of the ferromagnetic compound $[Ni^{II}(H_2O)_2\{Cr^{III}(CN)_6\}_{2/3}\square_{1/3}]\bullet xH_2O$ (75 K) is smaller than the one containing Cs^{I}, $[Cs^{I}Ni^{II}\{Cr^{III}(CN)_6\}]\bullet xH_2O$ (90 K) [26–29]. This is due to the average number of interactions between the paramagnetic sites that is larger in the absence of vacancies that can only be achieved when Cs^{I} is inserted. Finally, the presence of alkali ions may have large effect on the solubility of the compounds that, among other factors, highly impacts the synthesis of nanoparticles.

3.3 Synthetic Approaches to PBA Nanoparticles

3.3.1 Electrostatic Stabilization

The general scheme for the preparation of PB and most of the PBAs relies on mixing aqueous solutions of the two precursors and then recovering the precipitate that is subsequently filtered and dried. For a PBA, the two precursors may be $[M_N^{II}(H_2O)_6]Cl_2$ and $K_3[M_C^{III}(CN)_6]$ with M_N = V, Cr, Mn, Fe, Co, Ni or Cu and M_C = Cr, Fe or Co. The solubility of the different networks may be different and impacts the speed of precipitation and the size of the grains that form the powders and particularly the size and the dispersity of the nanoparticles.

Generally for PBAs, the relatively fast precipitation of the powder after mixing the two precursor solutions precludes the formation of large single crystals, particularly for the most insoluble networks. Only few compositions lead to single crystals large enough to be used for structural studies. Thus, in most cases, PBAs are obtained in powder form made of aggregated nanoparticles; the size and the dispersity of the nanoparticles is not controlled.

To, the best of our knowledge, the first reports concerning the preparation of PBA nanoparticles started at the end of the 1990s [30] and then developed quickly using different approaches (reverse micelles, organic or inorganic templates, etc...) that we will mention here. In 2006, a simple method based on the electrostatic stabilization in water and in the absence of any template agent was reported [31]. The method relies on the rapid precipitation of the objects in water and leads to weak size dispersion. It was found to depend on several chemical parameters that can be controlled by the chemist. Table 3.1 illustrates the effect of Cs^{I} and the relative stoichiometry of the two precursors for solutions with concentration close to 10^{-3} ML^{-1} on the stabilization of nanoparticles of one of the less soluble networks $Ni^{II}Cr^{III}$.

Table 3.1 Evolution of size of nanoparticles with time

$Cs^{I}Cl$	$[Ni^{II}(H_2O)]Cl_2$	$K_3[Cr^{III}(CN)_6]$	Size in nanometers after 10 min	Size in nanometers after 20 min	Size in nanometers after 60 min
0	1	0.66	50	300	Precipitate
0	1	1	10	15	15
1	1	1	6	6	6

In the absence of Cs^{I} and when the stoichiometry corresponds to that of the neutral network $[Ni^{II}\{Cr(CN)_6\}_{2/3}]$, the size of the objects evolves in solution until precipitation. The 300 nm particles obtained after 20 min is an average size and probably a large distribution is present in solution. The as-obtained powder has the composition expected for the neutral network without K^{I} ions. It is remarkable to notice that when the same experiment is carried out with solutions having the same concentration, the size of the objects stabilizes at 15 nm after 15 min and no precipitate is observed. The elemental analysis of the objects, that can be subsequently recovered, shows an almost equal amount of chromium and nickel with potassium ions to complete the charge balance. The same experiment was carried out in the presence of $Cs^{I}Cl$. In this case, objects with a size close to 6 nm form in solution within few seconds and no further evolution of the size was observed. A complete analysis of these objects was carried out and led to the conclusion that they are negatively charged and possess the following composition $Cs^{I}_{0.3}[Cs^{I}_{0.4}Ni^{II}\{Cr(CN)_6\}_{0.9}]$, with 0.3 Cs^{I} per unit formula as cations. Thus, the stabilization of the 6 nm nanoparticles in water is due to their negative charge that precludes their aggregation in solution. The transmission electronic microscopy imaging of the particles confirmed the size observed in solution by dynamic light scattering (DLS) and reveals that the objects possess a slight aspect ratio close to 1.1. This last information is important for the analysis of the magnetic behavior as we will see below. The same synthesis approach for other PBA using $[Cr^{III}(CN)_6]^{3-}$ and $[Fe^{III}(CN)_6]^{3-}$ as hexacyanometallate precursors leads also to nanoparticles with different sizes and dispersity depending on the kinetic of the substitution reaction and thus, among other parameters, on the lability of the $M_N^{II}(H_2O)_6^{2+}$ species and the degree of solubility of the network (see kinetic studies in ref. [9]). For example, for M_N = Ni and M_C = Fe and in the presence of Cs^I, 4.5 nm $Cs^INi^{II}Fe^{III}$ particles are obtained. While, it was difficult to prepare nanoparticles containing Mn^{II}, with a controlled size and dispersity, because of the large lability of $[Mn^{II}(H_2O)_6]^{2+}$ and the relatively high solubility of the related network in comparison to the Ni^{II} containing one.

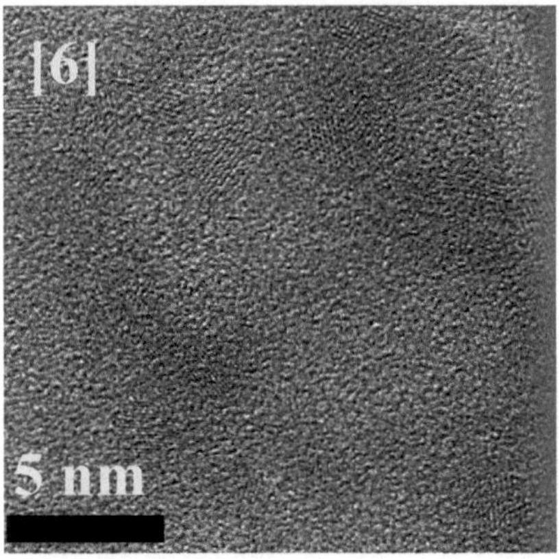

Figure 3.2 High-resolution transmission electronic microscopy image of a 6 nm $Cs^INi^{II}Cr^{III}$ nanoparticle.

3.3.2 Reverse Micelles Approach

Other approaches were developed; they allow preparing nanoparticles with small sizes that can be eventually tuned. The use of reverse micelles techniques is well known in the field of colloidal chemistry to confine the growth of extended networks and prepare nanoparticles [32]. It consists of stabilizing water droplets in oil (heptane for example) by a surfactant. The size of the droplets in which the synthesis takes place can be tuned by different chemical parameters. To the best of our knowledge, the first report of the preparation of PBA nanoparticles (Cu^{II}-hecyanoferrate(III)) was carried out using reverse micelles [30]. Then Mann and coworkers used the same approach and demonstrated that different PBA networks can be obtained as nanoparticles [33, 34]. The size of the objects could be tuned by changing the chemical parameters such as the water/surfactant ratio during the synthesis leading to objects of well-defined cubic shape in many cases. This approach was applied to the $Ni^{II}Cr_{2/3}$ network and allows preparing objects with an average of 3 nm size and relatively good dispersity (Fig. 3.3) [35].

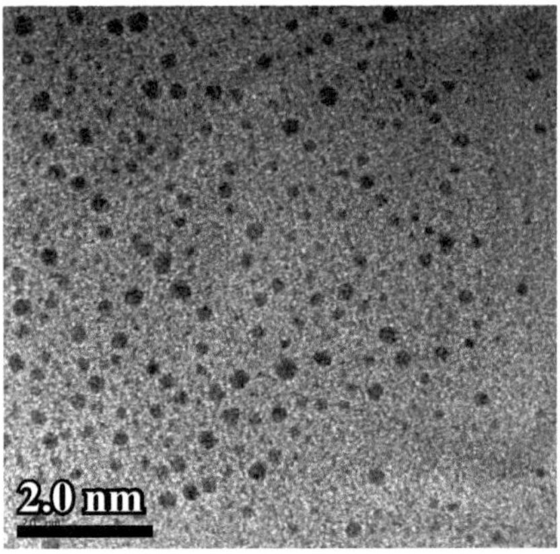

Figure 3.3 Transmission electronic microscopy image of 3 nm $Ni^{II}Cr_{2/3}$ nanoparticles (scale bar: 20 nm).

3.3.3 Confined Growth in Preformed Pores

Another approach that leads to sub-10 nm objects concerns the confined growth of the objects in mesoporous silica following the scheme depicted in Fig. 3.4 [21, 24, 36, 37]. SBA-15 and MCM-41 type silicas with pores of 7.5 and 5.3 nm for the former and 3.4 nm for the latter were first functionalized with the organic functionality $-(CH_2)_2C_5H_4N$ and then treated sequentially with a hexacyano- or octacyano-metallate and the $M_N^{II}(H_2O)_6^{2+}$

(M_N = Ni, Fe, Co) precursors. The objects possessing the same size as the silica pores were subsequently recovered by dissolving the silica network in HF [37].

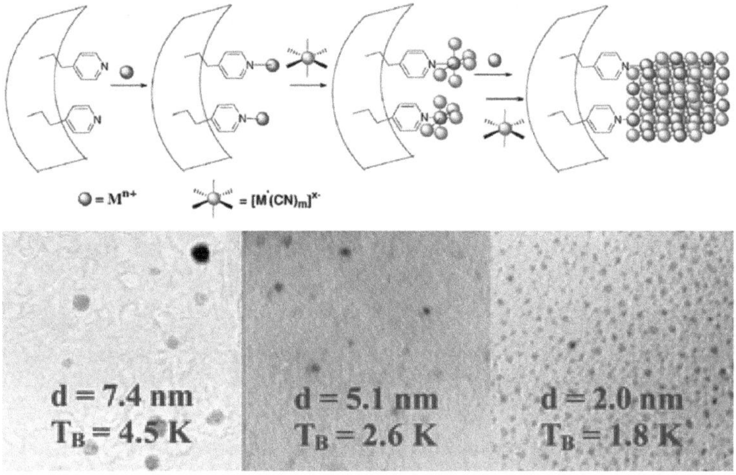

Figure 3.4 Synthetic scheme of the growth of PBA nanoparticles in the pores of functionalized mesoporous silica (top) and transmission electronic microscopy images of the $Ni^{II}Cr^{III}_{2/3}$ nanoparticles (bottom). ©Copyright (adapted from Ref. [37] with permission from the Centre National de la Recherche Scientifique (CNRS) and The Royal Society of Chemistry).

3.4 Preparation of Core–Shell Nanoparticles

The electrostatic approach leads to a dispersion of nanoparticles in aqueous solution without any templating agent that can be used as seeds for the preparation of core–shell objects. The approach consists in growing a shell network on the seeds, taking care to preclude any side nucleation of the shell network in solution.

The core and the shell may or may not have the same chemical composition. When the core and the shell are the same PBA network, particles of increased size can be obtained. This approach was used to grow the size of $Cs^{I}Ni^{II}Cr^{III}$ nanocrystals from 6 to more than 50 nm (Fig. 3.5) [38]. When the shell has a different composition than the seeds, particles made of two or possibly more networks can be prepared (Figs. 3.6 and 3.7) [39]. It is important to stress that the size of the objects can be controlled, in most cases, down to almost one molecular layer which corresponds to a thickness of 5 Å (length of the M_N^{II}–NC–M_C^{III} sequence). The control of the thickness of the shell network is important for the study of the magnetic behavior and particularly of anisotropy as we will see in the following sections. The TEM images in Figs. 3.6 and 3.7 reveals that he aspect ratio of the seeds' particles is conserved for the core–shell and the core–two-shells objects.

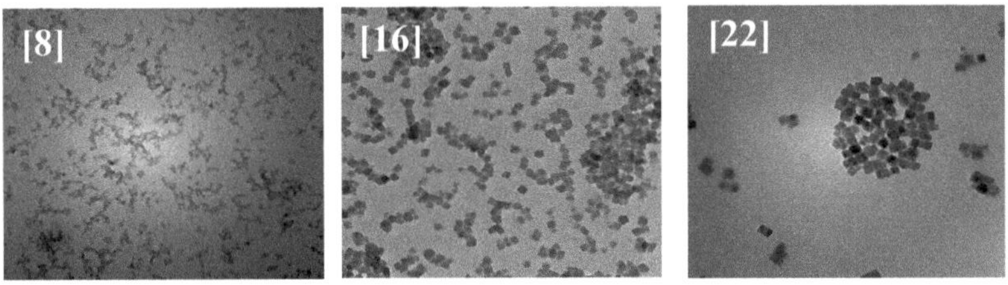

Figure 3.5 Transmission electronic microscopy images of $Cs^INi^{II}Cr^{III}$ nanoparticles (scale bars: 100 nm). ©Copyright adapted from Ref. [38].

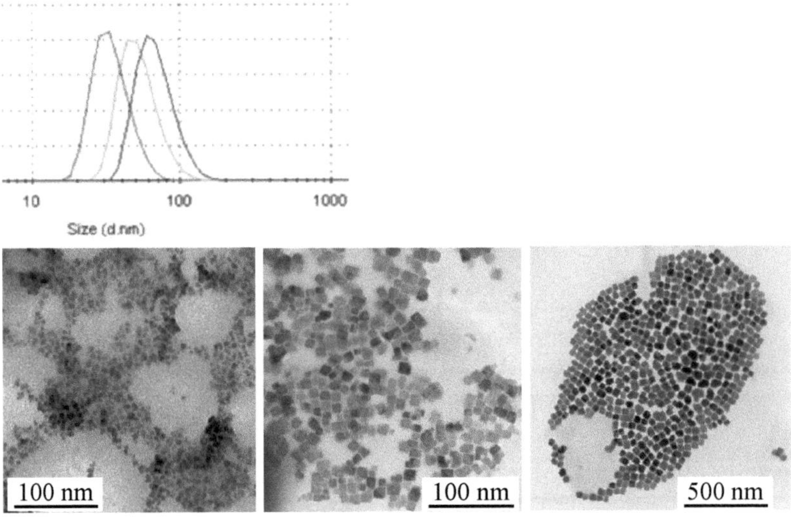

Figure 3.6 Dynamic light scattering (DLS) spectra showing the evolution of the nanoparticles' size (top) and transmission electronic microscopy images of the particles corresponding to (from left to right) those used as seeds $Cs^ICo^{II}Cr^{III}$ (12.1±2.9 nm), the core-shell $Cs^ICo^{II}Cr^{III}$@$Cs^IFe^{II}Cr^{III}$ (30.2±8.4 × 26.7±7 nm) and the core–two-shells $Cs^ICo^{II}Cr^{III}$@$Cs^IFe^{II}Cr^{III}$@$Cs^INi^{II}Cr^{III}$ (51±8.1 × 46.3±6.6 nm). ©Copyright adapted from Ref. [39].

Figure 3.7 High-angle annular dark-field scanning transmission electron microscopy (HAADF-STEM) image (left) of the $Cs^{I}Co^{II}Cr^{III}$@$Cs^{I}Fe^{II}Cr^{III}$@$Cs^{I}Ni^{II}Cr^{III}$ particles; zoom showing a difference of contrast between the core, the first shell and the second shell (middle) and high-resolution transmission electron microscopy (HRTEM) of one particle showing a difference in contrast for the three part of the object (right). ©Copyright adapted from Ref. [39].

The preparation of particles made from the 6 nm $Cs^{I}Ni^{II}Cr^{III}$ seeds on which a first shell of 2 × 1.5 nm of the same $Cs^{I}Ni^{II}Cr^{III}$ network and then a second shell of 2 × 1.5 nm of $Cs^{I}Co^{II}Cr^{III}$ were grown leads to 9 nm $Cs^{I}Ni^{II}Cr^{III}$ and then to 12 nm $Cs^{I}Ni^{II}Cr^{III}$@$Cs^{I}Co^{II}Cr^{III}$ particles, respectively. These objects can thus be described as 9 nm core $Cs^{I}Ni^{II}Cr^{III}$ surrounded by a shell of 1.5 nm thickness of $Cs^{I}Co^{II}Cr^{III}$. It is very difficult with such shell thickness to perform observation with transmission electronic microscopy to demonstrate the formation of such objects as for the largest ones shown in Fig. 3.7. However, as we will show below, magnetic studies may discriminate between a core–shell structure and particles made of a mixture in the volume of the two networks or a mixture of particles of the two networks in the sample. From the structural point of view, these core–shell architectures form with perfect epitaxy because an almost perfect matching between the cell parameters of the different networks is possible.

In the following section made of two parts. We will first briefly present the exchange-coupling mechanism that allows rationalizing the magnetic behavior of the bulk PBA compounds. In the second part, we will discuss the origin and the nature of the magnetic anisotropy of Ni^{II} and Co^{II} containing complexes necessary to understand the magnetic behavior of the nanoparticles made from networks containing these two metal ions.

3.5 Origin of the Magnetic Behavior of Bulk and Nanoscale Coordination Network

3.5.1 Mechanism of the Exchange Coupling in PBA Networks

First, it is worth reminding the electronic structure of the metal ions present in the nitrogen/water and carbon sites of the three-dimensional cubic PB and PBA structures. The geometry of the metal ions is octahedral. For the carbon site, the symmetry is O_h. While the coordination sphere of the other site can be either N_6 or $N_{6-z}O_z$ depending on the presence of vacancies; the symmetry is thus not O_h. We will, however, assume that, because the of cubic structure of the network, the local symmetry of the metal ions in the nitrogen/water site is O_h. The metal ions in the carbon sites are in the low spin state because of the large π back-bonding effect of the carbon side of the cyanide ligand, while the metal ions in the nitrogen or the mixed nitrogen/water sites are in the high spin state (except for Co^{III}) because the nitrogen side of cyanide has mainly a π-donor effect. The electronic structure of the low spin site is, thus, $t_{2g}^{n}e_{g}^{0}$, while for the high spin site it is t_{2g}^{n-m} e_{2g}^{m} where n is the number of d electrons. For example, Prussian blue has an $Fe^{II}(d^6)$ and $Fe^{III}(d^5)$ ions in the carbon and the mixed nitrogen/water sites, respectively leading to the $t_{2g}^{6}e_{g}^{0}$ (S = 0) and $t_{2g}^{3}e_{g}^{2}$ (S = 5/2) local electronic configurations.

Before discussing the mechanism of magnetic interaction in PBAs, it is worth presenting, briefly, the special case of PB. PB is famous because of its intense blue color due to the presence of an intervalence band around 700 nm corresponding to an optically activated electron transfer from low spin Fe^{II} to high spin Fe^{III}. PB is also a ferromagnet with a relatively low ordering temperature of 5.6 K [6, 40]. This behavior is due to the presence of a ferromagnetic exchange coupling between the paramagnetic Fe^{III} (S = 5/2) ions separated by around 10 Å in the solid. The origin of the ferromagnetic coupling has puzzled researcher for some time, because one may expect a weak antiferromagnetic coupling between the paramagnetic Fe^{III} ions due to a weak overlap between the magnetic orbitals (see below for a definition of the ferro- and antiferromagnetic coupling). It has been shown using molecules, as models of PB, that the ferromagnetic coupling is due, as for the color, to the mixed valence nature of the material [41, 42]. Assuming a trinuclear unit of the three-dimensional network, the wave function of the ground state can be described a mixture of the $Fe^{III}(S = 5/2)$—$Fe^{II}(S = 0)$—$Fe^{III}(S = 5/2)$ and the $Fe^{II}(S = 2)$—$Fe^{III}(S = 1/2)$—$Fe^{III}(S = 5/2)$ configurations. In the former, the central ion is diamagnetic and the coupling between the two paramagnetic Fe^{III} ions is either absent or weakly antiferromagnetic, while in the latter configuration and because the electron transfer, the central Fe^{III} is paramagnetic and is coupled to the two peripheral Fe^{II} and Fe^{III} ions. This coupling is ferromagnetic, which explains the overall ferromagnetic coupling in PB. Because the configuration obtained after electron transfer has a very weak contribution to the ground state, the observed coupling is not very large and the ordering temperature is rather low (5.6 K). The situation for PBAs is different because, in most cases, the two metal ions are paramagnetic separated by only 5 Å and their single electrons can directly interact through the cyanide bridge, which leads to an exchange coupling with larger magnitude.

The exchange-coupling phenomenon has a quantum mechanical origin and is governed by electrostatic interaction between the single electrons described by their orbitals. The result of these interactions may *in fine* lead to a situation where the electronic spins of two metal ions, bridged by a ligand, are either parallel or antiparallel, which corresponds to a ferromagnetic or antiferromagnetic interaction, respectively. One model that rationalizes the nature of the interaction is based on the localized Heilter–London approach and was developed by Kahn in order to give an orbital interpretation of the mechanism of exchange coupling in molecular systems useful for synthetic chemists [43]. Within this model and if one considers a binuclear species (Fig. 3.8), the orbitals describing the single electrons are those corresponding to the mononuclear fragments and are called "magnetic orbitals." The symmetry of the binuclear species is not O_h, it is, however, possible to consider an O_h local symmetry for the metal ions, i.e. t_{2g} and e_g orbitals as explained above

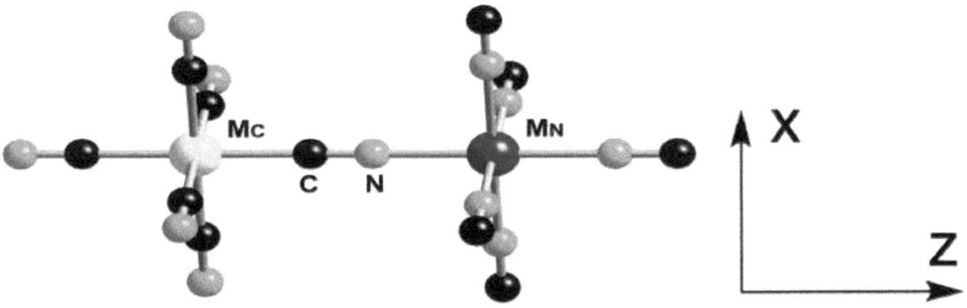

Figure 3.8 Schematic view of the binuclear species used for the analysis of the magnetic mechanism.

In such a case, the t_{2g} orbitals of the $M_C(CN)_6$ fragment are the molecular orbitals obtained from the interaction of the d_{xy}, d_{xz} and d_{yz} metallic orbitals and the ligand orbitals adapted in symmetry. One such molecular orbital (d_{xz}) is depicted in Fig. 3.9. For the $M_N(NC)_6$ fragment, the t_{2g} orbitals are similar; their weight is smaller on the nitrogen atoms linked to the metal ion then on the carbon atoms. The e_g orbitals are obtained from the $d_{x^2-y^2}$ and d_{z^2} metallic orbitals and the ligands orbitals adapted in symmetry as depicted in Fig. 3.9.

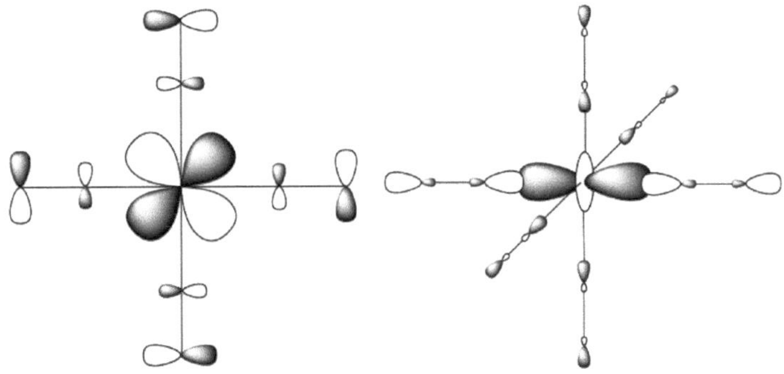

Figure 3.9 Scheme of the d_{xz} (left) and the d_{z^2} (right) magnetic orbitals corresponding to the $M_C(CN)_6$ and $M_N(NC)_6$ species, respectively.

If we consider the case of the $(NC)_6Cr^{III}$—CN—$Mn^{II}(NC)_6$ species; the three single electrons of Cr^{III} ($t_{2g}^3e_g^0$, S = 3/2) with orbital symmetry t_{2g} interact with the five single electrons of Mn^{II} ($t_{2g}^3e_g^2$, S = 5/2) among them three with orbital symmetry t_{2g} and two with orbital symmetry e_g (Fig. 3.10). The interaction between the single electrons of the same orbital symmetry (t_{2g}) leads to an antiferromagnetic contribution (J_{AF}) to the overall coupling, because the two sets of orbitals possess non-zero overlap integrals that tend to put the spin of their electrons antiparallel. While the (electrostatic) interaction between electrons described with orbitals of different symmetry (t_{2g}/e_g) leads to a ferromagnetic contribution to the coupling because the overlap integral is zero and the overlap density that is related to the extent of the exchange is relatively large (Fig. 3.9). The resulting exchange-coupling interaction in the case of the Cr^{III}/Mn^{II} network is thus difficult to predict on the basis of this simple qualitative arguments because it results from the sum of the ferro- and antiferromagnetic contributions as depicted in Fig. 3.10. Experimental results for the $Cs^{I}[Mn^{II}\{Cr^{III}(CN)_6\}] \cdot xH_2O$ PBA and on the molecular species $[Cr^{III}(CNMn^{II}L_5)_6](ClO_4)_9$, where L is a pentadentate ligand, show that the coupling is antiferromagnetic [28, 44, 45]. This leads to the conclusion that the antiferromagnetic contribution (J_{AF}: t_{2g}/t_{2g}) is the leading one. Substituting Mn^{II} by Ni^{II} ($t_{2g}^6e_g^2$, S = 1) should lead to a ferromagnetic interaction because the antiferromagnetic contribution is zero (Ni^{II} has no single electrons in the t_{2g} orbitals), as was experimentally observed [26, 46]. Using V^{II} ($t_{2g}^3e_g^0$, S = 3/2) or Cr^{II} ($t_{2g}^3e_g^1$, S = 3/2) leads to an antiferromagnetic exchange coupling larger than in the case of the Mn^{II} containing network because of the reduction of the ferromagnetic contribution in the case of Cr^{II} and its absence of V^{II}. PBA molecular magnets containing Cr^{II} orders at 240 K [47] and those with V^{III} and/or V^{II} orders at temperatures higher than room temperature [48] owing to this large antiferromagnetic exchange-coupling interaction.

When Mn^{II} is substituted with Co^{II} ($t_{2g}^5e_g^2$, S = 3/2), the number of single electrons in the t_{2g} orbitals is reduced from 3 to 1 leading to a reduction of the magnitude of the antiferromagnetic contribution. As a result, the ferromagnetic one becomes the leading one and the coupling is ferromagnetic. However, its magnitude is weaker than for the Ni^{II} containing network where the antiferromagnetic contribution is zero. This difference in the magnitudes of the exchange-coupling interaction between the Co^{II} and the Ni^{II} containing networks leads to a smaller ordering temperature for the former compared to the latter PBA ($T_C(Co^{II}Cr^{III})$ = 25 K, while $T_C(Ni^{II}Cr^{III})$ = 90 K) [28]. The magnetic properties of the nanoparticles based on these networks will also be different as it will be briefly explained below. Before analyzing the magnetic behavior of the PBA at the nanoscale, it is necessary to introduce briefly the concepts of magnetic anisotropy; we will focus on the cases of Ni^{II} and Co^{II} species because we will mainly study PBA nanoparticles based on these two ions.

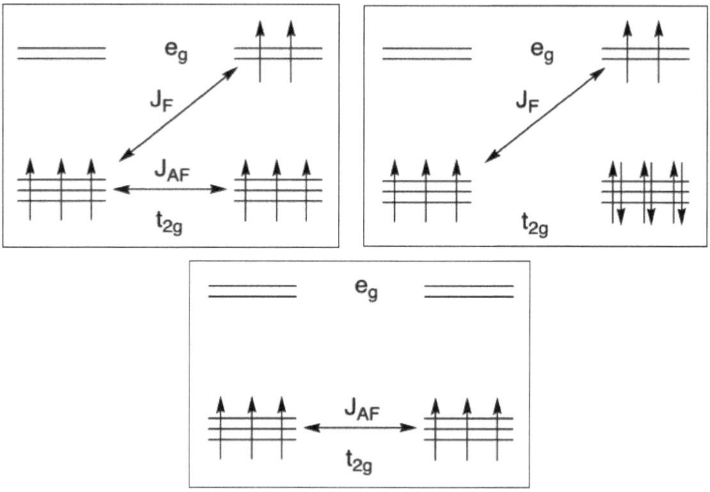

Figure 3.10 Contribution of the different exchange pathways to the overall exchange interaction: t_{2g}/t_{2g}, J_{AF} and t_{2g}/e_g, J_F.

3.5.2 Magnetic Behavior of Nanoscale Objects

The magnetic anisotropy of single domain nanoparticles can be understood within the frame of the Néel–Brown thermal activated model [49–51]. The single domain particles are considered as possessing a giant spin. Their relaxation is activated and follows the Arrhenius law that describes the temperature dependence of the relaxation time of the magnetization, $\tau = \tau_0 \exp(\frac{\Delta E}{k_B T})$, where τ_0 is a characteristic attempt time, k_B the Boltzmann constant and ΔE ($=K_{eff}V$) the average energy barrier; K_{eff} is the effective anisotropy constant and V the mean volume of the nanoparticles. The energy barrier and the blocking temperature of the nanoparticles are determined by analyzing their dynamic magnetic properties studying the thermal dependence of the alternative current (ac) magnetic susceptibility at different frequencies, ν (with $\tau = 1/2\pi\nu$), of the oscillating magnetic field. When ν is close to the blocking temperature T_B, the out-of-phase component χ'' of the magnetic susceptibility becomes temperature dependent. Its maximum corresponds to the blocking temperature at the applied frequency with the relaxation time τ.

If the objects are not isolated and feel dipolar magnetic interaction, the behavior is complex because the dipolar interaction may lead to collective behavior and/or affect the anisotropy of the surface of the objects. In such a case, the analysis of the dynamic behavior can be made using the approach proposed by Mydosh using the parameter Φ that allows discriminating between various magnetic behaviors, where $\Phi = (T_{max} - T_{min})/(T_{max}(\log\nu_{max} - \log\nu_{min}))$ with T_{max} and T_{min} being the temperatures of the maxima of the $\chi'' = f(T)$ curves for the two extreme applied frequencies ν_{max} and ν_{min}, respectively. Values of τ_0 and Φ larger than 10^{-13} s and 0.12, respectively for are expected for single magnetic domain nanoparticles without (or with very weak) dipolar interactions [52]. Lower values for τ_0 and Φ are found for (1) strongly interacting single domain particles or (2) disordered systems [51, 53–55].

3.5.3 Magnetic Anisotropy in Mononuclear Ni^{II} and Co^{II} Complexes

The objective of this section is to present the main results concerning the magnetic anisotropy of Co^{II} and Ni^{II} mononuclear complexes that will be used in order to analyze and understand the magnetic behavior of PBA nanoparticles because we will mainly focus on the magnetic anisotropy of nanoparticles containing these two metal ions. The magnetic anisotropy of the nanoparticles has its origin, at least partially, in the magnetic anisotropy of the metal ions composing the objects (magneto-crystalline anisotropy), in addition to shape anisotropy.

Magnetic anisotropy of molecular based materials is the consequence of the application of a magnetic field on a given materials formed from ensemble of magnetically isolated molecules. In such a case, the observed magnetic anisotropy can be related to that of the isolated molecules [56]. However, the phenomenon has its origin in the electronic structure of the molecule in zero applied magnetic field.

In the case of a Ni^{II} complex with octahedral symmetry (O_h) and in the absence of any applied magnetic fields, the ground spectroscopic term ($^3A_{2g}$) is perfectly isotropic and the m_S sub-levels 0 and ±1 of the $S = 1$ electronic spin are degenerate. In such a case and if we neglect spin-spin interaction, applying a magnetic field does not generate any magnetic anisotropy. The description of the wave functions by only the ground state is not strictly accurate; the contribution of the excited states must be considered. However, if the symmetry is O_h including the contribution of the excited state does not lift the degeneracy of the m_S sub-levels and no magnetic anisotropy can be observed. Only if the symmetry is lower, say D_{4h}, it is possible to show that the degeneracy of the m_S levels is lifted leading to $m_S = 0$ and ±1 with different energies; this is called zero-field-splitting (ZFS). It is responsible of the appearance of magnetic anisotropy. The origin of ZFS is the simultaneous effect of a structural distortion ($O_h \rightarrow D_{4h}$ for example) and spin-orbit coupling. Based on symmetry arguments, one notices that for O_h symmetry the three components of the orbital moment $\hat{L}$, $\widehat{L_x}$, $\widehat{L_y}$ and $\widehat{L_z}$, possess the same symmetry (T_{1g}), so that the coupling between the spin and orbital degrees of freedom does not break the symmetry of the spin wave functions and the $m_S = 0$ and ±1 sub-levels remain degenerate. While in the axial D_{4h} symmetry, the $\widehat{L_z}$ component of the orbital moment has a different symmetry from the other two components. The spin-orbit coupling is thus different for the axial direction (z) than for the equatorial one (xy), leading to a splitting of the energies of the $m_S = 0$ and ±1 sub-levels and thus to axial magnetic anisotropy (when a magnetic field is applied). It is important to stress that in the present case ($S = 1$ and D_{4h} symmetry), the ground spectroscopic term does not have orbital degeneracy so no orbital moment. The effect of spin-orbit coupling comes from the admixture of excited states that possess an orbital moment. In the frame of second order perturbation theory, the magnitude of ZFS, is thus proportional to the inverse of the energy difference between the ground state and the excited states coupled via the spin-orbit operator. If, in addition, a structural anisotropy is present in the xy plane, the spin-orbit coupling lifts the degeneracy of the $m_S = \pm 1$ sub-levels leading to a rhombic anisotropy. This is because the $\widehat{L}_x$ and $\widehat{L}_y$ components of the orbital moment have different symmetries and thus different energies if a distortion is present in the *xy* plane. It is possible to, phenomenologically, describe ZFS by two parameters *D* and *E*; the former is the axial parameter and the latter the rhombic one as depicted in Fig. 3.11. Because of the structure of the spin Hamiltonian used to describe ZFS, *D* is negative when the $m_S = \pm 1$ are the ground levels. These parameters are accessible experimentally by low temperature magnetometry studies and particularly by electron paramagnetic resonance (EPR) spectroscopy.

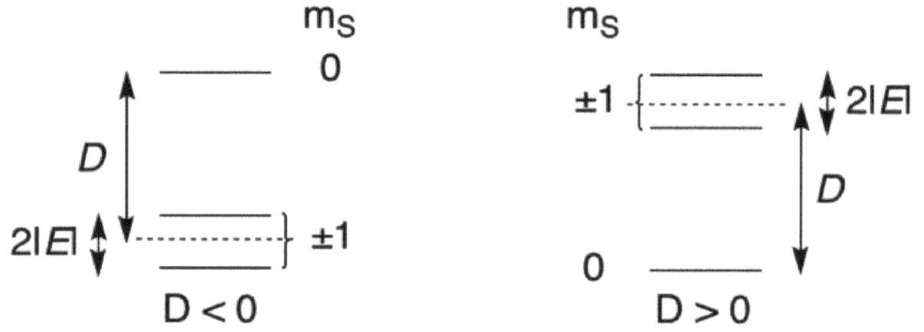

Figure 3.11 Zero-field splitting of the $S = 1$ levels.

If the ground levels are $m_S = \pm 1$ ($D < 0$), it costs less energy to align the electronic spin with the magnetic field applied along the principal symmetry axis leading to what is called an Ising type anisotropy; the magnetic moments can thus be blocked in one direction if the energy barrier between the two spin directions is larger than thermal energy ($E = 0$). While for positive *D* values, applying a weak magnetic field perpendicular to the principal symmetry axis aligns the spin in the plane where it has the possibility to rotate, so that if the field is cut no blocking of the magnetization occurs in a privileged direction.

The sign of the ZFS axial parameter *D* depends on the nature of the distortion; it will thus be different for compressed and elongated D_{4h} molecules. This is explained below.

In summary and for a $S = 1$ distorted molecule, the magnitude of ZFS depends on the inverse of the energy difference between the ground and the excited states coupled by the spin-orbit operator and the sign of the axial parameter depends on the nature of distortion. Up to now, we reasoned on the interaction between energy states, it is however possible translate these ideas using the electronic configuration of the d orbitals.

The energy splitting of the d orbitals for weakly distorted Ni^{II} complexes, obtained from ab initio calculations is depicted in Fig. 3.12 [56–58].

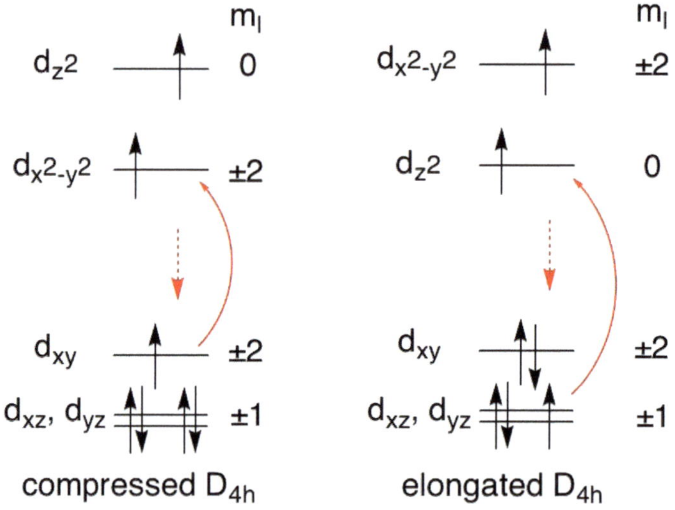

Figure 3.12 Splitting of the real d orbitals for compressed and elongated Ni^{II} octahedron (d^8, $S = 1$); the black arrows indicate the orientation of the spins in the different orbitals and the dashed red arrow indicate the excitation between two orbitals corresponding to the coupling between two Slater determinants.

For the compressed case, the d_{z^2} orbital has the highest energy while the $d_{x^2-y^2}$ is highest for the elongated case, as expected from simple crystal field theory. For ligands that have weak π-donor effect, the energy difference between the d_{xy} and the (d_{xz}, d_{yz}) orbitals is not the same for the elongated and compressed cases, but their relative energy is the same; they are closer for the compressed case. The ground and the first electronic configurations correspond to the electron filling shown in Fig. 3.12 for the compressed and elongated distortions. For the compressed case, the lowest energy excitation that has the main contribution involves the d_{xy} and the $d_{x^2-y^2}$ orbitals that possess the same orbital momenta values $m_l = \pm 2$. It is, thus, the $\widehat{l_z} \cdot \widehat{s_z}$ part of the spin-orbit operator that couples the ground and this excited state. So, the $m_S = \pm1$ sublevels that correspond to the $\widehat{s_z}$ operator are more stabilized than the $m_S = 0$ one leading to negative contribution to D. The spin-orbit operator cannot act on excitations involving orbitals with m_l values that differ by 2 in absolute value. Thus, for the elongated case, the excitation involving the $m_l = \pm2$ and $m_l = 0$ orbitals (the lowest in energy) cannot contribute to ZFS. The lowest excited state that it coupled to the ground state by the spin-orbit operator involves the excitation between the $m_l = \pm1$ (d_{xz}, d_{yz}) and the $m_l = 0$ (d_{z^2}) orbitals (Fig. 3.12). In this case, because the m_l values involved differ by ±1 and not by 0 as for the compressed case, it is the $\widehat{l_x}\cdot\widehat{s_x} + \widehat{l_y}\cdot\widehat{s_y}$ part of the spin-orbit operator that acts because it couples level with m_l values that differ by ±1. In this case, it is the $m_S = 0$ sublevel that is more stabilized than the $m_S = \pm 1$ leading to a positive D value [59, 60].

In summary, one expects, for a $S = 1$ system, a negative D value for a tetragonally compressed octahedron and to a positive D value for an elongated octahedron responsible of the Ising and planar magnetic anisotropy, respectively. The largest the distortion is, the largest the ZFS is and the largest the magnetic anisotropy is. In the case of PBA, the geometry around the Ni^{II} ions is very close to an octahedron. However, for nanoparticles, the coordination sphere of Ni^{II} ions present on their surface may be slightly distorted because water molecules replace cyanide for these ions. However, the distortion is not very large and magnetic anisotropy is not expected to be huge.

For Co^{II}, the analysis is similar but the main difference is that the excitations involve orbitals with different m_l values, as shown in Fig. 3.13, for the two types of distortion leading to positive D values for the two cases. The other important difference with Ni^{II} is that the energy gap between the ground and excited states is much smaller because the orbitals involved are closer in energy (Fig. 3.13). The result is a much larger ZFS and thus a larger magnetic anisotropy for Co^{II}, though an easy plane of magnetization is expected. It is worth noting that the wave functions describing the ground and the excited states are multideterminantal and our assumption of only one determinant is an approximation. However, when we analyze in details real cases, we find for Co^{II} positive D values in distorted octahedral geometry larger in magnitude than for Ni^{II}. It is, thus expected, that nanoparticles containing Co^{II} species at their surface present larger magnetic anisotropy than those containing only Ni^{II} ions.

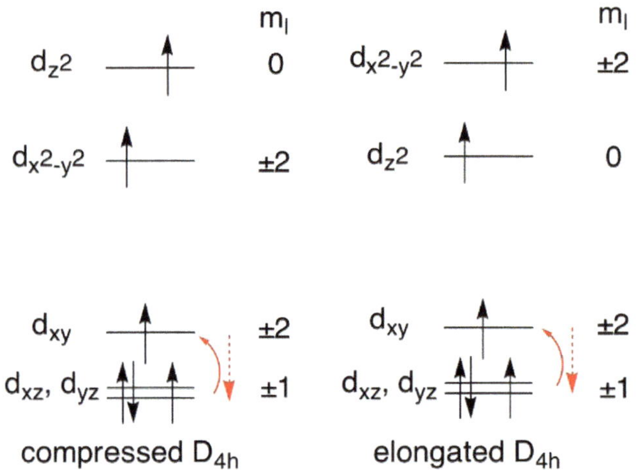

Figure 3.13 Splitting of the real d orbitals for compressed and elongated Co^{II} octahedron (d^7, $S = 3/2$); the black arrows indicate the orientation of the spins in the different orbitals and the dashed red arrow indicate the excitation between two orbitals corresponding to the coupling between two Slater determinants.

3.6 Magnetic Properties of Core Particles

3.6.1 PBA Nanoparticles Based on Hexacyanoferrate(III)

As stated above, cubic 4.5 nm negatively charged $Ni^{II}Fe^{III}$ nanoparticles are obtained by spontaneous electrostatic stabilization in aqueous solution in the presence of Cs^I; they have the formula $Cs_{0.36}[Cs^I_{0.46}Ni^{II}(H_2O)_{0.36}\{Fe^{III}(CN)_6\}_{0.94}\square_{0.06}]$. The density of vacancies is very low (6%) so that the coordination sphere of about only 1/3 of the Ni^{II} ions within the volume of the objects are linked to one water molecule; the Ni^{II} ions belonging to the surface (around 50% for 4.5 nm particles) have of course additional water molecules pointing out of the particles surface and are not included in the formula. The magnetic behavior of these objects was investigated when recovered with cetyltrimethylammonium (CTA) and with the organic polymer polyvinylpyrrolidone (PVP) leading to relatively concentrated and diluted samples. The length of the CTA alkyl chain is around 1 nm so that the average distance between the nano-objects is between 1.5 and 2 nm and in this case, one expects the presence of interparticle magnetic dipolar interactions, while for the PVP sample these interactions should be reduced. The investigation of the magnetic behavior shows a large dependence of the coercive field values and the blocking temperature with the dilution (Figs. 3.14 and 3.15) measured for three different samples: NPs in CTA (NP@CTA), NPs in PVP (NP@PVP) and NPs highly diluted in PVP (dil_NP@PVP).

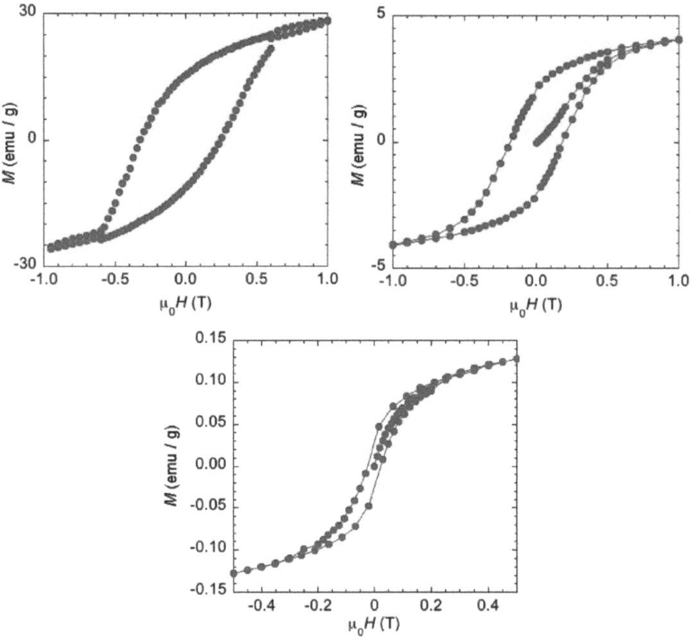

Figure 3.14 Hysteresis loops at T = 2 K of $Cs^{I}Ni^{II}Fe^{III}$ 4.5 nm particles embedded in CTA (left), PVP (middle) and highly dilutes in PVP (right).

The coercive field values decrease from 3300 to 1800 and then to 220 Oe upon dilution. The out-of-phase ac curves, $\chi'' = f(T)$, show a large decrease of the blocking temperature going from 14 to 6.5 and to 2.7 K for the highly diluted sample. The thermal dependence of χ'' was measured for the three samples at different frequencies, which allows the analysis of the dependence of the relaxation time with temperature and the extraction of the characteristic parameters τ_0 and ΔE. The τ_0 value for NP@CTA and NP@PVP were found equal to 10^{-38} and 10^{-16} s, which is not physically sound (must be larger than 10^{-13} s) with ΔE values of 1200 and 230 K. The corresponding Mydosh parameters, Φ, are smaller than 0.12 (0.03 and 0.07 for NP@CTA and NP@PVP samples, respectively). The particles embedded in PVP have larger Φ, lower energy barrier, lower blocking temperature and smaller coercive field. The highly diluted sample, dil_Np@PVP, has $\tau_0 = 1.2 \times 10^{-12}$ s and $\Phi = 0.09$, with an energy barrier of 66 K and a blocking temperature of 2.7 K. The value of the coercive field is 220 Oe. All these features are consistent with the presence of relatively large non-homogeneous dipolar interactions that are highly reduced in the highly diluted sample, where a magnetic behavior very close to that of single magnetic domain well-isolated objects is found.

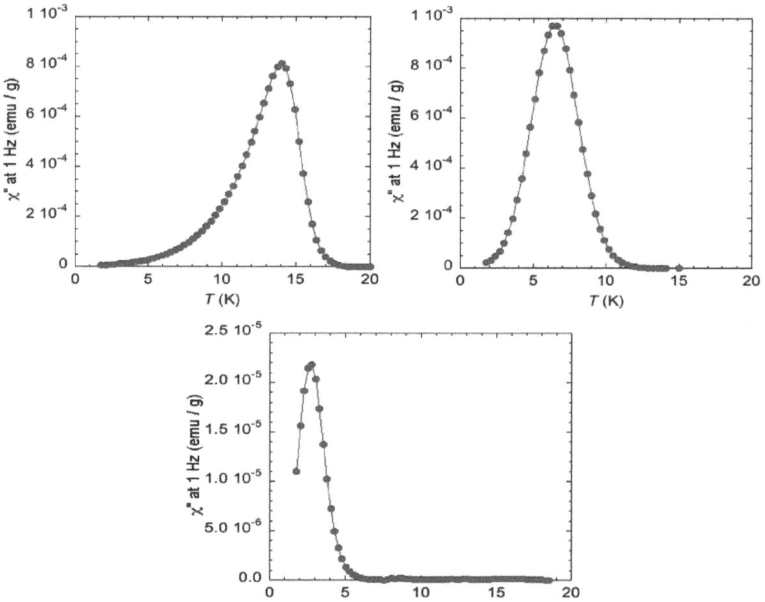

Figure 3.15 Thermal dependence of the out-of-phase component, χ'', of the magnetic susceptibility measured at $\nu = 1$ Hz of $Cs^{I}Ni^{II}Fe^{III}$ 4.5 nm particles embedded in CTA (left), PVP (middle) and highly dilutes in PVP (right).

It is interesting to compare the behavior of the $Cs^{I}Ni^{II}Fe^{III}$ particles with different degree of dilution to those grown ($Ni^{II}Fe^{III}_{0.46}$) and thus embedded within mesoporous silica of variable porosity whose preparation were presented above (Fig. 3.4). The blocking temperature of the three samples (NP@SiO_2 with sizes of 7.4, 5.1 and 2 nm) obtained from the maxima of the ZFCM curves was found equal to 4.5, 2.6 and 2.2 K, respectively. The 5.1 nm objects possess almost the same blocking temperature as the $Cs^{I}Ni^{II}Fe^{III}$ 4.5 nm ones highly diluted in PVP. However, the ac data are different. For the 5.1 nm NP@SiO_2 sample, the τ_0 value (7.7×10^{-15} s) is too small for single magnetic domain objects and the Mydosh parameter (Φ = 0.06) indicates either the presence of dipolar coupling or some kind of intrinsic disorder. Thanks to the organization of the pores within the SBA-15 type silica, their average distance and thus the separation between the NPs is known, which allows an estimation of their average magnetic dipolar interaction. The average interaction energy can be estimated to be equal to 0.1 K, which in the case of the 5.1 nm particles excludes a large effect of interparticle dipolar coupling. Thus, it is reasonable to assume that the observed behavior is due to the intrinsic properties of the objects. The main differences between the $Cs^{I}Ni^{II}Fe^{III}$ objects and those embedded in SiO_2 are the density of vacancies and the shape; the formers are cubic with almost no vacancies while the latter are spherical with 33% vacancies of $[Fe(CN)_6]^{3-}$. The relatively weak (around 5 cm^{-1}) exchange-coupling interaction for Ni^{II}—NC—Fe^{III} and the larger density of vacancies for the NP@SiO_2 leads a smaller magnetic correlation length resulting in a non-full alignment of the magnetic moment present at the surface with those in the volume (surface/volume ratio is around 50%). In addition, the spherical shape has the effect of decreasing the number of coordination bonds of the surface ions thus decreasing further their coupling with the volume ions. These features may be at the origin of the magnetic surface disorder of the objects embedded in the silica pores and thus the difference in the magnetic behavior between the two types of objects.

3.6.2 PBA Nanoparticles Based of Hexacyanochromate(III): Size Dependence

In the case of PBA $Cs^{I}Ni^{II}Cr^{III}$ nanoparticles, the exchange coupling between Cr^{III} and Ni^{II} through the cyanide bridge is close to 15 cm^{-1} [46], and the magnetic moment of a particle of only 6 nm size is relatively large (around 10^4 Bohr Magneton), magnetic dipolar interactions may thus be large enough to affect the magnetic behavior of the objects if they are not well isolated.

The study of the ac susceptibility of a series of $Cs^{I}Ni^{II}Cr^{III}$ nanoparticles with size ranging from 6 to 22 nm embedded in PVP shows that the attempt time τ_0 is too small and does not correspond to what is expected for isolated particles, but rather to objects feeling dipolar interaction. The nanoparticles were thus highly diluted in the PVP so that the average distance between the objects is larger than few hundreds of nanometers as shown in Fig. 3.16 [38].

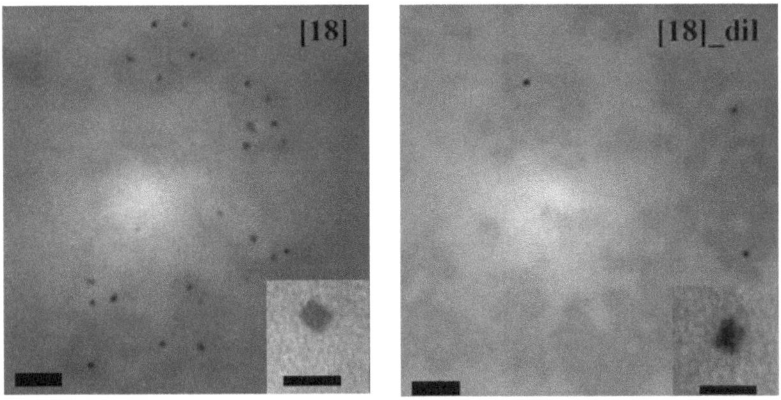

Figure 3.16 Transmission electronic microscopy images of 18 nm $Cs^{I}Ni^{II}Cr^{III}$ particles embedded in PVP weak dilution (left) and high dilution (right) (scale bar: 200 nm); insets correspond to the enlargement of one particle (scale bar: 50 nm). ©Copyright adapted from Ref. [38].

The $\chi'' = f(T)$ curves recorded at ν = 1 Hz show the presence of maxima at blocking temperatures ranging from 2.7 to 21 K for the largest objects (22 nm) (Fig. 3.17). It is worth noting that for the non-highly diluted 22 nm particles (same starting sample as the highly diluted one), the maximum of the $\chi'' = f(T)$ curve was found at T = 43 K; diluting objects shifted the blocking temperature from 43 to 21 K. The objects embedded in PVP feel, thus, large ferromagnetic coupling of magnetic dipolar origin.

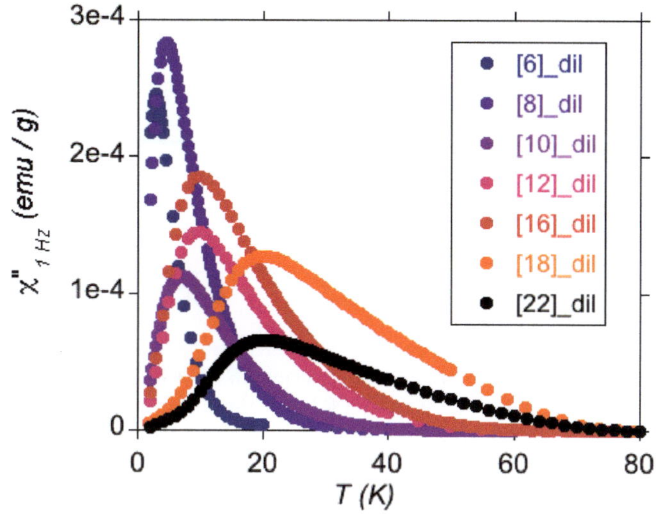

Figure 3.17 Thermal dependence of the out-of-phase component, χ'', of the magnetic susceptibility measured at ν = 1 Hz of the highly diluted $Cs^{I}Ni^{II}Cr^{III}$ particles with different sizes. ©Copyright adapted from Ref. [38].

The ac measurements performed at different frequencies for the diluted and the non-diluted samples allow extracting the values of the energy barriers (ΔE) that were plotted as a function of the volume (Fig. 3.18). For the non-diluted samples, a linear dependence with the volume was found, while for the diluted ones the dependence is not linear. This is consistent with the fact that for non-diluted samples the nanoparticles' anisotropy is mainly due to the volume, while it is not in the case for the diluted samples. We, thus, assumed that the magnetic anisotropy of the isolated objects has a contribution from their surface following a phenomenological dependence $\Delta E = K_{\text{eff}}V = K_V V + K_S S$ relating the effective anisotropy constant K_{eff} and the size of the particles d: $\Delta E/V = K_{\text{eff}} = K_V + 6K_S/d$ with K_V and K_S the anisotropy constants per volume and surface units, respectively [61]. The fit of the data leads to $|K_V| = (0.5 \pm 0.1) \times 10^3$ J/m^3 and $|K_S| = (3.5 \pm 0.6) \times 10^{-6}$ J/m^2.

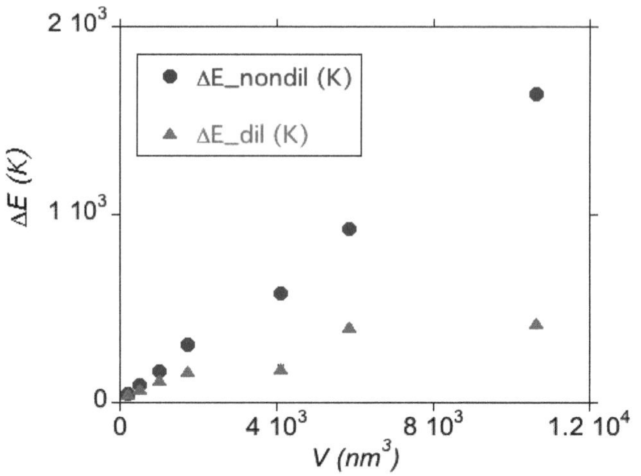

Figure 3.18 Plot of the anisotropy energy ΔE as a function of the volume of the particles V for the weakly (●) and the highly (▲) diluted samples ©Copyright adapted from Ref. [38].

For the $Cs^{I}Ni^{II}Cr^{III}$ PBA nanoparticles, the Cr^{III} ions are magnetically isotropic. The magneto-crystalline anisotropy is, thus, mainly due to the anisotropy of the Ni^{II} ions (and to exchange anisotropy). It was demonstrated by X-ray absorption spectroscopy (XAS) and X-ray magnetic circular dichroism (XMCD) studies on the 6 nm nanoparticles that the surface Ni^{II} ions have a compressed tetragonal geometry, due to a coordination sphere corresponding to $Ni^{II}(NC)_5(OH_2)$, with the water molecule pointing out of the surface [62]. The local anisotropy in the bulk originates mainly from the very slightly distorted $Ni^{II}(NC)_6$ sites present just below the surface and from some Ni^{II} ions having coordinated water molecules. Indeed, considering a cubic PBA particle of formula $[Cs^{I}_{0.4}Ni^{II}(H_2O)_{0.6}\{Cr(CN)_6\}_{0.9}\square_{0.1}]^{0.3-}$, on average only one Ni^{II} out of ten is coordinated to a water molecule in the volume of the object and 40% of the Ni^{II} ions (surface ions) have at least one water molecule pointing out of the particles' surface. There are 4 Ni^{II} per nm^3 in the bulk and 2 per nm^2 at the surface of the objects. This leads to the local anisotropy constant $|K_V| = (1.0 \pm 0.2) \times 10^{-3}$ meV per Ni^{II} and $|K_S| = (1.2 \pm 0.2) \times 10^{-2}$ meV per Ni^{II}. The local anisotropy of the Ni^{II} ions at the surface is, thus, one order of magnitude higher than the one in the bulk. As particles of 6 and 22 nm display, respectively 40% and 10% of atoms at the surface, the contribution of surface anisotropy to the whole anisotropy is about 88% and 61%, respectively. This clearly shows that surface anisotropy is the main source of anisotropy in those particles, which is still considerable for nanoparticles with the largest sizes. And this surface anisotropy is due to the coordination of water molecules that induces a structural distortion and ZFS as explained above. The intriguing result from XAS is that the surface Ni^{II} were modeled by a tetragonally *compressed* geometry, which is not expected from water molecules that normally have a weaker ligand field than the nitrogen end of the cyanide. This compressed geometry may thus be due to the presence of the organic polymer in which the particles are embedded. Finally, since the Ni^{II} have compressed geometry, the axial anisotropy parameter D is negative with an Ising type local anisotropy so that the local moments are expected to be perpendicular to the surface of the particles. The PBA nanoparticles are slightly elongated and this elongation remains when their size increases (Fig. 3.6). Their aspect ratio is close to 1.1. A calculation of the shape anisotropy in the case of the $Cs^{I}Ni^{II}Cr^{III}$ leads to a value (500 J/m^3) very close to the anisotropy experimentally found (550 J/m^3). Thus, shape anisotropy in addition to surface anisotropy are responsible of the presence of a coercive field that was shown to increase upon decreasing the size and increasing dilution (Fig. 3.19). This is consistent with an increase of the contribution of surface anisotropy for smaller and well-isolated particles. Shape anisotropy tends to align the easy axis or the particle along its largest axis, while local anisotropy of the surface Ni^{II} ions is perpendicular to the surface and thus perpendicular to the largest axis for four of the faces. The competition between the two components is the result of the effective anisotropy experimentally observed. Experiments on a single layer of nanoparticles and on isolated ones give information on the resulting orientation of the magnetic moments of the particles (see below).

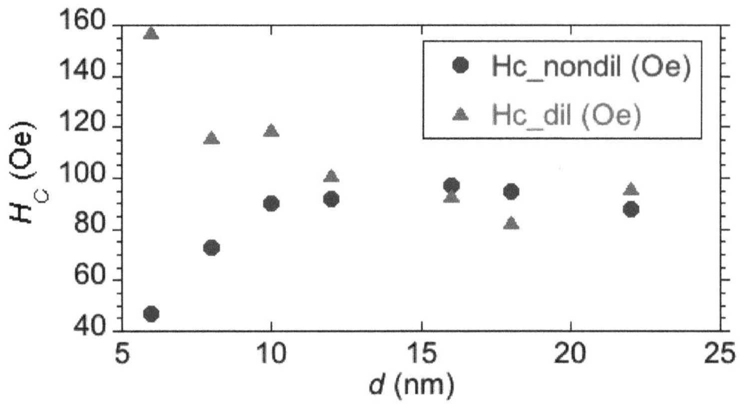

Figure 3.19 Plot of the coercive field H_C at 2 K, as a function of the nanoparticles size for the weakly (●) and the highly (▲) diluted samples. ©Copyright adapted from Ref. [38].

3.7 Core–Shell Nanoparticles

3.7.1 Magnetic Properties of Core–Shell $Cs^{I}Ni^{II}Cr^{III}$@$Cs^{I}Co^{II}Cr^{III}$ Nanoparticles

As stated above, the magnetic anisotropy of Co^{II} octahedral complexes with a tetragonally distorted geometry is expected to be much larger than the equivalent Ni^{II} ones. In addition, Co^{II} ions are expected to have an easy plane of magnetization (positive D value). Nanoparticles containing Co^{II} species at their surface ($Cs^{I}Ni^{II}Cr^{III}$@$Cs^{I}Co^{II}Cr^{III}$: 8 + 2 × 1 = 10 nm) were prepared and compared to $Cs^{I}Ni^{II}Cr^{III}$ particles of the same size (10 nm). In order to demonstrate the effect of the surface ions, the magnetic behavior of the pure $Cs^{I}Ni^{II}Cr^{II}$ and the core–shell objects were compared to those of alloy particles made from a mixture of Co^{II} and Ni^{II} in the volume of the objects ($Cs^{I}Ni^{II}_{x}Co^{II}_{1-x}Cr^{III}$) [63]. The main difficulty, here, is to ensure that the core–shell and the alloy-like objects possess the assumed architecture and do not consist of mixtures of pure Ni^{II}- and pure Co^{II}-containing objects. Because of the relatively small size of the objects, TEM imaging cannot discriminate between pure and core–shell (or alloy-like) objects; neither infrared spectroscopy nor X-ray powder diffraction do. Actually, the best technique is magnetism. The temperature dependence of the ZFCM curves show that the blocking temperature (maximum of the curve) of the core–shell objects $Cs^{I}Ni^{II}Cr^{III}@Cs^{I}Co^{II}Cr^{III}$ (28.4 K) is larger than those of the pure $Cs^{I}Ni^{II}Cr^{III}$ and the alloy-like $Cs^{I}Ni^{II}{}_{x}Co^{II}_{1-x}Cr^{III}$ particles (12.7 and 21.3, respectively) demonstrating the formation of the core–shell objects (Fig. 3.20). The hysteresis loops are consistent with the trend of the blocking temperatures (Fig. 3.20); the particles with the highest blocking temperature possess the largest coercive field. The coercive field of $Cs^{I}Ni^{II}Cr^{III}$ nanoparticles is the lowest reflecting the relatively weak anisotropy of Ni^{II}. The coercive fields of the alloy-like, $Cs^{I}Ni^{II}{}_{x}Co^{II}_{1-x}Cr^{III}$, and the core-shell, $Cs^{I}Ni^{II}Cr^{III}@Cs^{I}Co^{II}Cr^{III}$, objects, are equal to 530 and 2360 Oe, respectively. For 10 nm core–shell particles with a shell of 1 nm of $Cs^{I}Co^{II}Cr^{III}$ network, the percentage of Co^{II} ions is 47 among them 26 are at the surface layer, while for the alloy-like objects the percentage of Co^{II} ions present at the surface is only 12% assuming a statistical distribution of Ni^{II} and Co^{II} ions. The magnetic moment of the alloy-like particles is the same as for the core–shell ones because they have the same composition and the same size. The mean volume V is thus the same. The difference in the blocking temperature, that is proportional to their energy barrier ($\Delta E = K_{\mathrm{eff}}V$), is due to the difference in their anisotropy constant that is larger for the core–shell objects than for the allow-like ones' due to the larger number of anisotropic Co^{II} ions at the surface for the former than for the latter.

The hysteresis loop of the core–shell objects (Fig. 3.20) does not show any break and the ZFCM curve has one maximum and not two that would correspond to the core and to the shell. This is consistent with an intimate coupling between the shell and the core magnetization that rotate together in the presence of the magnetic field, and not to the sum of two independent components, as a result of a ferromagnetic coupling at the interface between the two networks. This is an indirect evidence of the formation of the core–shell objects and not to a mixture of particles of the pure networks during the synthetic process. In this case, shape anisotropy and local anisotropy are not in competition, which probably contribute to the larger overall anisotropy of the Co^{II}-containing objects.

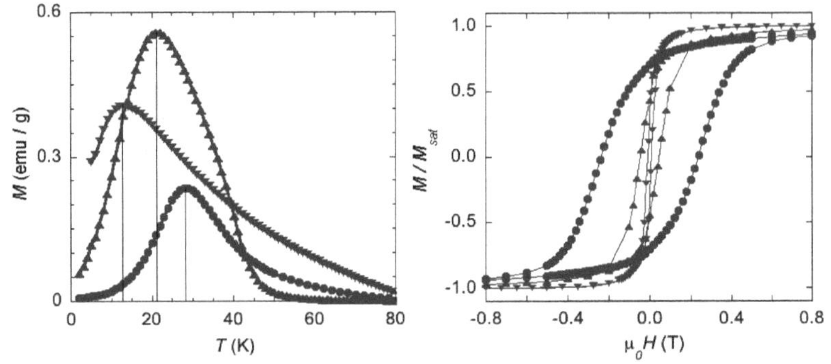

Figure 3.20 Zero-Field Cooled Magnetization curves (with an applied field of 30 Oe) for pure $Cs^{I}Ni^{II}Cr^{III}$ (▼), alloy-like $Cs^{I}Ni^{II}_{x}Co^{II}_{1-x}Cr^{III}$ (▲) and core–shell $Cs^{I}Ni^{II}Cr^{III}$@$Cs^{I}Co^{II}Cr^{III}$ (●) particles showing maxima at 13, 28 and 21 K, respectively (left) and normalized hysteresis loops at T = 2 K. Adapted from Ref. [63] with permission from The Royal Society of Chemistry.

3.7.2 Other Core–Shell Particles: Preparation and Magnetic Behavior

In order to reveal the effect of surface anisotropy and preclude the contribution of volume anisotropy, one possible approach is to use a diamagnetic core in combination with ferromagnetic shells. This is theoretically possible using diamagnetic nanoparticles as the core network. However, it was not possible to prepare diamagnetic nanoparticles with a small size (less than 10 nm) and narrow distribution on which a ferromagnetic network could be grown.

A smart approach consists in using a void as a core. This can be carried out by the same seeds-growth approach and then dissolving the core network. Such approach leads, however, to shell-particles containing a large amount of vacancies necessary to permit the dissolution of the core, but may have an impact on their magnetic behavior.

Prussian blue nanoparticles containing a void of about 130 nm and a shell of 30 nm were prepared by chemically etching the objects [64]. The comparison of the magnetic behavior of filled and hollow 190 nm particles shows a difference in their remnant magnetization while their coercive field remains the same. The larger remnant magnetization is due to an increase of intraparticle dipolar interaction between the interior parts of the shell that allows a better alignment of the spins with the magnetic field. The fact that the coercive field did not change is not surprising because of the relatively thick shell (30 nm) where mainly volume anisotropy is present.

Mn^{II}-containing PBA networks are more soluble than the related Ni^{II} or Co^{II} ones. $Rb^{I}Mn^{II}Fe^{III}$@$Rb^{I}Co^{II}Fe^{III}$@$Ni^{II}Cr^{III}$ were, thus, prepared using the seeds-growth approach and then the core was dissolved by leaving the objects in water at 45°C for 45 min leading to the hollow particles □@$Rb^{I}Co^{II}Fe^{III}$@$Ni^{II}Cr^{III}$ that have a 40 nm void, a first shell of 11 nm and second shell of 12 nm [65]. The magnetic studies show a behavior corresponding to the superposition of the $Rb^{I}Co^{II}Fe^{III}$ and the $Ni^{II}Cr^{III}$ properties with critical temperatures of 20 and 75 K, respectively. Simple estimation of the volume of $Ni^{II}Cr^{III}$ metal ions contributing to the outer 12 nm shell leads to the conclusion that it corresponds to particles of about 90 nm size. Since the single domain size of the $Ni^{II}Cr^{III}$ objects was found to be around 22 nm, the outer shell will behave as a bulk network from the magnetic point of view with, thus, the same critical temperature as the bulk as experimentally observed. This is also the case for the intermediate $Rb^{I}Co^{II}Fe^{II}$ 11 nm shell (whose size corresponds to an equivalent particle of 50 nm) that will, thus, have the behavior of the bulk network also as observed. So, the hollow □@$Rb^{I}Co^{II}Fe^{III}$@$Ni^{II}Cr^{III}$ particles have a magnetic behavior corresponding to the superposition of the two networks.

It is possible to grow PBA shells onto another nanomaterial such as gold. 133 nm particles with a gold core of 21 nm and two subsequent shells of 16 and 40 nm of the $K^{I}Ni^{II}Fe^{III}$ and $K^{I}Ni^{II}Cr^{III}$ networks were prepared (Au^{0}@$K^{I}Ni^{II}Fe^{III}$@$^{I}Ni^{II}Cr^{III}$:21@2 × 16@2 × 40 nm). Then gold was dissolved by HF leading to the hollow objects □@$K^{I}Ni^{II}Fe^{III}$@$K^{I}Ni^{II}Cr^{III}$ [66]. Since the outer shell has a much larger volume than the inner one (around 25 times larger), its magnetic behavior dominates the magnetic properties of the particles. This system is quite interesting because it can serve as a test for the validity of the model developed for the $Cs^{I}Ni^{II}Cr^{III}$@$Cs^{I}Co^{II}Cr^{III}$ (8 + 2 × 1 = 10 nm) particles where enhanced anisotropy was assumed to be due to the surface ion; hollow particles with an outer shell not thicker than few nanometers would be interesting to investigate for this purpose.

3.8 Magnetic Properties of PBA Nanoparticles Assembled as Single Layers on Substrates: From Assembly to Single Object Behavior

3.8.1 Nanoparticles Assembled on Functionalized Si(100) Surface

Ultimately, it is desired to investigate the magnetism of single nanoparticles, which can be motivated by a fundamental understanding of the interplay between surface and shape anisotropy. One approach to do so is assemble the particles on substrates with a control of their density. The PBA objects possess two important features that can be exploited to do so: (1) they are stable in solution in the absence of templating agents with the surface M_C^{III} ions terminated by CN ligands and (2) they are negatively charged. These two features allow their assembly on adequately prepared substrates either by electrostatic interaction and/or by coordination bonds.

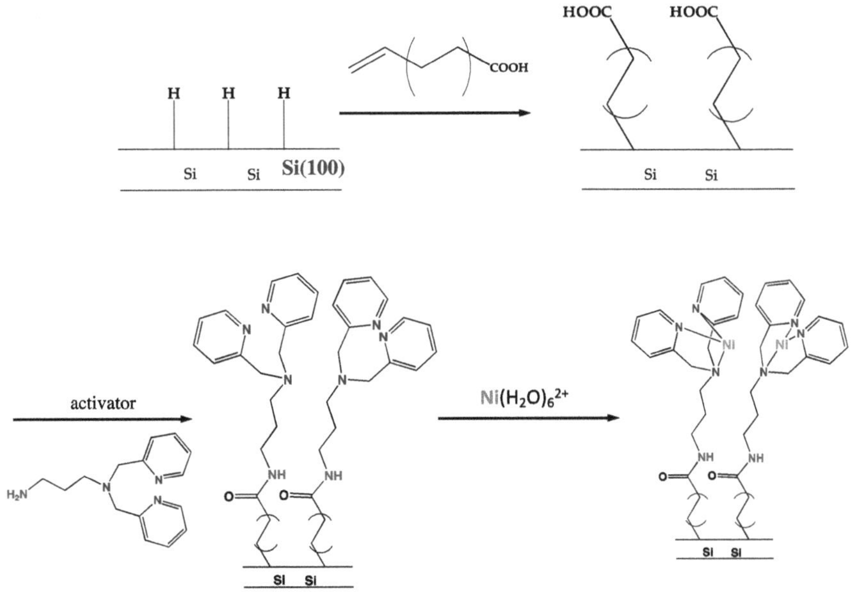

Figure 3.21 Synthetic scheme of the functionalization of Si(100) substrate.

A first approach consists in using a functionalized Si(100) substrate following the scheme presented in Fig. 3.21: (1) a Si(100) substrate with a native SiO_2 layer is etched by HF leading to H-terminated Si(100) surface, (2) the as-prepared surface is functionalized by an undecanoic/decane:1/100 mixture, (3) a peptidic bond is realized by reacting the dangling carboxylate groups with an amine terminated tridentate ligand and (4) coordination of Ni^{II} was performed by immerging the substrate in aqueous solution of $[Ni^{II}(H_2O)_6]^{2+}$ [67].

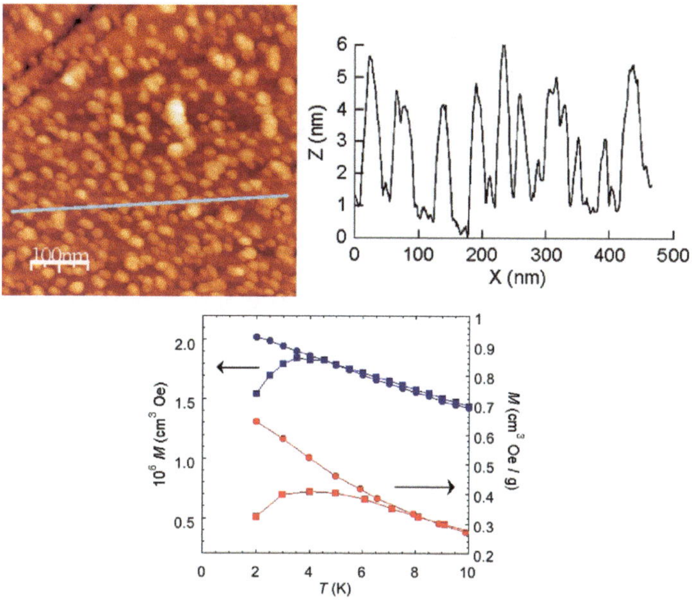

Figure 3.22 AFM image of the monolayer of the 6 nm $Cs^INi^{II}Cr^{III}$ nanoparticles (left) and topography along a 0.5 μm line (middle) and FC (●) and ZFC (■) magnetization curves for the nanoparticles diluted in PVP (blue) and grafted onto Si(100) (red) (right). Adapted with permission from Fleury, B., Volatron, F., Catala, L., Brinzei, D., Rivire, E., Huc, V., David, C., Miserque, F., Rogez, G., Baraton, L., Palacin, S., Mallah, T. (2008). Grafting a monolayer of superparamagnetic cyanide-bridged coordination nanoparticles on Si(100), *Inorg. Chem.*, **47**, pp. 1898–1900. Copyright (2008) American Chemical Society.

Soaking the as-prepared substrate in an aqueous solution containing the 6 nm $Cs^INi^{II}Cr^{III}$ nanoparticles leads to relatively isolated objects as shown by atomic force microscopy (AFM) imaging (Fig. 3.22). The magnetic properties in the form of FCM and ZFCM curves were measured for the single layer of particles and compared to those embedded in PVP. The blocking temperatures corresponding to the maximum of ZFCM are almost the same (3.5 K) demonstrating that the objects survive the grafting process and that they are well isolated on the surface. It is, unfortunately, not possible to have further information on the magnetic behavior and particularly the magnetic anisotropy of the objects from magnetometry studies. One possibility is to perform angle-dependent XMCD studies or to use local probes as shown in the following sections.

3.8.2 Nanoparticles Assembled as Stripes on Functionalized Si(100)

The magnetic properties of the 6 nm $Cs^{I}Ni^{II}Cr^{III}$ objects were studied when organized as stripes on a patterned surface. Focused ion beam (FIB) with a low dose of Ga^{+} source was used to perform stripes at the functionalized Si(100) surface (Fig. 3.21). The substrate was rinsed and then soaked in to the aqueous solution containing the nanoparticles leading to a surface patterned with 2-micrometers stripes on which a single layer of the NPs is assembled [68]. The NPs density within the stripes is large so that dipolar interactions are present leading to a collective behavior with a critical temperature close to that of the bulk (90 K).

The magnetic behavior was investigated locally using a Scanning Hall probe microscopy (SHPM) technique that directly probes the magnetic stray fields at surfaces. For T = 20 K ($<T_C$), a representative series of pictures of the grafted nanoparticles on a scan area of 7 × 7 μm^2, was collected (Fig. 3.23). Specifically, a hysteresis-type cycle is performed by varying the applied field within H = ±10 kOe. Figure 3.23 shows, for comparison, the hysteresis loop of $Cs^{I}Ni^{II}Cr^{III}$ particles when measured by conventional magnetometry on a massive powder specimen at 20 K. We first note that, for the same conditions of height of the tip and value of H, the color scales in Fig. 3.23b–h indicate stronger stray fields than that observed for T = 100 K in a previous experience. Although the contours are relatively smooth

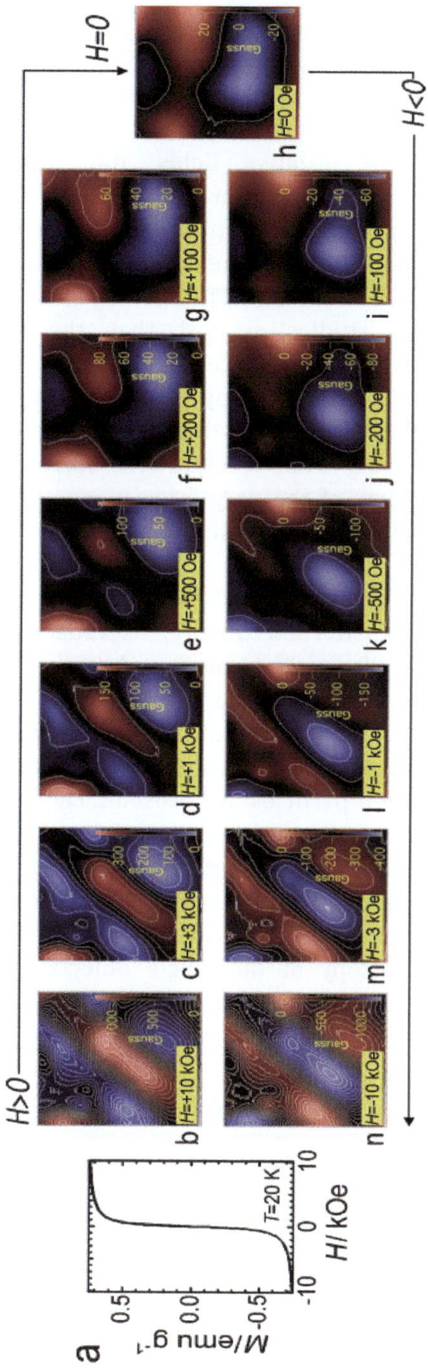

Figure 3.23 (a) Magnetization curve M versus H taken on a powder sample of $Cs^{I}Ni^{II}Cr^{III}$ nanoparticles. (b–n) Applied-field evolution of the magnetic domains of the particles grafted on patterned subtract. These images were collected at $T = 20$ K with the tip in the lift mode ($h = 350$ nm). The scan area is 7×7 μm^2. ©Copyright adapted from Ref. [68].

because of the height of the Hall probe from the sample, the main magnetic features are evident. Oblique stripes are neatly visible for $H > 1$ kOe (b–d). As H vanishes (e–g) down to $H = 0$ (h), magnetic spots appear, suggesting the presence of random magnetic domains. At negative fields, the spots rearrange themselves (i–k) to again obtain the stripe-like shape for $H < 1$ kOe (l–n), though with opposite direction with respect to positive H. For $|H| > 1$ kOe the intensity of the stray field increases without significant modification of the image.

The topographic studies suggest that the particles are elongated along the planar surface as was observed by TEM imaging. A particle shape anisotropy thus promotes a hard axis of magnetization perpendicularly to the surface. As a consequence, the dipolar field arising from interparticle correlations should lie in the plane. In view of the very large magnetic moment per particle and the relatively high surface coverage, the dipolar field is significantly large. At zero applied fields, one would therefore expect the presence of magnetic domains uncorrelated with respect to the patterned texture, as indeed observed in Fig. 3.23h. As the applied field is increased, both the anisotropy and the dipolar field are overcome, and the magnetic moments of the nanoparticles tend to align along the magnetic field direction. In this situation, strong stray fields are detected solely in the region where the particles are present. The analysis of the data together with micromagnetic calculations lead to the conclusion that, at 20 K, in addition to the out-of-plane magnetic moment that develops along the applied magnetic field, one needs to consider an in-plane component with a M_{in}/M_{out} ratio of 0.9. The shape anisotropy of the single objects plays thus an important role in the overall behavior of a collection of the particles.

3.8.3 Magnetic Behavior of Single PBA Nanoparticles

For this experiment, negatively charged $K^{I}Ni^{II}Cr^{III}$ particles with an average size around 22 nm obtained by spontaneous stabilization were used [69]. Their average size is close to the single magnetic domain one, which allows the investigation of the magnetic behavior of single objects above and below the critical size. A slightly different approach, from the one presented above, was used to position the nanoparticles on different location of the substrate. A silicon substrate was locally oxidized by an AFM tip leading to SiO_2 protuberance that were then functionalized by protonated aminopropyltriethoxysilane (APTES) [70]. Then the substrate was immerged in the solution containing the particles. Electrostatic interaction allows positioning the particles only on the protonated APTES locations.

Low temperature Magnetic Force Microscopy (LT-MFM) was used to perform the local magnetic characterization of the particles and investigate their magnetic reversal upon applying the magnetic field [69]. Well-isolated NPs are present on the surface. However, couples of NPs or small aggregates can be found; they are not always easy to differentiate due to the small size of the NPs compared with the size of the tip apex (≈50 nm). As an example, single isolated NPs and dimers of NPs can be observed. The height profiles point out, that the single isolated NP has 19.7 ± 2 nm height and the two attached NPs has heights of 20.8 ± 2.1 and 22.2 ± 2.2 nm.

We focus on the case of an isolated object. The MFM tip was previously aligned in the +*z* direction and the magnetic moment of the NP was fully saturated in the same direction by applying an external magnetic field of +10 kOe.

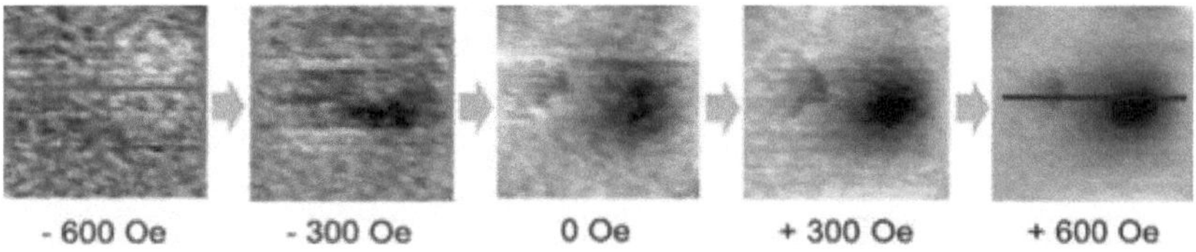

Figure 3.24 Set of magnetic force images of a single $K^INi^{II}Cr^{III}$ particle (image sizes: 114 × 114 nm^2) at B_{ext} of −600 Oe, −300 Oe, 0 Oe, +300 Oe, and +600 Oe, from left to right. ©Copyright adapted from Ref. [69].

A set of magnetic images (Fig. 3.24) were recorded sweeping the external magnetic field from −600 to +600 Oe. The image shows bright contrast at B_{ext} = −600 Oe, reflecting magnetic repulsion between tip and NP, so the magnetization of the NP is pointing to the −*z* direction (opposite to the MFM tip). At B_{ext} = −300 Oe, almost the same bright contrast is observed, so the magnetization reversal of the NP has not taken place yet. Once the external field is increased up to 0 Oe, the resultant magnetic moment of the NP presents dark contrast. The magnetic moments of tip and NP are now pointing in the same direction (attractive interaction). So, the magnetic reversal occurs at a field between −300 and 0 Oe. Normally, the reversal must occur at a positive field (the coercive field of the particle). But the presence of the stray field of the tip, when oriented in opposite direction to the external field (negative applied field), shifts negatively the reversal field of the particle. If we assume that the tip stray field is around +300 Oe, the reversal field will indeed be positive. This is consistent with the magnetic measurements that give a value of ±170 Oe for the coercive field of the particles dispersed in PVP. The intensity of the dark contrast increases as B_{ext} reaches positive values and at B_{ext} = +600 Oe, the NP seems to have been fully reversed. This experiment on a 20 nm single object assumed to be isolated shows that the coercive field is very close to an assembly of particles embedded in PVP whether highly diluted or not. This leads to the conclusion that the magnetization reversal of a 20 nm object should not be highly affected by dipolar interaction so that single objects may be addressed even though they are part of a relatively dense assembly of objects. Other similar experiments were performed on couples of particles and lead to the conclusion that the coercive field of one of the objects in influenced by the other one due to an antiferromagnetic coupling between the two objects. This is in contrast with the experiment performed on a dense assembly of objects within single layer (Section 8.4) where a ferromagnetic coupling (due to dipolar interaction) was observed. The difference is probably due to the collective influence of dipolar interaction. When the NPs are assembled in a bi-dimensional layer the collective effects forces the local magnetization of the single objects to lie within the layer to minimize dipolar energy. While for a couple of particles, if the easy axis of each particle is perpendicular to the surface, the dipolar energy is minimized when their magnetic moments are antiparallel (antiferromagnetic coupling).

3.9 Conclusion

The ability to stabilize perfectly well-crystallized PBA nanoparticles in aqueous solution in the form negatively charged objects opened the way to prepare objects with tunable size and a size control down to the molecular layer, using the seeds-growth approach. Using the same approach, epitaxial growth leads to core–multishell objects possessing perfectly well-defined core/shell and shell/shell interfaces. The molecular nature the objects excludes any interpenetration of the two different networks at the level of the interface because the two coordination sites (M_C and M_N) are electronically different precluding their exchange (Ni^{II} or Co^{II} are not stable in the octahedral C site). This is why these objects have perfectly well-defined interface, more difficult to achieve in oxides' based nanoparticles. All these characteristics are required for a thorough study of their magnetic properties and particularly the effect of the surface molecules on their magnetic anisotropy. Another important aspect that results from the molecular nature of these objects is the knowledge of the coordination sphere of the metal ions at the objects' surface; the M_C metal ions are octahedral with 6 carbon atoms and no exception is possible and the M_N metal ions have at least one water molecule and are distorted. These characteristics allowed us to reduce the number of unknown parameters and to relate the local anisotropy of the surface ions to that of model molecules. The study of the magnetic anisotropy of a single layer of nanoparticles and of isolated objects reveal the interplay between local anisotropy, the anisotropy of the objects with the contribution of the shape and collective behavior governed by dipolar interactions. One interesting perspective is the design of core–multishell objects with alternation of ferromagnetic and diamagnetic networks with a tunable thickness of the diamagnetic spacer, which allows investigating the magnetic coupling between two networks within a single object. Such objects should be chemically accessible.

Acknowledgments

The authors thank the Institut Universitaire de France (IUF) for financial support.

References

1. Frisch, J. L. (1710). Notitia caerulei berolinensis nuper inventi, *Miscellanea Berolinensia Ad Incrementum Sci.*, **1**, pp. 377–378.

2. Robin, M. B. (1961). The color and electronic configurations of Prussian blue, *Spectrochim Acta,* **17**, pp. 1095–1095.
3. Ludi, A., Gudel, H. U. (1968). Structure of $Co_3(Co(Cn)_6)_2$ and $Cd_3(Co(Cn)_6)_2$, Hydrates, *Helv. Chim. Acta,* **51**, pp. 2006–2016.
4. Brown, D. B., Shriver, D. F. (1969). Structures and solid-state reactions of Prussian blue analogs containing chromium manganese iron and cobalt, *Inorg. Chem.,* **8**, pp. 37–42.
5. Ron, G., Ludi, A., Engel, P. (1973). Crystal-structures of cadmium hexacyanocobaltate(III) and cadmium hexacyanocobaltate(III) hydrate, *Chimia,* **27**, pp. 77–79.
6. Buser, H. J., Ludi, A., Fischer, P., Studach, T., Dale, B. W. (1974). Neutron-diffraction study of Prussian blue, $Fe_4[Fe(Cn)_6]_{3.14}D_2o$, *Z. Phys. Chem. Neue. Fol.,* **92**, pp. 354–357.
7. Beall, G. W., Milligan, W. O., Korp, J., Bernal, I. (1977). Crystal-structure of $Mn_3[Co(Cn)_6]_{2.12}h_{2o}$ and $Cd_3[Co(Cn)_6]_{2.12}H_2O$ by neutron and x-ray-diffraction, *Inorg. Chem.,* **16**, pp. 2715–2718.
8. Buser, H. J., Schwarzenbach, D., Petter, W., Ludi, A. (1977). Crystal-structure of Prussian blue—$Fe_4[Fe(Cn)_6]_3 \cdot xH_2O$, *Inorg. Chem.,* **16**, pp. 2704–2710.
9. Catala, L., Mallah, T. (2017). Nanoparticles of Prussian blue analogs and related coordination polymers: From information storage to biomedical applications, *Coord. Chem. Rev.,* **346**, pp. 32–61.
10. Zhao, Y. Y., Zhang, W., Hu, M. (2017). Well-defined cyanometallate coordination-polymer nanoarchitectures realized by wet-chemical manipulation, *Chemnanomat,* **3**, pp. 780–789.
11. Griebler, W. D., Babel, D. (1982). X-ray and magnetic studies of perovskite-related cyano compounds $CsMIIMIII(Cn)_6$, *Z. Naturforsch. B,* **37**, pp. 832–837.
12. Ellis, D., Eckhoff, M., Neff, V. D. (1981). Electrochromism in the mixed-valence hexacyanides. 1. Voltammetric and spectral studies of the oxidation and reduction of thin-films of Prussian blue, *J. Phys. Chem.,* **85**, pp. 1225–1231.
13. Itaya, K., Shibayama, K., Akahoshi, H., Toshima, S. (1982). Prussian-blue-modified electrodes: An application for a stable electrochromic display device, *J. Appl. Phys.,* **53**, pp. 804–805.
14. DeLongchamp, D. M., Hammond, P. T. (2004). High-contrast electrochromism and controllable dissolution of assembled Prussian blue/polymer nanocomposites, *Adv. Funct. Mater.,* **14**, pp. 224–232.

15. Shiozaki, H., Kawamoto, T., Tanaka, H., Hara, S., Tokumoto, M., Gotoh, A., Satoh, T., Ishizaki, M., Kurihara, M., Sakamoto, M. (2008). Electrochromic thin film fabricated using a water-dispersible ink of Prussian blue nanoparticles, *Jpn. J. Appl. Phys.*, **47**, pp. 1242–1244.

16. Matsui, J., Kikuchi, R., Miyashita, T. (2014). A trilayer film approach to multicolor electrochromism, *J. Am. Chem. Soc.*, **136**, pp. 842–845.

17. Lee, K. M., Tanaka, H., Takahashi, A., Kim, K. H., Kawamura, M., Abe, Y., Kawamoto, T. (2015). Accelerated coloration of electrochromic device with the counter electrode of nanoparticulate Prussian blue-type complexes, *Electrochim. Acta*, **163**, pp. 288–295.

18. Sato, O., Iyoda, T., Fujishima, A., Hashimoto, K. (1996). Photoinduced magnetization of a cobalt-iron cyanide, *Science*, **272**, pp. 704–705.

19. Moulin, C. C. D., Villain, F., Bleuzen, A., Arrio, M. A., Sainctavit, P., Lomenech, C., Escax, V., Baudelet, F., Dartyge, E., Gallet, J. J., Verdaguer, M. (2000). Photoinduced ferrimagnetic systems in Prussian blue analogues $C_x(I)Co_4[Fe(CN)_{(6)}](y)$ (C^I = alkali cation). 2. X-ray absorption spectroscopy of the metastable state, *J. Am. Chem. Soc.*, **122**, pp. 6653–6658.

20. Escax, V., Bleuzen, A., Moulin, C. C. D., Villain, F., Goujon, A., Varret, F., Verdaguer, M. (2001). Photoinduced ferrimagnetic systems in Prussian blue analogues (CxCo4)-Co-I[Fe(CN)(6)](y) (C-I = alkali cation). 3. Control of the photo- and thermally induced electron transfer by the [Fe(CN)(6)] vacancies in cesium derivatives, *J. Am. Chem. Soc.*, **123**, pp. 12536–12543.

21. Moore, J. G., Lochner, E. J., Ramsey, C., Dalal, N. S., Stiegman, A. E. (2003). Transparent, superparamagnetic K(x)(I)Coy(II)[Fe-III(CN)(6)]1-Silica nanocomposites with tunable photomagnetism, *Angew. Chem. Int. Ed.*, **42**, pp. 2741–2743.

22. Hozumi, T., Hashimoto, K., Ohkoshi, S. (2005). Electrochemical synthesis, crystal structure, and photomagnetic properties of a three-dimensional cyano-bridged copper-molybdenum complex, *J. Am. Chem. Soc.*, **127**, pp. 3864–3869.

23. Fornasieri, G., Bleuzen, A. (2008). Controlled synthesis of photomagnetic nanoparticles of a Prussian blue analogue in a silica xerogel, *Angew. Chem. Int. Ed.*, **47**, pp. 7750–7752.

24. Aouadi, M., Fornasieri, G., Briois, V., Durand, P., Bleuzen, A. (2012). Chemistry of cobalt(II) confined in the pores of ordered silica monoliths: From the formation of the monolith to the CoFe Prussian blue analogue nanocomposite, *Chem. Eur. J.*, **18**, pp. 2617–2623.

25. Ohkoshi, S. I., Tokoro, H. (2012). Photomagnetism in cyano-bridged bimetal assemblies, *Acc. Chem. Res.*, **45**, pp. 1749–1758.
26. Gadet, V., Mallah, T., Castro, I., Verdaguer, M., Veillet, P. (1992). High-Tc molecular-based magnets—a ferromagnetic bimetallic chromium(III) nickel(II) cyanide with Tc = 90 K, *J. Am. Chem. Soc.*, **114**, pp. 9213–9214.
27. Mallah, T., Ferlay, S., Auberger, C., Helary, C., LHermite, F., Ouahes, R., Vaissermann, J., Verdaguer, M., Veillet, P. (1995). Hexacyanometalates: Molecular precursors for high-spin molecules and high-Tc molecule-based magnets, *Mol. Cryst. Liq. Crys. A*, **273**, pp. 141–151.
28. Verdaguer, M., Bleuzen, A., Marvaud, V., Vaissermann, J., Seuleiman, M., Desplanches, C., Scuiller, A., Train, C., Garde, R., Gelly, G., Lomenech, C., Rosenman, I., Veillet, P., Cartier, C., Villain, F. (1999). Molecules to build solids: High-Tc molecule-based magnets by design and recent revival of cyano complexes chemistry, *Coord. Chem. Rev.*, **190**, pp. 1023–1047.
29. Ruiz, E., Rodriguez-Fortea, A., Alvarez, S., Verdaguer, M. (2005). Is it possible to get high T-c magnets with Prussian blue analogues? A theoretical prospect, *Chem. Eur. J.*, **11**, pp. 2135–2144.
30. Moulik, S. P., De, G. C., Panda, A. K., Bhowmik, B. B., Das, A. R. (1999). Dispersed molecular aggregates. 1. Synthesis and characterization of nanoparticles of $Cu_{-2}[Fe(CN)_{(6)}]$ in H_2O/AOT/n-heptane water-in-oil microemulsion media, *Langmuir*, **15**, pp. 8361–8367.
31. Brinzei, D., Catala, L., Louvain, N., Rogez, G., Stephan, O., Gloter, A., Mallah, T. (2006). Spontaneous stabilization and isolation of dispersible bimetallic coordination nanoparticles of CsxNi[Cr(CN)(6)](y), *J. Mater. Chem.*, **16**, pp. 2593–2599.
32. Taleb, A., Petit, C., Pileni, M. P. (1997). Synthesis of highly monodisperse silver nanoparticles from AOT reverse micelles: A way to 2D and 3D self-organization, *Chem. Mater.*, **9**, pp. 950–959.
33. Vaucher, S., Dujardin, E., Lebeau, B., Hall, S. R., Mann, S. (2001). Higher order construction of molecule-based magnets, *Chem. Mater.*, **13**, pp. 4408–4410.
34. Vaucher, S., Fielden, J., Li, M., Dujardin, E., Mann, S. (2002). Molecule-based magnetic nanoparticles: Synthesis of cobalt hexacyanoferrate, cobalt pentacyanonitrosylferrate, and chromium hexacyanochromate coordination polymers in water-in-oil microemulsions, *Nano Lett.*, **2**, pp. 225–229.

35. Catala, L., Gacoin, T., Boilot, J. P., Riviere, E., Paulsen, C., Lhotel, E., Mallah, T. (2003). Cyanide-bridged Cr-III-Ni-II superparamagnetic nanoparticles, *Adv. Mater.,* **15**, pp. 826–+.

36. Clavel, G., Guari, Y., Larionova, J., Guerin, C. (2005). Formation of cyano-bridged molecule-based magnetic nanoparticles within hybrid mesoporous silicaw, *New J. Chem.,* **29**, pp. 275–279.

37. Folch, B., Guari, Y., Larionova, J., Luna, C., Sangregorio, C., Innocenti, C., Caneschi, A., Guerin, C. (2008). Synthesis and behaviour of size controlled cyano-bridged coordination polymer nanoparticles within hybrid mesoporous silica, *New J. Chem.,* **32**, pp. 273–282.

38. Prado, Y., Mazerat, S., Riviere, E., Rogez, G., Gloter, A., Stephan, O., Catala, L., Mallah, T. (2014). Magnetization reversal in (CsNiCrIII)-Cr-II(CN)(6) coordination nanoparticles: Unravelling surface anisotropy and dipolar interaction effects, *Adv. Funct. Mater.,* **24**, pp. 5402–5411.

39. Catala, L., Brinzei, D., Prado, Y., Gloter, A., Stephan, O., Rogez, G., Mallah, T. (2009). Core-multishell magnetic coordination nanoparticles: Toward multifunctionality on the nanoscale, *Angew. Chem. Int. Ed.,* **48**, pp. 183–187.

40. Ito, A., Suenaga, M., Ono, K. (1968). Mossbauer study of soluble Prussian blue insoluble Prussian blue and Turnbulls blue, *J. Chem. Phys.,* **48**, pp. 3597–3600.

41. Mayoh, B., Day, P. (1976). Charge-transfer in mixed-valence solids. 8. Contribution of valence delocalization to ferromagnetism of Prussian blue, *J. Chem. Soc. Dalton Trans.*, pp. 1483–1486.

42. Rogez, G., Marvilliers, A., Riviere, E., Audiere, J. P., Lloret, F., Varret, F., Goujon, A., Mendenez, N., Girerd, J. J., Mallah, T. (2000). A mixed-valence mixed-spin Prussian-blue-like heptanuclear complex, *Angew. Chem. Int. Ed.,* **39**, pp. 2885–2887.

43. Kahn, O. (1993). *Molecular Magnetism* (VCH Publishers, Inc., New York), p. 145.

44. Mallah, T., Ferlay, S., Scuiller, A., Verdaguer, M. (1996). A rationale molecular approach to high-spin molecules and molecular magnets, *NATO Adv. Sci. I C-Mat.,* **484**, pp. 597–614.

45. Scuiller, A., Mallah, T., Verdaguer, M., Nivorozkhin, A., Tholence, J. L., Veillet, P. (1996). A rational route to high-spin molecules via hexacyanometalates: A new mu-cyano CrIIIMn6(II) heptanuclear complex with a low-lying S = 27/2 ground state, *New J. Chem.,* **20**, pp. 1–3.

46. Mallah, T., Auberger, C., Verdaguer, M., Veillet, P. (1995). A heptanuclear $CrNi_6$ complex with a low-lying S = 15/2 ground-state, *J. Chem. Soc. Chem. Commun.*, **1**, pp. 61–62.

47. Mallah, T., Thiebaut, S., Verdaguer, M., Veillet, P. (1993). High-T(C) molecular-based magnets—ferrimagnetic mixed-valence chromium(III)-chromium(II) cyanides with T(C) at 240-Kelvin and 190-Kelvin, *Science*, **262**, pp. 1554–1557.

48. Ferlay, S., Mallah, T., Ouahes, R., Veillet, P., Verdaguer, M. (1995). A room-temperature organometallic magnet based on Prussian blue, *Nature*, **378**, pp. 701–703.

49. Néel, L. (1949). Théorie du traînage magnétique des ferromagnétiques en grains fins avec applications aux terres cuites, *Ann. Geophys.*, **5**, pp. 99–136.

50. Néel, L. (1954). Anisotropie magnétique superficielle et surstructures d'orientation, *J. de Phys. et Le Radium*, **15**, pp. 225–239.

51. Brown, W. F., Jr. (1963). Thermal fluctuations of a single-domain particle, *Phys. Rev.*, **130**, pp. 1677–1686.

52. Dormann, J. L., Cherkaoui, R., Spinu, L., Nogues, M., Lucari, F., D'Orazio, F., Fiorani, D., Garcia, A., Tronc, E., Jolivet, J. P. (1998). From pure superparamagnetic regime to glass collective state of magnetic moments in gamma-Fe_2O_3 nanoparticle assemblies, *J. Magn. Magn. Mater.*, **187**, pp. L139–L144.

53. Aharoni, A. (1973). Relaxation time of superparamagnetic particles with cubic anisotropy, *Phys. Rev. B*, **7**, pp. 1103–1107.

54. Bessais, L., Ben Jaffel, L., Dormann, J. L. (1992). Relaxation time of fine magnetic particles in uniaxial symmetry, *Phys. Rev. B*, **45**, pp. 7805–7815.

55. Singh, V., Seehra, M. S., Bonevich, J. (2009). AC susceptibility studies of magnetic relaxation in nanoparticles of Ni dispersed in silica, *J. Appl. Phys.*, **105**, pp. 07B518–3.

56. Cahier, B., Perfetti, M., Zakhia, G., Naoufal, D., El-Khatib, F., Guillot, R., Riviere, E., Sessoli, R., Barra, A. L., Guihery, N., Mallah, T. (2017). Magnetic anisotropy in pentacoordinate Ni-II and Co-II complexes: Unraveling electronic and geometrical contributions, *Chem. Eur. J.*, **23**, pp. 3648–3657.

57. Charron, G., Malkin, E., Rogez, G., Batchelor, L. J., Mazerat, S., Guillot, R., Guihery, N., Barra, A. L., Mallah, T., Bolvin, H. (2016). Unraveling sigma and pi effects on magnetic anisotropy in cis-NiA(4)B(2) complexes: Magnetization, HF-HFEPR studies, first-

principles calculations, and orbital modeling, *Chem. Eur. J.*, **22**, pp. 16848–16860.

58. El-Khatib, F., Cahier, B., Lopez-Jorda, M., Guillot, R., Riviere, E., Hafez, H., Saad, Z., Guihery, N., Mallah, T. (2018). A bis-binuclear Ni-II complex with easy and hard axes of magnetization: Complementary experimental and theoretical insights, *Eur. J. Inorg. Chem.*, pp. 469–476.
59. Gomez-Coca, S., Cremades, E., Aliaga-Alcalde, N., Ruiz, E. (2013). Mononuclear single-molecule magnets: Tailoring the magnetic anisotropy of first-row transition-metal complexes, *J. Am. Chem. Soc.*, **135**, pp. 7010–7018.
60. Ruamps, R., Batchelor, L. J., Maurice, R., Gogoi, N., Jimenez-Lozano, P., Guihery, N., de Graaf, C., Barra, A. L., Sutter, J. P., Mallah, T. (2013). Origin of the magnetic anisotropy in heptacoordinate Ni-II and Co-II complexes, *Chem. Eur. J.*, **19**, pp. 950–956.
61. Bodker, F., Morup, S., Linderoth, S. (1994). Surface effects in metallic iron nanoparticles, *Phys. Rev. Lett.*, **72**, pp. 282–285.
62. Prado, Y., Arrio, M. A., Volatron, F., Otero, E., Moulin, C. C. D., Sainctavit, P., Catala, L., Mallah, T. (2013). Magnetic anisotropy of cyanide-bridged core and core-shell coordination nanoparticles probed by X-ray magnetic circular dichroism, *Chem. Eur. J.*, **19**, pp. 6685–6694.
63. Prado, Y., Dia, N., Lisnard, L., Rogez, G., Brisset, F., Catala, L., Mallah, T. (2012). Tuning the magnetic anisotropy in coordination nanoparticles: Random distribution versus core-shell architecture, *Chem. Commun.*, **48**, pp. 11455–11457.
64. Hu, M., Furukawa, S., Ohtani, R., Sukegawa, H., Nemoto, Y., Reboul, J., Kitagawa, S., Yamauchi, Y. (2012). Synthesis of Prussian blue nanoparticles with a hollow interior by controlled chemical etching, *Angew. Chem. Int. Ed.*, **51**, pp. 984–988.
65. Risset, O. N., Knowles, E. S., Ma, S. Q., Meisel, M. W., Talham, D. R. (2013). RbjMk[Fe(CN)(6)](l) (M = Co, Ni) Prussian blue analogue hollow nanocubes: A new example of a multilevel pore system, *Chem. Mater.*, **25**, pp. 42–47.
66. Maurin-Pasturel, G., Long, J., Guari, Y., Godiard, F., Willinger, M. G., Guerin, C., Larionova, J. (2014). Nanosized heterostructures of Au@ Prussian blue analogues: Towards multifunctionality at the nanoscale, *Angew. Chem. Int. Ed.*, **53**, pp. 3872–3876.
67. Fleury, B., Volatron, F., Catala, L., Brinzei, D., Rivire, E., Huc, V., David, C., Miserque, F., Rogez, G., Baraton, L., Palacin, S., Mallah, T. (2008).

Grafting a monolayer of superparamagnetic cyanide-bridged coordination nanoparticles on Si(100), *Inorg. Chem.*, **47**, pp. 1898–1900.

68. Ghirri, A., Candini, A., Evangelisti, M., Gazzadi, G. C., Volatron, F., Fleury, B., Catala, L., David, C., Mallah, T., Affronte, M. (2008). Magnetic imaging of cyanide-bridged Co-ordination nanoparticles grafted on FIB-patterned Si substrates, *Small*, **4**, pp. 2240–2246.
69. Pinilla-Cienfuegos, E., Kumar, S., Manas-Valero, S., Canet-Ferrer, J., Catala, L., Mallah, T., Forment-Aliaga, A., Coronado, E. (2015). Imaging the magnetic reversal of isolated and organized molecular-based nanoparticles using magnetic force microscopy, *Part. Part. Syst. Charac.*, **32**, pp. 693–700.
70. Coronado, E., Forment-Aliaga, A., Pinilla-Cienfuegos, E., Tatay, S., Catala, L., Plaza, J. A. (2012). Nanopatterning of anionic nanoparticles based on magnetic Prussian-blue analogues, *Adv. Funct. Mater.*, **22**, pp. 3625–3633.

Chapter 4

Strategies for Developing Nanocomposites Containing Cyano-Bridged Coordination Polymers

Giulia Fornasieri and Anne Bleuzen
ICMMO, Univ Paris Sud, Université Paris-Saclay, 91405 Orsay, France
giulia.fornasieri@u-psud.fr

Cyano-bridged coordination polymers have attracted much attention over the past 20 years because of the diversity and the tunability of their chemical structures leading to a wide range of electronic properties. This versatility makes Prussian blue-type materials candidates for applications in many fields, such as electronics [1–4], catalysis [5–7], separation [8], sensing [9, 10], biomedical applications [11], energy storage [12–14], and even nuclear decontamination [11, 15].

Nevertheless, the integration of such functional objects into real applications depends on an additional processing step allowing the control of their shape, size, and organization in (or at the surface of) a solid matrix. Several works report on the study of Prussian blue analogue (PBA) isolated nanoparticles

Prussian Blue-Type Nanoparticles and Nanocomposites: Synthesis, Devices, and Applications
Edited by Yannick Guari and Joulia Larionova

ISBN 978-981-4800-05-1 (Hardcover), 978-0-429-02474-0 (eBook)
www.panstanford.com

[16, 17]. Different synthetic strategies have been considered to control the size of the objects, such as the precipitation in reverse micelle media [18, 19] or in ionic liquids [20], the polymer-assisted co-precipitation [21–23], the template-free route [24, 25] based on the electrostatic stabilization of the nanoparticles as well as the precipitation in a hard confining matrix such as alumina, silica, and lamellar hydroxides. This chapter reviews the progress in the elaboration of PBA-containing nanocomposites; thus, it will focus in the confined precipitation of PBA in a preformed matrix and will not concern the synthesis of free individual nanoparticles.

4.1 Confined Precipitation of Prussian Blue Analogue: An Overview

Cyano-bridged coordination polymers are synthesized in aqueous solution by the reaction between a Lewis base cyanometallate complex anion $[M'(CN)_x]^{p-}$ (x = 6,8) and a Lewis acid M^{q+} solvated metal ion. The reaction consists in the substitution of the aqua ligands by the bridging cyano ligands, with an exchange rate that depends on the water-exchange rate of $[M(H_2O)_6]^{q+}$ and is usually very fast. The precipitation of Prussian blue analogues, obtained by simply mixing the two aqueous solutions, is generally almost instantaneous. Consequently, the synthesis of PBA in the porosity of a matrix faces two main problems that are external precipitation and pore obstruction from an untimely contact between the two species. In order to overcome these issues, several strategies that are reviewed in the following have been developed.

When we consider the works reporting on PBA-containing nanocomposites we can enumerate a large variety of matrices with different chemical nature, such as carbon materials (graphene, carbon nanotubes), biopolymers (alginate, chitosan, apoferritin, ...), synthetic organic polymers (nafion, polycarbonate, polyelectrolyte multilayer films, ...), and last but not least inorganic materials (alumina, silica, lamellar hydroxides, ...). The morphologies of matrices greatly vary too, with powders, beads, films, monoliths and with porosities going from the nanometer to the micrometer and that can be disordered or organized.

The elaboration schemes are not strictly dependent on the nature of the matrix; all of them intend to prevent the uncontrolled precipitation of PBA outside the host matrix. They can be grouped in three approaches:

(1) Introduction of two "inert" precursors and triggering of the precipitation by an electrochemical reaction.
(2) Step-by–step growth of PBA nanoparticles by anchorage of one precursor on the matrix followed by sequential impregnations with the two precursors.
(3) Two-step process consisting in the successive introduction of the two PBA precursors in the matrix.

4.2 Electrochemical Route

This approach consists in the immersion of the matrix in a solution containing unreactive species, such as M^{3+} and $[M'(CN)_6]^{3-}$ ions, pair of precursors known not to spontaneously precipitate as the corresponding PBA. An electrochemical process produces the selective reduction of one precursor to trigger the formation of the coordination polymer in the porosity of the matrix. The confinement is very effective; nevertheless, this technique can produce only a very limited number of compounds. Almost all works report on the synthesis of Prussian blue nanotubes and nanowires, though the precipitation of V(IV)Fe(II) [26] and Fe(II)Cr(III) [27, 28] PBA are reported.

The electrochemical route can be implemented on macroscopic matrices such as alumina [26–32] and polycarbonate membranes [33, 34] or silica mesoporous films [35], but also on dispersed carbon nanotubes [36], graphene nanosheets [37], and polystyrene microspheres [38].

The porous alumina templates are fabricated in-situ by a two-step anodizing process on an aluminum foil [39]. This template is known to present oxygen vacancies and to be positively charged; then $[M(CN)_6]^{3-}$ could be attracted by the electric charge of the pores walls and can be reduced to $[Fe(CN)_6]^{4-}$ through an electro-assisted redox process triggering the precipitation of $Fe_4[Fe(CN)_6]_3 \cdot xH_2O$ (Fig. 4.1). Morphology of PB depends on the electrodepositing process, alternative current (AC) electrodeposition produces dense single crystals nanowires (with diameters from 50 nm to 200 nm) growing from the bottom [26, 27, 29, 31] (Figs. 4.2a,b) while direct current (DC) cycling creates nanotubes (with external diameter of 200 nm) see Fig. 4.2c [30]. During DC electrodeposition nanotubes are assumed to nucleate in the pores walls and expand toward the center of the alumina pores gradually giving nanowires with longer deposition times.

The precipitation of PB in a polycarbonate membrane instead of alumina is used to synthesize free PB nanotubes and nanowires (Fig. 4.3) [33]. After precipitation of PB, the organic membrane can be easily dissolved with chloroform while elimination of alumina necessitates a strong basic treatment that can damage PB objects.

Mesoporous silica films, exhibiting regular arrays of mesoporous channels oriented perpendicular to the surface [40], have been considered as template. Electro-assisted self-assembly with CTAB is a powerful method to produce very appealing crack-free ordered silica with pores diameter of 3 nm. Unfortunately the mesopores walls, synthesized with hydroxyl ions as catalyst, are negatively charged and a resistance to mass transport of the $[Fe(CN)_6]^{3-}$ anions in such confined environment has been observed limiting the precipitation of PB in the porosity [35].

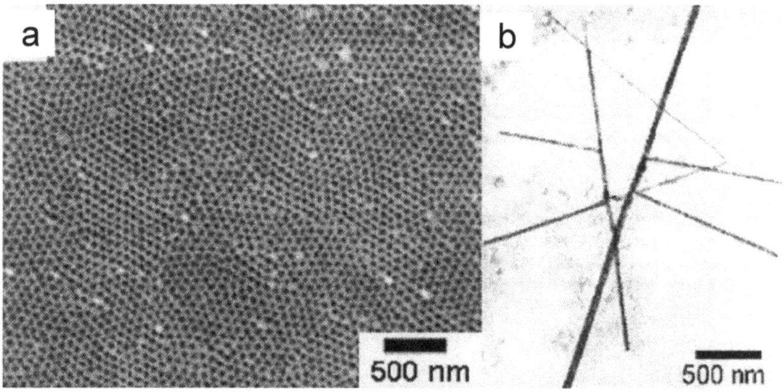

Figure 4.1 (a) SEM image of porous alumina template. (b) TEM image of dispersed Prussian blue nanowires. Adapted from [32], with the permission of AIP Publishing.

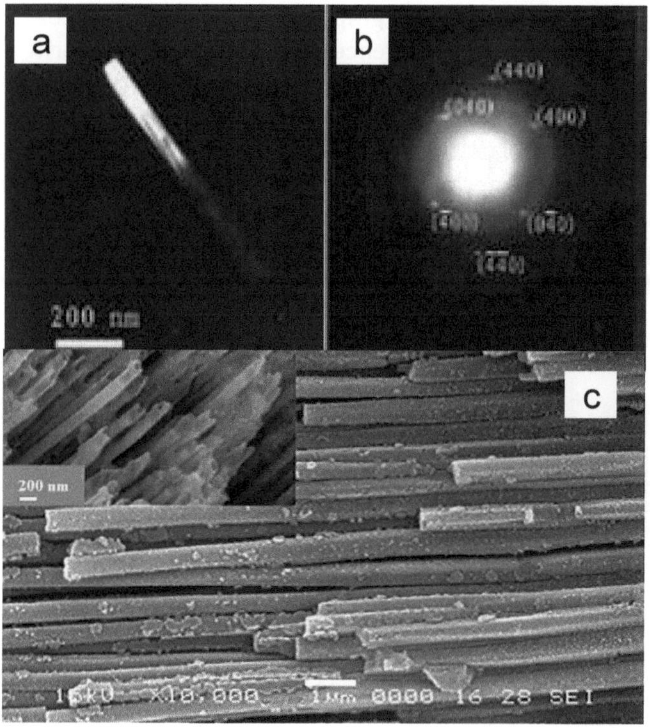

Figure 4.2 (a) Dark field image of a single Prussian blue nanowire, (b) corresponding selected area electron diffraction image (c) SEM image of highly ordered PB nanotube arrays after selective etching of polycarbonate matrix (inset: top view of PB nanotubes). Adapted from [29] and [30].

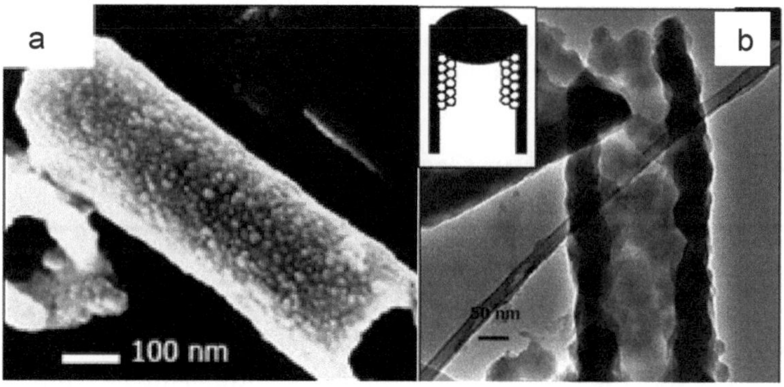

Figure 4.3 (a) SEM image of Prussian blue nanotube. (b) TEM image of a PB nanotube obtained after 30 min electrodeposition. Adapted from [33].

Carbon nanotubes and graphene nanosheets can spontaneously reduce Fe(III) to Fe(II). Hence PB/carbonaceous hybrid material can be prepared by simply mixing Fe^{3+} and $[Fe(CN)_6]^{3-}$ solutions in the presence of graphene or nanotubes aqueous dispersion. The PB nanoparticles are formed on the surface of the carbonaceous objects, giving 40 nm PB nanoparticles randomly deposited on multiwalled carbon nanotubes [36] (Fig. 4.4a) or 20 nm PB NPs on graphene nanosheets [37] with no isolated PB particles.

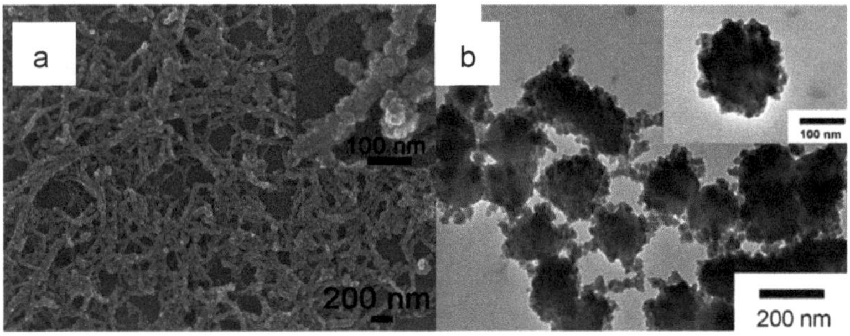

Figure 4.4 (a) SEM image of PB/carbon nanotube hybrids. (b) TEM image of strawberry-like PB/polystyrene microspheres. Adapted from [36] and [38].

A last example of PB composite by electrochemical route consists in the precipitation of 15 nm PB nanoparticles on the surface of 125 nm polystyrene (PS) spheres giving strawberry-like composites (Fig. 4.4b) [38]. The microspheres of polystyrene are beforehand functionalized by imidazolium groups, a mild reducing agent, and the precipitation of PB is obtained by dispersion of the microsphere in a Fe^{3+} and $[Fe(CN)_6]^{3-}$ solution.

4.3 Step-by-Step Growth

The anchorage of one precursor on the matrix is the starting point for the nucleation of PBA nanoparticles on the surface by sequential impregnations with the two precursor solutions. This method is very laborious but has the benefit to fully control the growth of the coordination network producing well-defined isolated nano-objects dispersed in the matrix. This approach has been applied to materials with various chemical natures:

alumina membranes [41, 42], biopolymers (alginate [43] and chitosan [44–46]), and silica [47–53].

For alumina nanocomposites the template is the alumina membrane already presented in the section on electrochemical route (Section 4.2). The positively charged surface can electrostatically interact with the negatively charged $[M(CN)_6]^{3-}$ entities, then the PBA nanoparticles growth is obtained by applying the solutions of the precursors sequentially to the template membranes with washings between every dipping. Literature reports on the elaboration of PB/alumina [41] and CoFe PBA/alumina nanocomposites [42]. Nanoparticles are formed and adhere on the inner wall of the template and can coalesce to create a film if a sufficient number of cycles are employed (10 or 12 cycles for alumina with diameters of pores of 200 nm).

Two types of biopolymers have been considered: chitosan of animal origin and the plant-derived alginate.

Chitosan is a biopolymer derived from the partial deacetylation of chitin and thus possessing free amino groups allowing coordination of metal ions. It is water soluble and can be processed in various forms: powders, pastes, films, gels. The drying of gel beads under supercritical CO_2 conditions lead to macroporous beads. The intrapore growth of the cyano-bridged network of PBA is obtained by an initial chemical anchorage of the divalent metal salts M^{2+} to the $-NH_2$ functionalities of porous chitin beads followed by consecutive repeated coordinations of the cyanometallate precursors $[M'(CN)_6]^{3-}$ (Fig. 4.5a). The precursors are solubilized in methanol and every impregnation lasts 48 h followed by washings and drying under vacuum. This method has been applied to a large variety of PBA, with anchoring Ni^{2+}, Cu^{2+}, Fe^{2+}, Co^{2+}, and Mn^{2+} cations followed by coordination by $[Fe(CN)_6]^{3-}$ and $[Cr(CN)_6]^{3-}$ [45]. The consecutive impregnations with Fe^{2+}, pyrazine, and $[Ni(CN)_4]^{2-}$ produces the growth of nanoparticles of the 3D spin crossover coordination polymer [Fe(pyrazine){$Ni(CN)_4$}] [46]. The step-by-step growth of PBA gives well-dispersed nanoparticles with sizes of 2–3 nm with an amount of metal ions into the host chitosan beads of around 6–8 wt% (Fig. 4.5b).

The solubilization in water of the as-obtained nanocomposites gives stable colloidal solutions containing nanoparticles with a residual chitosan shell.

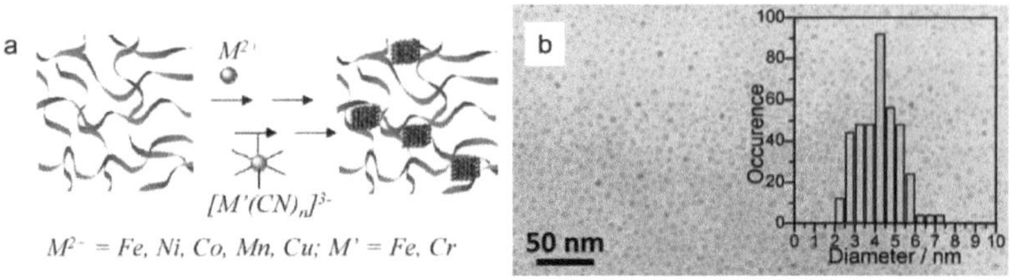

Figure 4.5 (a) Schematic representation of the preparation of cyano-bridged polymers into chitosan beads. (b) TEM image of NiFe PBA/chitosan after solubilization in aqueous solution. Adapted from [45].

Alginate is a polysaccharide constituted by block copolymer chains with β-D-mannurate and α-L-guluronate residues containing carboxylic groups (Fig. 4.6a). Divalent cations form with alginate ionotropic gels M^{q+}/alginate, which can be prepared in various shapes, such as beads, films, membranes and monoliths. Different gels can be prepared by putting into contact the aqueous sodium alginate solution with an aqueous solution containing M^{q+} cations (Ni^{2+}, Cu^{2+}, Fe^{2+}, Mn^{2+}, and Eu^{3+}). A gel is formed upon coordination by carboxylate functions with M^{q+} cations homogeneously dispersed in the three-dimensional matrix. After gel formation water is exchanged with acetonitrile. The polycyano metallate network is formed by successive coordination of cyanometallates $[M'(CN)_m]^{3-}$ ($M' = Fe^{3+}$; Cr^{3+} with $m = 6$ and Mo^{5+} with $m = 8$) unities followed by impregnation with M^{q+} solution in acetonitrile. Each impregnation takes 48 h followed by washings. After two growth cycles the nanocomposites are thoroughly washed and dried under super-critical CO_2 to obtain aerogels. This method gives coordination polymer nanoparticles with spherical shape, non-aggregated and having a size from 3 to 7 nm (Fig. 4.6b). The atomic ratio M/M′ calculated by EDS is higher than expected. This deviation suggests that not all M^{q+} sites presented in the ionotropic gels are accessible to coordination by cyanometallate precursors but around 50% remain coordinated to carboxylate in order to maintain the gel structure [43].

Colloidal crystals of silica can be employed as mold for the precipitation of PBAs. Thus the synthesis of a cobalt hexacyanoferrate coordination polymer is achieved by alternately immersing the opals in $CoCl_2$ and $K_3[Fe(CN)_6]$ aqueous solutions. Infiltration is allowed to proceed for 30 min and is followed by quick washings with water [47]. The replication process gives a framework of CoFe PBA with 5–10 nm-thick walls with a homogenous texture, suggesting that it is constructed from a single domain of crystalline $K_xCo_4[Fe(CN)_6]_y$ (Figs. 4.7a,b). Weak interactions between Co^{2+} and silica surfaces are the starting point for the precipitation of single crystals of the coordination polymer within the interstitial spaces during the first stage of the coprecipitation. These nanocrystals grow in preference to the formation of new nuclei during further additions of the reactants.

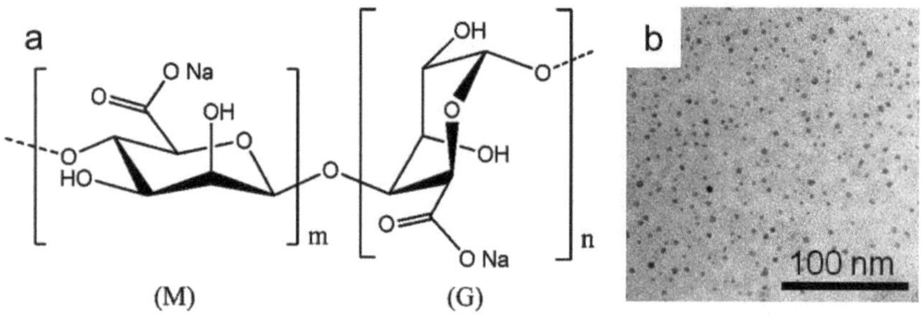

Figure 4.6 (a) Chemical structure of mannurate (M) and guluronate (G) residues of Na-alginate. (b) TEM image of NiFe PBA/alginate nanoparticles after solubilization in aqueous solution. Adapted from [43].

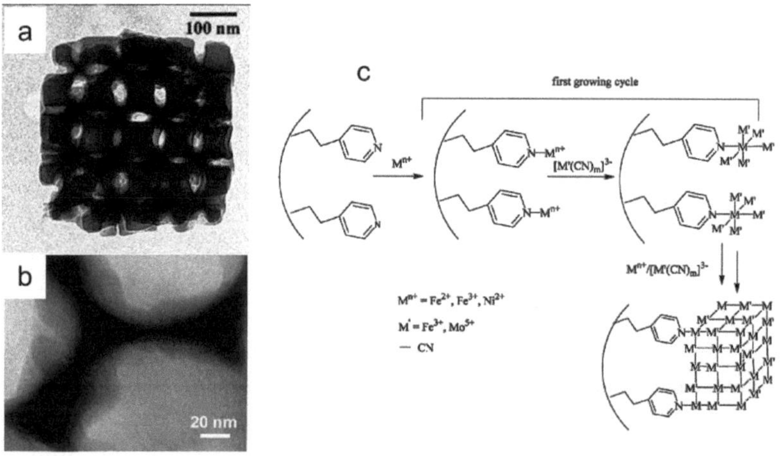

Figure 4.7 (a) TEM image of fractured sample of CoFe PBA inverse opal. (b) High magnification image showing continuous wall structure of crystalline CoFe PBA (adapted from [47]). (c) Schematic representation of the intrapore growth of cyano-bridged polymers into mesoporous silica (adapted from [49]).

Grafting of $-(CH_2)_2C_5H_4N$ groups on mesostructured silica provides chemical anchoring of M^{q+} cations on silica surface and allows controlled growth of PBA nanoparticles by sequential impregnations (Fig. 4.7c). This approach has been applied to a large variety of PBA, with Ni^{2+}, Fe^{2+}, Co^{2+}, and Fe^{3+} cations followed by coordination by $[Fe(CN)_6]^{3-}$, $[Co(CN)_6]^{3-}$, and $[Mo(CN)_8]^{3-}$ [48–50]. At each step the hybrid silica powder is poured in the methanolic solution containing the precursor and stirred overnight. Hosts with different porosities, from 3.4 to 7.5 nm, have been considered.

The precipitation of polycyanometallate nanoparticles can plug the porous channels and a maximum of three growth cycles can be implemented. Further treatment leads to saturation and may provoke the external precipitation in the bulk [49]. This method produces spherical nanoparticles with a diameter slightly smaller than the pore channel.

CuFe PBA nanoparticles can be synthesized in a mesoporous silica matrix by grafting of Cu^{2+} cations on an azido-functionalized silica by click chemistry (by a copper-catalyzed azide-alkyne cycloaddition reaction) followed by sequential infiltrations to grow the coordination polymer (Fig. 4.8). After five consecutive treatments, 6 nm spherical nanoparticles are obtained [53].

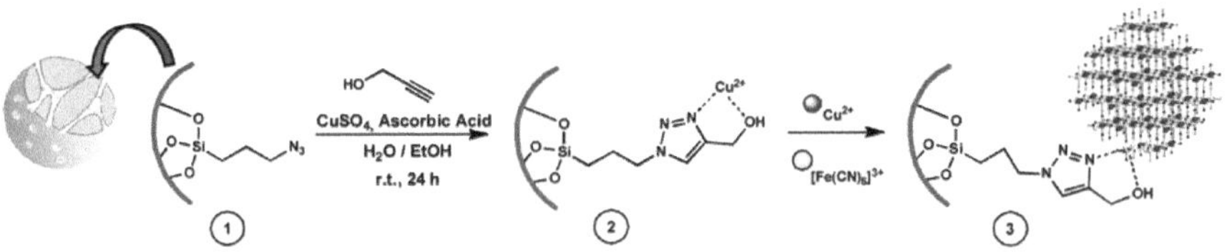

Figure 4.8 Precipitation of CuFe PBA nanoparticles in mesoporous silica by click chemistry. Adapted from [53].

A last example of this approach consists in the growth of PBA in chiral nematic organosilica thick films. The surface of the matrix is functionalized with bipyridium ligands and $[Fe(CN)_5NH_3]^{3-}$ is grafted by treatment with an aqueous solution containing the corresponding sodium salt. Silica films are consecutively immersed into aqueous solution of Fe^{3+} or Zn^{2+}, water, aqueous solution of $[Fe(CN)_6]^{4-}$ and water for four cycles.

Growth is not completely controlled as PBA microcrystals are deposited on the outer surface [51].

4.4 Two-Step Process

This strategy is facile and less time consuming nevertheless it can lead to a perfect control of the final nanocomposite.

This method gathers very different substrates going from organic films (made of bio-derived chitosan [54] or synthetic Nafion [55]) to silica [56–58] monoliths passing through lamellar hydroxides [59, 60].

CoFe PBA has been synthesized using chitosan as a morphological structure-directing agent [54]. First, a cast film is obtained by depositing an acetic acid solution of chitosan containing Co^{2+} cations on a glass slide. Chitosan amide and hydroxyl groups can form coordinate bonds with Co^{2+} ions dispersing these species in the matrix. Upon exposure to a hexacyanoferrate solution the films are transformed into composites containing dispersed CoFe PBA nanoparticles. The size of the objects depends on the chitosan concentration and can be varied from 2.8 to 5.5 nm. The particles are spatially isolated within the biopolymer matrix, suggesting that nucleation in the film is confined to localized regions, and that aggregation of the clusters is constrained by the cocooning effect of the surrounding chitosan matrix.

Nanoparticles of CoCr PBA can be prepared in the porosity of Nafion117® membranes [55]. This perfluorinated polymer consists of main chains of hydrophobic perfluorocarbon and side chains of hydrophilic sulfonic acid that aggregate forming nanometric cavities. First Nafion is immersed into a $K_3[Cr(CN)_6]$ aqueous solution; $[Cr(CN)_6]^{3-}$ and K^+ move into the cavities. After drying, the membrane is immersed into $CoCl_2$ solution; Co^{2+} ions diffuse into the cavities forming CoCr PBA spherical nanoparticles (8 nm) (Fig. 4.9). It is important to notice that when the order of impregnations is inversed (first Co^{2+} then $[Cr(CN)_6]^{3-}$) thin films of CoCr PBA are deposited on both sides of the Nafion membrane.

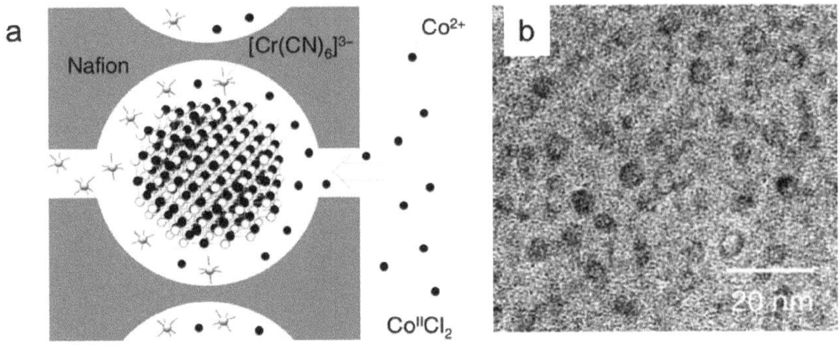

Figure 4.9 (a) Schematic representation of the precipitation of PBA in a cavity of a Nafion membrane. (b) TEM image of a PBA/Nafion nanocomposite. Adapted from [55].

Layered nanocomposites are elaborated by confined growth of 2D layers of cyano-bridged coordination polymers into layered double hydroxides (LDH). LDH are made of positively charged mixed metal hydroxides layers $[M(II)_{3-x}M(III)_x(OH)_6]^{x+}$ whose charge is counterbalanced by exchangeable anions A^{n-} (in general NO_3^-) located in the interlayer domains. The intercalation is performed in two steps. The first step consists in the intercalation of $[M'(CN)_6]^{3-}$ between the layers by anionic exchange of NO_3^- by immersing the pristine LDH in a saturated aqueous solution of the corresponding salt. The second step is the formation of the cyano-bridged metallic species between the LDH sheets by successive treatment with an excess of $M(NO_3)_2$. This approach has been executed on $[Mg_2Al(OH)_6](NO_3)$ and $[Zn_2Al(OH)_6](NO_3)$ matrices to intercalate 2D NiFe and CoFe layers in the first [61] and 2D NiCr layers in the second [59]. The intercalation of $[M'(CN)_6]^{3-}$ anions between the brucite-like sheets leads to an augmentation of the basal spaces from 8.8 Å to about 11 Å, then the inorganic host acts as a templating agent, preventing the formation of a 3D polymeric PBA system and restricting its growth in two dimensions (Fig. 4.10).

Transition metal layered simple hydroxides (LSH) instead of LDH produce 3D PBA nanoparticles. Thus, a copper layered hydroxide $[Cu_2(OH)_3(OAc)]\cdot H_2O$ can be used to produce a cubic CuFe PBA (Fig. 4.11a) [60]. The 3D cyano metallate network is produced by treatment with a $K_4[Fe(CN)_6]$ aqueous solution giving isolated nanoplatelets by pseudomorphic replication of the hydroxide; LSH acts as a metal reservoir controlling the growth of the secondary molecular phase. In LSH the AcO^- counter anions are an integral part of the layered structure, the exchange by $[Fe(CN)_6]^{4-}$ produces the collapse of the lamellar structure of the starting LSH giving cubic nanostructures packed in the layers grown from pseudomorphic transformation of copper hydroxyl acetate into CuFe PBA crystallites (Fig. 4.11b).

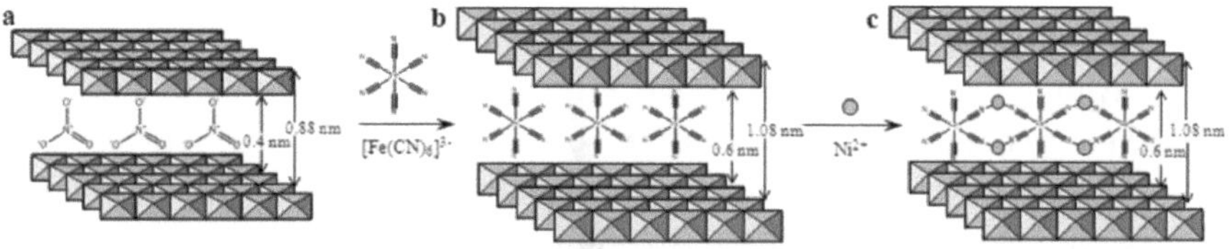

Figure 4.10 Schematic representation of the precipitation of PBA in the interlayer space of LDHs. Adapted from [61].

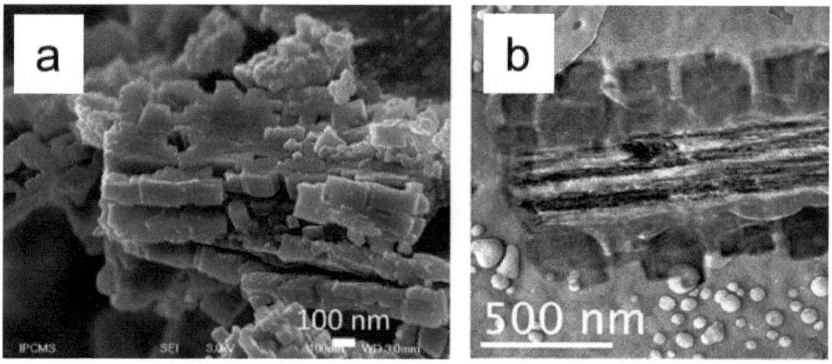

Figure 4.11 CuFe PBA particles produced by treatment of $Cu_2(OH)_3(OAc)\cdot H_2O$ LSH. (a) SEM image, (b) TEM image of microtomed sample (cross section) showing the residual LSH (center) decorated by cuboids of CuFe PBA. Adapted from [60].

CoFe PBA nanoparticles can be elaborated in a disordered silica matrix [62]. The porous network of sol-gel silica, obtained by hydrolysis-condensation of a silicon alkoxide, is used as a confining medium to elaborate PBA nanoparticles. The direct incorporation of CoFe PBA precursors in a silica sol results in phase segregation due to the different rates of polymerization (PBA precipitation is much faster than silica gelation). The elaboration of a homogeneous nanocomposite can be achieved by decorrelating the two condensation processes. A Co^{2+}-containing silica sol is produced by polymerization of tetraethyl orthosilicate (TEOS) in acidic conditions and solubilization of $Co(NO_3)_2 \cdot 6H_2O$. The gelation is achieved by addition of a KOH solution that produces simultaneously the formation of a tetrahedral Co(II) complex with a coordination sphere composed of hydroxide ligands and anionic SiO^- groups of the silica matrix. This negatively charged and low electrophilic Co(II) complex does not react directly with $[Fe(CN)_6]^{3-}$ that can be introduced in the medium without precipitation of CoFe PBA. After gelation and drying of the system, the precipitation of CoFe PBA nanoparticles is triggered by simple lowering of the pH. Preformed silica xerogel is an effective confining medium affording the formation of CoFe PBA nanoparticles (40 nm) made of the assembly of several crystallographic domains, each of about 8 nm.

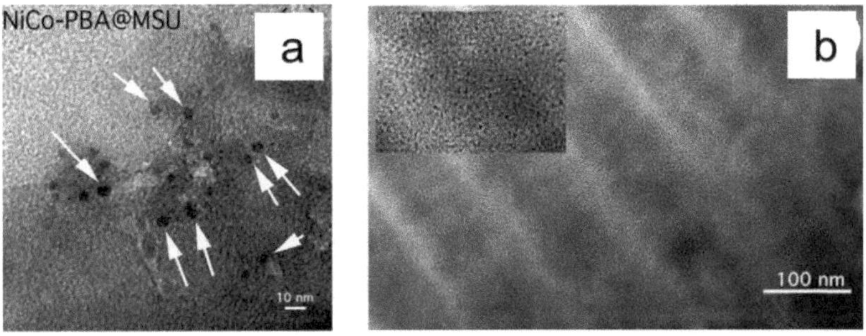

Figure 4.12 TEM images of (a) NiCo-PBA@MSU containing 0.8 mol% PBA, (b) NiCo-PBA@MSU containing about 3.6 mol% PBA. Adapted from [57].

Different works report on nanocomposites with ordered mesoporous silica matrices.

Basic polymerization of TEOS with cetyltrimethylammonium bromide (CTAB) as structure-directing agent gives a hexagonal ordered mesoporous (cylindrical channels of about 2 nm) silica powder, named MCM-41. CTA^+ cations, in electrostatic interaction with silica matrix, can be easily exchanged by Co^{2+} cations by soaking the powder in a $Co(NO_3)_2$ methanol solution. Precipitation of CoFe PBA nanoparticles is achieved by slurry the Co^{2+}-containing silica in a $K_3[Fe(CN)_6]$ aqueous solution [58].

PBA nanoparticles can also be prepared within a mesoporous MSU-type silica. Synthesis of MSU silica is achieved via a two-step process where stable hybrid micelles, made of non-ionic surfactants surrounded by a diffuse layer of silicate oligomers interacting with the PEO chains, are created first. These "hybrid micelles" can be concentrated by simple evaporation giving monolithic silica gel with 3D wormhole or 2D hexagonal structure [63]. The introduction of metal ions (Co^{2+}, Ni^{2+}, Fe^{2+}) in the initial system leads to a homogeneous distribution of species in the final material. $K_3[M'(CN)_6]$ (M' = Fe or Co) is further added to the concentrated solution to give monolithic nanocomposite containing the corresponding PBA [57, 64]. At low concentration (0.8% mol) well dispersed 4 nm particles are obtained whereas higher concentrations (from 1.1 to 5.2% mol) give 10 to 30 nm particles with less ordered silica (Fig. 4.12).

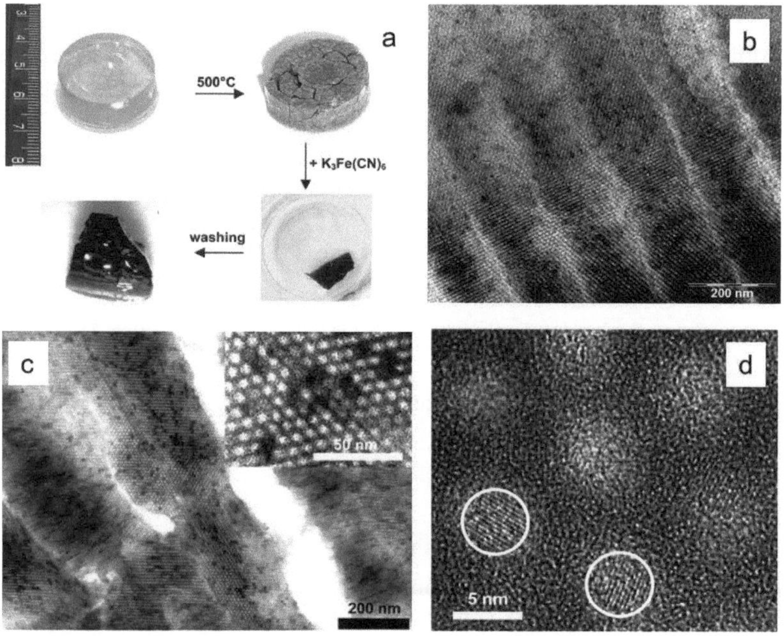

Figure 4.13 (a) elaboration scheme of CoFe PBA/silica nanocomposites, (b) TEM image of nanocomposite containing 1% mol RbCoFe PBA and 3D-cubic mesostructure, (c) TEM image of nanocomposite containing 1% mol RbCoFe PBA and 2D-hexagonal mesostructure, (d) HRTEM image showing PBA nanocrystals. Adapted from [56, 67] and [68].

A last example of the two-step process makes use of SBA15-type silica monoliths.

The facile synthetic procedure is versatile and allows the elaboration of monoliths with various and well-ordered nanostructures (2D hexagonal, 3D cubic, lamellar) containing transition metal M^{2+} cations in various amounts (until 23.5% mol) [56, 65]. The synthesis of the monoliths is obtained by hydrolysis and condensation of a silicon alkoxide solubilized with Pluronic P123, $(EO)_{20}(PO)_{70}(EO)_{20}$ as structure-directing agent. The polymerization is initiated by the addition of an acidified aqueous solution of $M(NO_3)_2$. The M^{2+} cations are initially incorporated as octahedral complex into the ordered copolymer phase. The thermal treatment for opening the porosity produces the deprotonation of these complexes to give tetrahedral lowly condensed species (similar to those of ref 62) in interaction with silica walls [66]. The formation of PBA nanoparticles is attained by impregnation with a $K_3[M'(CN)_6]$ aqueous acidic solution (Fig. 4.13a). The ordered porosity effectively confines the precipitation of PBA giving single-crystal monodisperse nanoparticles (Fig. 4.13b,c,d) [67]. CoFe, NiFe, NiCr PBA nanoparticles with different geometries (spherical, elongated, interconnected superparticles, ...) can be obtained [68].

References

1. Sato, O., Iyoda, T., Fujishima, A., Hashimoto, K. *Science* 1996, 272, 704–705.
2. Taguchi, M., Yamada, K., Suzuki, K., Sato, O., Einaga, Y. *Chem Mater* 2005, 17, 4554–4559.
3. DeLongchamp, D. M., Hammond, P. T. *Chem Mater* 2004, 16, 4799–4805.
4. Matsui, J., Kikuchi, R., Miyashita, T. *J Am Chem Soc* 2014, 136, 842–845.
5. Goberna-Ferrón, S., Hernández, W. Y., Rodríguez-García, B., Galán-Mascarós, J. R. *ACS Catal* 2014, 4, 1637–1641.
6. Sebastian, J., Srinivas, D. *Chem Commun* 2011, 47, 10449.
7. Dharman, M. M., Yu, J.-I., Ahn, J.-Y., Park, D.-W. *Green Chem* 2009, 11, 1754.
8. Jin, W., Toutianoush, A., Pyrasch, M., Schnepf, J., Gottschalk, H., Rammensee, W., Tieke, B. *J Phys Chem B* 2003, 107, 12062–12070.

9. Ohkoshi, S., Arai, K., Sato, Y., Hashimoto, K. *Nat Mater* 2004, 3, 857–861.
10. Zhao, G., Feng, J.-J., Zhang, Q.-L., Li, S.-P., Chen, H.-Y. *Chem Mater* 2005, 17, 3154–3159.
11. Long, J., Guari, Y., Guérin, C., Larionova, J. Dalton Trans 2016, 45, 17581–17587.
12. Lee, H.-W., Wang, R. Y., Pasta, M., Woo Lee, S., Liu, N., Cui, Y. *Nat Commun* 2014, 5, 5280.
13. Wessells, C. D., Huggins, R. A., Cui, Y. *Nat Commun* 2011, 2, 550.
14. Li, C. H., Nanba, Y., Asakura, D., Okubo, M., Talham, D. R. *RSC Adv* 2014, 4, 24955.
15. Olatunji, M. A., Khandaker, M. U., Mahmud, H. N. M. E., Amin, Y. M. *RSC Adv* 2015, 5, 71658–71683.
16. Larionova, J., Guari, Y., Sangregorio, C., Guérin, C. *New J Chem* 2009, 33, 1177.
17. Catala, L., Mallah, T. *Coord Chem Rev* 2017, 346, 32–61.
18. Vaucher, S., Fielden, J., Li, M., Dujardin, E., Mann, S. *Nano Lett* 2002, 2, 225–229.
19. Catala, L., Gacoin, T., Boilot, J.-P., Rivière, é., Paulsen, C., Lhotel, E., Mallah, T. *Adv Mater* 2003, 15, 826–829.
20. Clavel, G., Larionova, J., Guari, Y., Guérin, C. *Chem Eur J* 2006, 12, 3798–3804.
21. Uemura, T., Ohba, M., Kitagawa, S. *Inorg Chem* 2004, 43, 7339–7345.
22. Catala, L., Gloter, A., Stephan, O., Rogez, G., Mallah, T. *Chem Commun* 2006, No. 9, 1018.
23. Pajerowski, D. M., Frye, F. A., Talham, D. R., Meisel, M. W. *New J Phys* 2007, 9, 222–222.
24. Brinzei, D., Catala, L., Louvain, N., Rogez, G., Stéphan, O., Gloter, A., Mallah, T. *J Mater Chem* 2006, 16, 2593–2599.
25. Brinzei, D., Catala, L., Rogez, G., Gloter, A., Mallah, T. *Inorg Chim Acta* 2008, 361, 3931–3936.
26. Zhou, P., Xue, D., Luo, H., Yao, J., Shi, H. *Nanotechnology* 2004, 15, 27–31.
27. Yao, J., Xue, D., Zhou, P. *J Electrochem Soc* 2006, 153, C608.
28. Gao, B., Yao, J., Xue, D. *J Magn Magn Mater* 2010, 322, 2505–2508.
29. Zhou, P., Xue, D., Luo, H., Chen, X. *Nano Lett* 2002, 2, 845–847.
30. Xian, Y., Hu, Y., Liu, F., Xian, Y., Feng, L., Jin, L. *Biosens Bioelectron* 2007, 22, 2827–2833.

31. Luo, H., Chen, X., Zhou, P., Shi, H., Xue, D. *J Electrochem Soc* 2004, 151, C567.
32. Zhou, P. H., Xue, D. S. *J Appl Phys* 2004, 96, 610–614.
33. Ravindran, S., Singh, K. V., Andavan, G. T. S., Ozkan, M., Gao, Y., Hu, E., Ozkan, C. S. *Nanotechnology* 2006, 17, 714–718.
34. Qu, F., Shi, A., Yang, M., Jiang, J., Shen, G., Yu, R. *Anal Chim Acta* 2007, 605, 28–33.
35. Goux, A., Ghanbaja, J., Walcarius, A. *J Mater Sci* 2009, 44, 6601–6607.
36. Zhai, J., Zhai, Y., Wen, D., Dong, S. *Electroanalysis* 2009, 21, 2207–2212.
37. Jin, E., Lu, X., Cui, L., Chao, D., Wang, C. *Electrochim Acta* 2010, 55, 7230–7234.
38. Mao, H., Song, J., Zhang, Q., Liu, D., Gong, N., Li, Y., Wu, Q., Verpoort, F., Song, X.-M. *Nanotechnology* 2013, 24, 215601.
39. Masuda, H., Fukuda, K. *Science* 1995, 268, 1466–1468.
40. Walcarius, A., Sibottier, E., Etienne, M., Ghanbaja, J. *Nat Mater* 2007, 6, 602–608.
41. Johansson, A., Widenkvist, E., Lu, J., Boman, M., Jansson, U. *Nano Lett* 2005, 5, 1603–1606.
42. Shi, Y., Zhou, B., Wu, P., Wang, K., Cai, C. *J Electroanal Chem* 2007, 611, 1–9.
43. Tokarev, A., Agulhon, P., Long, J., Quignard, F., Robitzer, M., Ferreira, R. A. S., Carlos, L. D., Larionova, J., Guérin, C., Guari, Y. *J Mater Chem* 2012, 22, 20232.
44. Guari, Y., Larionova, J., Molvinger, K., Folch, B., Guérin, C. *Chem Commun* 2006, No. 24, 2613–2615.
45. Folch, B., Larionova, J., Guari, Y., Molvinger, K., Luna, C., Sangregorio, C., Innocenti, C., Caneschi, A., Guérin, C. *Phys Chem Chem Phys* 2010, 12, 12760.
46. Larionova, J., Salmon, L., Guari, Y., Tokarev, A., Molvinger, K., Molnár, G., Bousseksou, A. *Angew Chem Int Ed* 2008, 47, 8236–8240.
47. Vaucher, S., Dujardin, E., Lebeau, B., Hall, S. R., Mann, S. *Chem Mater* 2001, 13, 4408–4410.
48. Clavel, G., Guari, Y., Larionova, J., Guérin, C. *New J Chem* 2005, 29, 275–279.
49. Folch, B., Guari, Y., Larionova, J., Luna, C., Sangregorio, C., Innocenti, C., Caneschi, A., Guérin, C. *New J Chem* 2008, 32, 273–282.
50. Folch, B., Larionova, J., Guari, Y., Datas, L., Guérin, C. *J Mater Chem* 2006, 16, 4435–4442.

51. Wang, P.-X., Zamarion, V. M., Hamad, W. Y., MacLachlan, M. J. *Dalton Trans* 2015, 44, 14724–14731.

52. Delchet, C., Tokarev, A., Dumail, X., Toquer, G., Barré, Y., Guari, Y., Guerin, C., Larionova, J., Grandjean, A. *RSC Adv* 2012, 2, 5707.

53. Turgis, R., Arrachart, G., Delchet, C., Rey, C., Barré, Y., Pellet-Rostaing, S., Guari, Y., Larionova, J., Grandjean, A. *Chem Mater* 2013, 25, 4447–4453.

54. Collins, A. M., Mann, S., Hall, S. R. *Nanoscale* 2010, 2, 2370.

55. Kosaka, W., Tozawa, M., Hashimoto, K., Ohkoshi, S. *Inorg Chem Commun* 2006, 9, 920–922.

56. Durand, P., Fornasieri, G., Baumier, C., Beaunier, P., Durand, D., Rivière, E., Bleuzen, A. *J Mater Chem* 2010, 20, 9348.

57. Alamri, H., Ballot, N., Long, J., Guari, Y., Larionova, J., Kleinke, K., Kleinke, H., Prouzet, E. *Chem Mater* 2014, 26, 875–885.

58. Vo, V., Van Minh, N., Lee, H. I., Kim, J. M., Kim, Y., Kim, S. J. *Mater Res Bull* 2009, 44, 78–81.

59. Coronado, E., Martí-Gastaldo, C., Navarro-Moratalla, E., Ribera, A. *Inorg Chem* 2010, 49, 1313–1315.

60. Lang, M., Delahaye, E., Foix, D., Ihiawakrim, D., Ersen, O., Leuvrey, C., Grenèche, J.-M., Rogez, G., Rabu, P. *Eur J Inorg Chem* 2016, 2016, 2030–2038.

61. Layrac, G., Tichit, D., Larionova, J., Guari, Y., Guérin, C. *J Phys Chem C* 2011, 115, 3263–3271.

62. Fornasieri, G., Bleuzen, A. *Angew Chem Int Ed* 2008, 47, 7750–7752.

63. de Zárate, D. O., Bouyer, F., Zschiedrich, H., Kooyman, P. J., Trens, P., Iapichella, J., Durand, R., Guillem, C., Prouzet, E. *Chem Mater* 2008, 20, 1410–1420.

64. Mouawia, R., Larionova, J., Guari, Y., Oh, S., Cook, P., Prouzet, E. *New J Chem* 2009, 33, 2449.

65. Delahaye, E., Moulin, R., Aouadi, M., Trannoy, V., Beaunier, P., Fornasieri, G., Bleuzen, A. *Chem Eur J* 2015, 21, 16906–16916.

66. Aouadi, M., Fornasieri, G., Briois, V., Durand, P., Bleuzen, A. *Chem Eur J* 2012, 18, 2617–2623.

67. Fornasieri, G., Aouadi, M., Durand, P., Beaunier, P., Rivière, E., Bleuzen, A. *Chem Commun* 2010, 46, 8061.

68. Moulin, R., Delahaye, E., Bordage, A., Fonda, E., Baltaze, J.-P., Beaunier, P., Rivière, E., Fornasieri, G., Bleuzen, A. *Eur J Inorg Chem* 2017, 2017, 1303–1313.

Chapter 5

Electrochemical Energy Storage in Prussian Blue Batteries and Capacitors

Neal Kuperman, Prasanna Padigi, Gary Goncher, Joseph Thiebes, David Evans, and Raj Solanki

Department of Physics, Portland State University, 1719 SW 10th Ave., Portland, OR 97201, USA

nwk@pdx.edu

The proliferation of portable electronic devices, the appeal of electrical vehicles, and a need for load leveling storage in renewable energy power grids have highlighted the need for high energy density storage with electrochemically rechargeable batteries. Among the more promising storage systems that have emerged involve utilizing Prussian blue (PB) and its analogues as cathodes materials for high capacity, rechargeable batteries and pseudocapacitors. The open structure of PB lattice allows for efficient discharging/charging of such devices via intercalation/de-intercalation of ions and rapid charge exchange via redox reactions. In the following section, we discuss the unique structure of the 300-year-old pigment and how it can solve our current energy storage needs.

Prussian Blue-Type Nanoparticles and Nanocomposites: Synthesis, Devices, and Applications
Edited by Yannick Guari and Joulia Larionova

ISBN 978-981-4800-05-1 (Hardcover), 978-0-429-02474-0 (eBook)
www.panstanford.com

5.1 Introduction

The first published report of synthesis of Prussian blue (iron (III) hexacyanoferrate (II)) dates back to the early 18th century and to this day, it continues to attract scientific curiosity that arises not only from its zeolitic structure but also because it can be oxidized or reduced (redox) electrochemically [1, 2]. Reflecting the period of its discovery, its original ingredients included animal blood and its stoichiometry and structure were unknown for a long time, but lack of this knowledge did not stop its utilization. Shortly after its first synthesis, its startling blue color generated commercial applications, mostly as a pigment for paintings [1, 3]. Today, this pigment is produced under various trade names on industrial scale for a large number of applications, including paints, printing inks, blue prints, and "blue" or indigo for laundry [4, 5]. Its affinity to sequester metal ions has led to its application as an antidote for ingestion of heavy metal and radioactive isotopes, where it is administered orally [6, 7]. It has also been used as biosensor electrodes and has been investigated for electrochromic and electrocatalysis applications [2, 8, 9]. Prussian blue's zeolitic structure that allows for ion intercalation, its stability in acidic solutions, and ability to reversibly switch between the redox states makes it an attractive material for fabricating batteries and capacitors for energy storage. These applications are further examined in detail as follows.

5.2 Prussian Blue Structure

Although this pigment has been around for over 300 years, its structure was not determined until much later due to difficulty in characterizing its composition. This difficulty was due to the nanoscale size of the pigment particles that adsorbed impurity ions. The Keggin–Miles model was first to correctly identify the polymeric nature of PB [10]. This model also identified two forms of PB, "soluble" and "insoluble." These terms do not mean actual solubility, but the tendency for PB to form colloidal suspensions in the presence of some salts. The insoluble form of PB refers to the presence of iron(III) ions in the interstitial locations instead of water or potassium ions. It was not until 1970s that systematic studies utilizing X-ray and neutron diffraction helped to establish its cubic lattice structure, and confirmed the composition as $Fe_4[Fe(CN)_6]_3 \cdot xH_2O$ (x = 14–16) [11, 12]. The neutron data also provided information on location of the water molecules in the lattice: at nitrogen coordinated to Fe(III) and as interstitials, uncoordinated to metal centers.

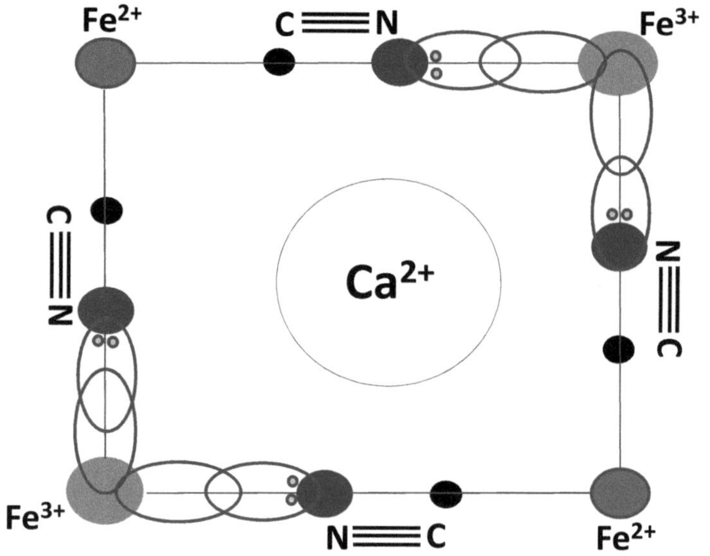

Figure 5.1 Cross-section view of ironhexacyanoferrate complex.

PB has a 3-dimensional, open, zeolite-like structure. The cubic framework frame consists of cyanide ions that act as bridges linking the Fe^{2+} and Fe^{3+} ions forming the Fe(III)–N–C–Fe(II) sequence. The Fe(II)–C–N–Fe(III) sequence has average bond distances of Fe(II)–C = 0.20 nm, Fe(III)–N = 0.198 nm, and C–N = 0.11 nm [11]. The open cubic structure allows for intercalation of cations into the interstitial sites, including water [10]. A side view of this face-centered-cubic (fcc) crystal structure is shown in Fig. 5.1, with a Ca ion intercalated in the interstitial space. The strong bonds between the carbon end of the cyanide ligand and the partially filled d-orbitals of the Fe^{2+} ions, as well as the bonding between the nitrogen end of the cyanide ligand and the partially filled d-orbitals of the Fe^{3+} ions makes PB a very stable material. The coexistence of more than one oxidation state of the same element, Fe(III) and Fe(II), in PB characterizes it as a mixed valence compound. It is the electronic interaction between the two oxidation states of iron that gives it the blue color, makes it a semiconductor, and a molecular ferromagnet at low temperatures [12].

5.3 Synthesis of Prussian Blue

PB can be prepared by simply mixing aqueous solutions of salts of $[Fe^{II}(CN)_6]^{4-}$ (such as potassium ferrocyanide) and Fe(III) (such as ferric nitrate or ferric chloride), or alternatively, salts of $[Fe^{III}(CN)_6]^{3-}$ and Fe(II), resulting in precipitation of nanoscale particles in an aqueous colloidal suspension [13]. These particles are not soluble in water, which leads to rapid precipitation, resulting in varying degrees of incorporation of water molecules at lattice defects [14]. To produce PB which does not become suspended in aqueous solution, an excess of Fe(III) or Fe(II) ions, respectively, can be added to the reaction mixture [13]. Because these particles are insoluble, precipitation is rapid, particularly in bulk synthesis, which can lead to poor crystallinity and amorphous agglomerations [15]. It is therefore desirable to slow the reaction. The size and crystallinity of the particles can vary with the synthesis conditions, which will affect its appearance, and its physical and electrochemical properties [16]. For example, it has been found that simultaneous dropwise addition of the two precursors to a common solution produces better crystallinity by slowing the reaction and allowing for more ordered lattice vacancies and incorporation of cations [15]. Using hydrothermal methods and introducing an excess of metal ions to the reaction mixture have also been shown to change the reaction rate and improved crystallinity [16]. A scanning electron micrograph (SEM) of the PB nanoscale crystals is shown in Fig. 5.2.

The crystalline structure of the $Fe_4[Fe(CN)_6]_3$ crystals can be characterized by using powder X-ray diffractometry (PXRD). A typical PB XRD scan performed over an angular range of 10–70° 2θ at room temperature is shown in Fig. 5.3.

The XRD spectrum of PB analyzed by Rietveld analysis using the XRD peak indexing software DicVol under the FULLPROF suite is shown in Fig. 5.3, where the unit cell lattice symmetry was identified as cubic belonging to the Pm3m space group [17, 18]. The unit cell parameters were determined to be $a_0 = b_0 = c_0 = 10.13$ Å and all the three angles ($\alpha = \beta = \gamma = 90°$) were found to be equal. A cubic structure model of PB is shown in Fig. 5.4.

Figure 5.2 SEM image of as prepared PB. Nanocubes range between 20 to 200 nm.

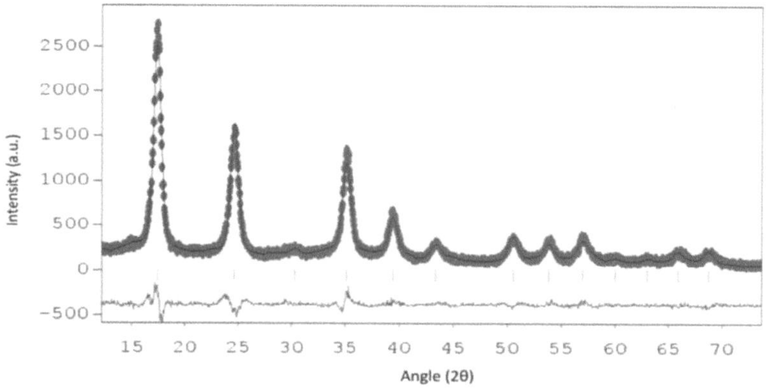

Figure 5.3 Powder XRD pattern for annealed PB. The experimental X-ray spectrum is shown in red and the Rietveld fit with the solid black line. The blue line shows the deviation between the two spectra.

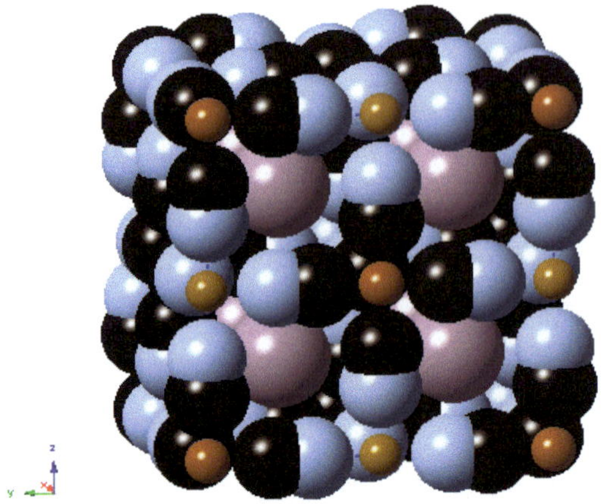

Figure 5.4 Cubic lattice structure of PB. Fe^{2+} and Fe^{3+} ions are shown as rust color spheres, and the large black/blue ligands between iron atoms are CN^- ions. Water molecules and other ions in solution can intercalate into the voids in the cubic lattice; in this diagram intercalated K^+ ions are shown in purple.

5.3.1 Prussian Blue Analogues

Substitution of one of the iron atoms in PB with another transition metal creates structures with new properties called Prussian blue analogues (PBAs). These materials can have quite different electrochemistry from PB, even though the lattice structure and dimensions are quite similar to it. These analogues include the signature metal–CN–metal bonds.

5.4 Prussian Blue Rechargeable Batteries

In a typical intercalation-type rechargeable battery, the host ion shuttles between the anode and the cathode. Let us consider a potassium ion (K^+) battery that consists of a graphite anode and a PB cathode, as illustrated in Fig. 5.5. During the charging process, the K^+ ions migrate from the cathode to the anode through the electrolyte and the electrons flow through the external circuit from the cathode to the anode. A reverse reaction occurs in the discharge process where the K^+ ion intercalates into the PB lattice. To achieve this "rocking-chair" dynamics requires a careful selection of cathode and anode pairs in order to maintain the required cell voltage and to realize a reasonable energy density without unduly increasing the weight or volume of the cell.

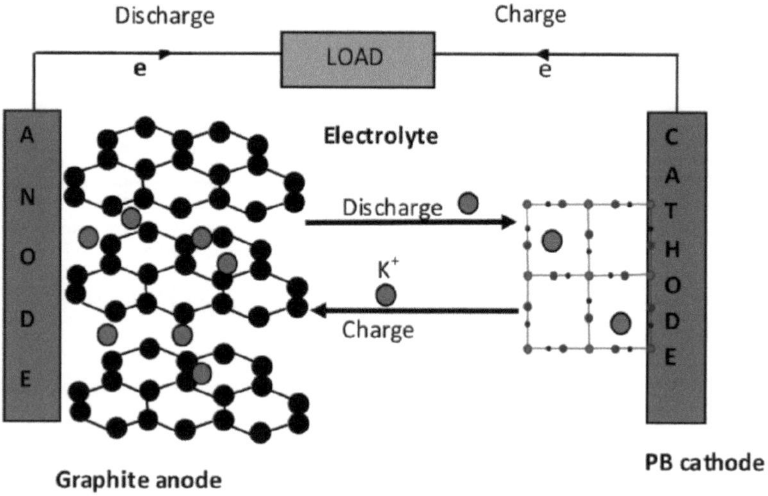

Figure 5.5 Charge/discharge processes in a K^+ ion cell consisting of K insertion compounds at both anode and cathode. Note that the anode and cathode are defined when used in the galvanic cell mode.

The open framework structure of PB previously mentioned provides the channels for intercalation/de-intercalation of cations into the interstitial sites ideally over thousands of cycles. Moreover, the redox capability of these materials provides a low cost option for application in electrochemical energy storage fields. Although, currently both PB and its analogues are being investigated for battery applications, the focus of this chapter will be on PB.

In intercalation-type rechargeable batteries, the cathode material undergoes charging through the process of oxidation. This involves donation or removal of electrons from the lattice and de-intercalation of cations from the interstitial space. In the case of PB, during the process of charging, as illustrated in Fig. 5.6, the lattice undergoes oxidation by removal of K^+ ions from the interstitial sites, and the Fe^{2+} ions present in the lattice undergo oxidation by donating an electron, resulting in PB turning into Prussian Green (PG) which is a fully oxidized state of PB. During the process of discharging, PB accepts an additional

K^+ ion and an electron, and reduces Fe^{3+} ions in the lattice to Fe^{2+} by adding the electron to the lattice.

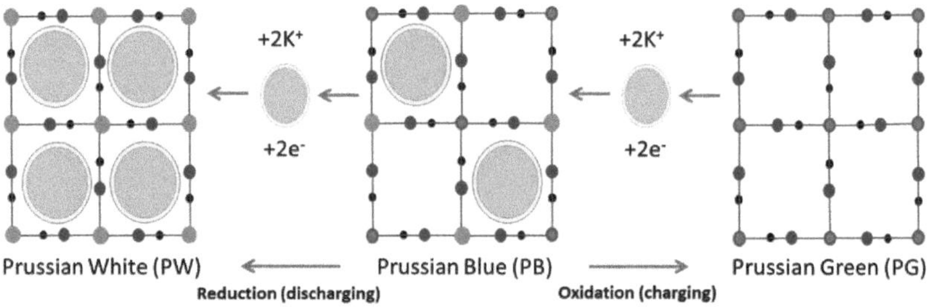

Figure 5.6 Concept of charging and discharging in PB as the cathode material.

5.5 Intercalating Species

In all intercalation type batteries, guest ions are reversibly inserted and extracted from the interstitial sites of the unit cell of both the cathode and the anode materials during the process of charging and discharging the battery. The guest ions are charged species that include Li^+, Na^+, K^+, Ca^{2+}, and Mg^{2+}. The charge present on the guest ions influences their rate of diffusion into the interstitial spaces of the cathode and anode materials due to the electrostatic interactions between the guest ions and the host lattice. Hence, multivalent ion diffusion in and out of the electrode framework has proved to be difficult because of the increased electrostatic interaction with the electrode's host lattice. The increased charge concentration can lead to slow kinetics and poor cyclability. However, the large pores and open framework in PB have resulted in relatively good performance of divalent and trivalent ion batteries as well [19]. For this type of battery, to utilize a higher fraction of the theoretical capacity, the guest ions would have to reversibly diffuse as far as possible, or more preferably all the way to the last unit cell present in every grain of the cathode and anode material. To achieve this would require minimum electrostatic interactions of the ions with the host lattice.

A technique known as cyclic voltammetry (CV) is employed to ascertain the ability of a metal hexacyanoferrate compound to undergo redox, and also shows the reversibility of the process. CV involves subjecting the working electrode or the cathode to a voltage sweep with respect to a reference electrode, and measuring the current during the voltage sweep.

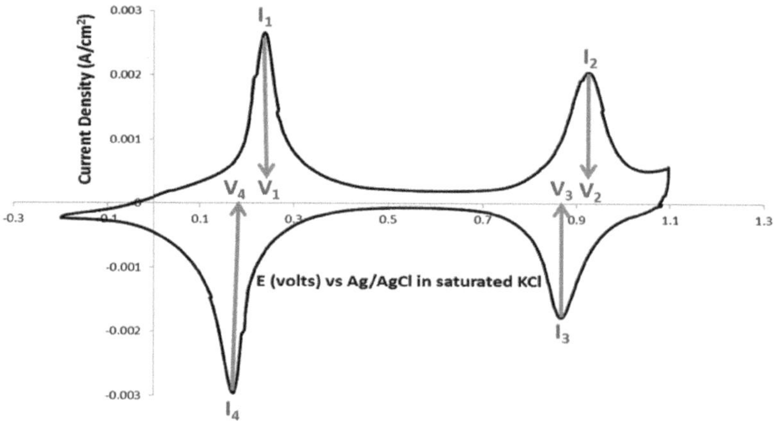

Figure 5.7 CV curve for PB electrode in 1M KNO_3 aqueous electrolyte at a scan rate of 0.5 mV/s.

The voltage sweep range is defined by two extreme points where the direction of the sweep of the voltage is reversed. Figure 5.7 shows a typical CV curve for a PB electrode. Such a plot allows one to study the kinetics involved in the intercalation of the ions as well as electron transfer associated with the intercalation of ions within the lattice of the active material.

The CV curve shown in Fig. 5.7 is characterized by the presence of oxidation peaks and their corresponding reduction peaks in the reverse sweep. The process of oxidation is associated with the removal of K^+ ions from the interstitial sites and concurrent removal of electrons from the lattice. This is characterized by peak oxidation voltages V_1, V_2, and corresponding peak oxidation currents I_1, I_2. The process of reduction is associated with the insertion of K^+ ions into the interstitial sites and additions of electrons to the lattice in order to maintain charge neutrality. This is characterized by peak reduction voltages V_3, V_4, and corresponding peak reduction currents I_3, I_4. The reversibility of the insertion and extraction of K^+ ions is reflected by the difference between the peak oxidation and reduction voltages

V_1, V_4 and V_2, V_3, respectively. The smaller the difference between the oxidation and reduction potentials, the higher the reversibility associated with the redox process, i.e., insertion and removal of K^+ ions [20].

The mobility of the ions in the interstitial sites is a diffusion limited process. The process of diffusion is dependent on the amount of electrostatic interaction between the intercalating ions and the host lattice. A sharp peak indicates low resistance associated with the insertion, diffusion, and extraction of ions and electrons from the lattice, due to reduced electrostatic interactions between the intercalating ions and crystal lattice. Similarly, a broad peak indicates increased resistance to insertion, diffusion, and extraction of ions and electrons from the lattice, due to enhanced electrostatic interactions between the intercalating ions and crystal lattice. Hence, the CV technique not only provides a means to observe and quantify the presence and reversibility of the redox processes associated with an electrode material, but also reflects the intensity of electrostatic interactions between the interacting ions and the host lattice of the electrode material.

Following initial characterization of the cathode using CV, the cathode is next subjected to constant current charge-discharge cycling to determine the amount of charge it can store per unit weight, using a technique called galvanic cycling. In galvanic cycling, an upper cut-off voltage and lower cut-off voltage are defined and the working electrode or cathode is subjected to a constant current. The voltage of the cathode is monitored with respect to a reference electrode such as Ag/AgCl.

During the charging cycle, as the voltage of the cathode reaches the upper cut-off voltage the polarity of the current reverses and the cathode is subjected to the process of discharging. When the voltage reaches the lower cut-off value, the process of discharging ends by again reversing current polarity, marking the beginning of the next charging cycle. This is illustrated in Fig. 5.8, where the charging process involves the extraction of Ca^{2+} ions from the interstitial sites of the PB cathode and the oxidation or extraction of equivalent number electrons from its lattice. The discharging process involves the insertion of Ca^{2+} ions into the interstitial sites and the reduction or addition of equivalent number of electrons to the cathode lattice.

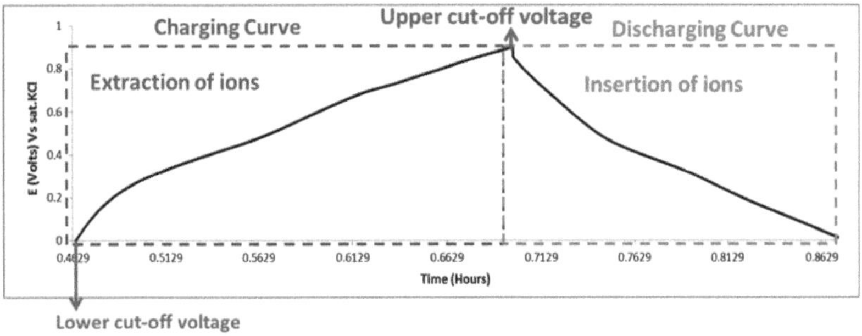

Figure 5.8 Charge-discharge profile of PB electrode with Ca^{2+} as the intercalating ion.

The magnitude of electrostatic interaction between the intercalating ions and the host lattice determines the resistance encountered by the diffusing ions. This is reflected by the shape of the charge and discharge profile. In some cases, it is possible to reduce the electrostatic interaction by addition of a small quantity of deionized (DI) water. It is postulated that the DI water produces a hydration sheath around the guest cations. This hydration sheath provides an improved shielding of the electrostatic interactions between positive charge possessed by the cations from the net negative charge possessed by the host lattice, enabling enhanced diffusion of the guest species into the interstitial spaces of the host lattice and access to the redox sites, thereby increasing the gravimetric capacity utilization [6, 8, 11, 21, 22].

5.6 Enhancing Storage Capacity of PB and PBAs

Recently, the interest in PB and PBAs as cathode materials for secondary batteries has grown as the desire to explore non-lithium ion technology increases. Many of the early PBA cathodes failed to deliver capacities above 100 mAh/g despite their high theoretical capacities of close to 170 mAh/g. Tabuchi et al. observed that by drying as synthesized PB powder it was possible to reduced the water content of the material and achieve higher capacities [23]. In 2012, Goodenough et al. synthesized a range of PBAs via co-precipitation with the iron bonded to the nitrogen end of the cyanide substituted with one of the following transition metals: Mn^{2+}, Fe^{2+}, Co^{2+}, Ni^{2+}, Cu^{2+}, or Zn^{2+} [24]. Although redox activity was observed in all the materials, they experienced poor coulombic efficiencies and a maximum reversible capacity of 70 mAh/g. Low reversible specific capacities were observed in some of the other early PBA cathodes synthesized via rapid coprecipitation without any modification or annealing [24–27]. Goodenough proposed that the poor performance was a result of water in the PBA crystal structure.

As already discussed, PB and PBAs synthesized using the typical co-precipitation method possess lattice defects, usually in the form of $Fe(CN)_6$ vacancies, which can be replaced by coordinated water [28–32]. Additional water may reside within or on the surface of the PB lattice. Using thermogravimetric analysis (TGA), inductive coupled plasma mass spectroscopy (ICPMS), and other techniques researchers have investigated the water content of materials synthesized via co-precipitation without modification [33, 34]. Results vary among materials, but water content in PB has been reported to be as high as 30% by weight [31]. Vacancies, and coordinated and interstitial water reduce the number of redox active sites, limit the achievable specific capacity, destabilize the lattice, and can cause side reactions with water. Additionally, it has been shown that the macroscopic crystal structure impacts the overall battery performance of a PBA [16]. Therefore, developing methods to optimize a PBA's structure or produce materials with a more uniform and pristine lattice is of great importance for battery applications.

In order to better control the co-precipitation reaction, Yang et al. adopted a citrate-assisted synthesis to produce low defect $NaCo[Fe(CN)_6]$ [32]. With the addition of trisodium citrate as a chelating agent, the reaction rate and crystallization process were slowed, allowing for more uniform growth. This resulted in an 11% reduction of lattice defects and a 52% decrease in water content when compared to a control synthesized without citrate ions. The citrate assisted PBA displayed a higher specific capacity (150 mAh/g) than the control (115 mAh/g). The high capacity and coulombic efficiency was attributed to the more pristine framework of the citrate assisted PBA, with suppressed defects.

Guo et al. proposed a single precursor synthesis in which PB, denoted high quality PB (HQ-PB) was synthesized with a single $NaFe(CN)_6$ precursor in an acidic environment [33]. The resulting PB nanocubes were on the order of 300–600 nm in size. Low quality PB (LQ-PB) synthesized under the unmodified co-precipitation technique, used as a control, were on the order of 20 nm. The crystal structures of the two materials were identical, however the 6% by weight of vacancies in the HQ–PB was much lower than the LQ-PB which had 32% by weight. By limiting the number of defects and reducing the water content, the HQ–PB exhibited superior electrochemical properties with a high specific capacity (170 mAh/g), no apparent capacity loss over 150 cycles, and a coulombic efficiency of near 100%.

A highly crystalline PB/graphene oxide composite (PB/GO) was synthesized in an acidic environment by the mixing of $Na_4[Fe(CN)_6]$ in the presence of Fe_2O_3/GO [35]. The Fe_2O_3/GO regulated the release of Fe ions and allowed for a slow and controlled growth. The resulting product had few vacancies and limited water molecules. The size of the PB/GO crystals was larger (300–500 nm) than the PB synthesized with GO (100–200 nm), which also suffered from substantial defects. The GO scaffolding provided the additional benefit of reducing structural imperfections and increasing the electronic conductivity of the material, resulting in high reversible capacity and good cyclability as a cathode in Na^+ ion batteries.

When comparing performance to standard lithium ion cathode materials, the theoretical capacities of PBAs fall short ($LiCoO_2$ = 274 mAh/g vs. PB = 180 mAh/g). However, due to structural instabilities upon extraction of more than half of the available lithium ions, the achievable capacity for $LiCoO_2$ is near 140 mAh/g [36]. This puts PBAs on par with $LiCoO_2$ with regards to specific capacity and energy storage. Table 5.1 shows a number of different PB and PBA cathodes tested with numerous cations. Some of the cations that have shown relatively high reversible capacities over a large number of cycles include Li^+, Na^+, K^+, Mg^{2+}, Zn^{2+}, and Ca^{2+}. This highlights the robust electrochemical properties, high structural stability, and versatility of PB and PBAs.

Table 5.1 Comparing PB and PBA battery performance

Electrode composition	Ion	Specific reversible capacity (mAh/g) @ Current Density (mA/g)	Capacity retention	Ref.
$KFe^{III}[Fe^{III}(CN)_6]$/multiwalled carbon nanotubes/activated carbon/polyvinylidene fluoride (PVDF)	K^+	123.08 mAh/g @ 111 mA/g	100% after 20 cycles at 388 mA/g	[16]
$K_{0.6}Ni_{1.2}[Fe(CN)6]_3$ $3.6H_2O$/amorphous carbon/graphite/PVDF	K^+	59 mAh/g @ c/6	100% after 1000 cycles	[37]
$K_{0.6}Ni_{1.2}[Fe(CN)_6]_3 3.6 H_2O$/amorphous carbon/graphite/PVDF	Na^+	59 mAh/g @ c/6	100% after 5000 cycles at 8.3C rate	
$Na_{1.85}Co[Fe(CN)_6]_{0.99}\cdot \square_{0.01}\cdot 1.9H_2O$/conducting graphite/polytetrafluoroethylene (PTFE)	Na^+	150 mAh/g @ 10 mA/g	90% capacity retention over 200 cycles	[32]
$Na_{1.92}Fe_2(CN)_6$/Ketjen black/PTFE	Na^+	160 mAh/g @ 10 mA/g	80% after 750 cycles	[38]
$KFe^{III}[Fe^{III}(CN)_6]$/carbon nanotunes/PTFE	Li^+	138 mAh/g @ 25 mA/g	69.5% after 50 cycles	[39]
$K_{0.07}Zn[Fe(CN)_6]_{0.69}$/Super P/PVDF	Zn^{2+}	69.1 mAh/g @ 60 mA/g	93% after 100 cycles at 300 mA/g	[40]
$K_{0.1}Cu[Fe(CN)_6]_{0.7}\cdot 3.6H_2O$/acetylene black/PVDF	Mg^{2+}	50 mAh/g @ 100 mA/g	74% at 1000 mA/g	[41]
$CaCoFe(CN)_6$/multiwalled carbon nanotunes/activated carbon/PVDF	Mg^{2+}	52.83 mAh/g @ 25 mA/g	95% after 100 cycles at 100 mA/g	[42]

$CaCoFe(CN)_6$/multiwalled carbon nanotunes/activated carbon/PVDF	Ca^{2+}	82 mAh/g @ 25 mA/g	95% after 100 cycles at 100 mA/g	
$K_2BaFe(CN)_6$/multiwalled carbon nanotunes/activated carbon/PVDF	Ca^{2+}	55.8 mAh/g @ 12.5 mA/g	100% after 30 cycles at 12.5 mA/g	[43]
$K_2Fe^{II}[Fe^{III}(CN)_6]$/multiwalled carbon nanotunes/activated carbon/PVDF	Ca^{2+}	150 mAh/g @ 25 mA/g	85% after 100 cycles at 125 mA/g	[17]

5.7 Prussian Blue Supercapacitors

5.7.1 Electric Double Layer Capacitors (EDLCs)

A supercapacitor or an ultracapacitor is a device for charge storage via the electrical double layer near high surface area electrodes in an appropriate electrolyte. These devices are often referred to as electrical double layer capacitors (EDLCs) after the mechanism of charge storage. A voltage is applied to electrodes separated by an ion-permeable separator, and ions near the electrodes form layers of positive and negative charge near the electrode-electrolyte interface as shown in Fig. 5.9. The small separation between these layers of ions (0.3–0.8 nm) produces very large capacitance values compared to typical electrolytic capacitors. EDLCs typically have very high surface area electrodes (>1000 m^2/g) such as activated carbon in order to store a significant amount of charge.

The advantages of this type of energy storage are the high charge/discharge rates and the very large number of reversible charge storage cycles (typically millions of cycles) that can be sustained before degradation. The primary drawbacks to this type of energy storage are the low voltage of approximately 2.5 V that can be sustained across the double layer, and the small storage capacity compared to batteries.

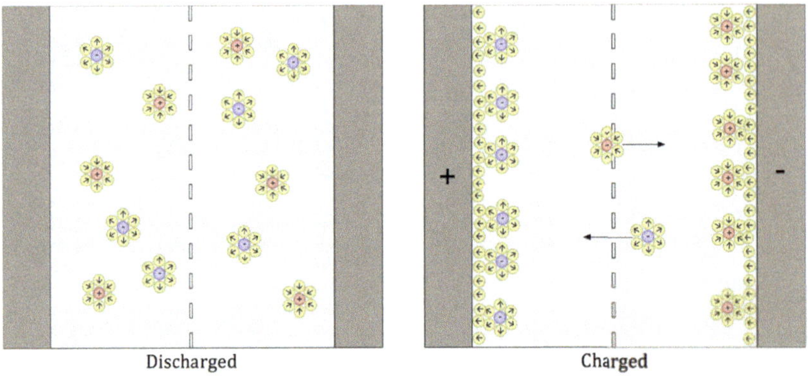

Figure 5.9 Schematic of an electric double layer capacitor (EDLC). Polar solvent molecules are shown in yellow, positive ions in red, and negative ions in blue with solvation spheres surrounding them. In the charged state ions are separated from the electrodes by a layer of polarized solvent molecules (inner Helmholz layer).

In a supercapacitor the electrodes can be identical (symmetric cell) or different (asymmetric cell) with an ion conductive separator between them. Aqueous electrolytes are typically used for high conductivity in order to obtain high power. Typical breakdown voltage for aqueous electrolytes is about 1 V, so device voltages are low. Porous electrodes are used for high surface area and thus high capacitance, but the type pore size is also important so there is not a direct linear relationship between surface area and capacitance. Tuning pore size of carbon electrodes can improve characteristics of supercapacitors by making more of the electrode surface accessible to the electrolyte [44].

5.7.2 Pseudocapacitors

Adding a layer of a redox-active material to one or both of the electrodes of an EDLC can substantially increase the total energy stored in the capacitor. If the layer is kept very thin, the redox reactions are fast surface reactions. The power density of the cell can thus be kept very high, approaching that of the double layer capacitor by itself, which is limited by ion mobility in the electrolyte.

The term pseudocapacitance is used to describe the behavior of these materials where charge storage occurs via an electron-transfer mechanism rather than charge accumulation in the double layer. Pseudocapacitors exhibit charge storage by two mechanisms: ions in the electrical double layer as for EDLCs, and charge transfer across the double layer. The electrochemical signature of the electrodes in a pseudocapacitor should be capacitive, with a linear relationship between amount of charge stored and the potential. The galvanostatic charge-discharge curves for these devices are thus triangular in shape.

There is some controversy about which material systems should properly be called pseudocapacitors, and which should be called batteries [5]. Batteries exhibit Faradaic behavior, where a significant plateau in potential is observed during constant current charging and discharging. Certain metal oxides and polymers exhibit true pseudocapacitive behavior when coated on carbon capacitor electrodes [3, 30]. In these cases very high rates of charging/discharging are observed, and very long lifetimes (number of charge/discharge cycles) are recorded.

Pseudocapacitive materials used to improve energy storage in EDLCs include RuO_2 and MnO_2. RuO_2 was used as early as 1971 as a pseudocapacitor, and has a high maximum theoretical specific capacitance of 1450 F/g; RuO_2 nanoparticles dispersed at 10% loading on activated carbon have shown experimental capacitance as high as 1340 F/g [3]. Since the weight loading is low, however, the areal capacitance is low. MnO_2 has also been used as a pseudocapacitor material with specific capacitance of 200–250 F/g. Although it has a high theoretical capacity of 1233 F/g, its low conductivity means it must be used in very thin layers with capacitance much lower than theoretical.

Other metal oxides such as $Ni(OH)_2$, Co_3O_4, and mixed metal oxides of Ni, Co, and Mn have also been used as pseudocapacitors [3]. While some of these oxides ($NiCo_2O_4$, for example) have exhibited very high average capacitance, there are clearly pseudocapacitive and battery-type contributions to their energy storage within the potential range used [3, 5]. Often this means there is degradation of the material with cycling (low cycling life) and limited power compared to EDLCs.

5.7.3 Prussian Blue Pseudocapacitors

As already discussed, PB with its large lattice constant allows easy intercalation/de-intercalation of ions. This potentially allows for more rapid insertion of ions into the lattice, allowing greater power density. The absence of large dimensional changes may also allow many charge/discharge cycles without much degradation in performance. Moreover, oxidation-reduction of both metal ions occurs within a potential range accessible in aqueous electrolytes.

PB supercapacitors were made by coating a slurry of colloidal PB Blue, activated carbon, and PVDF in N-methyl-2-pyrrolidone onto carbon paper electrodes [44]. Testing was done in a three-electrode cell with the PB electrode, a graphite counter electrode, and an Ag/AgCl reference electrode in 1M solutions of various ions at a pH of 2. Galvanostatic cycling of the electrodes have produced results that varied from pseudocapacitive behavior to Faradaic behavior depending on the ion used (Fig. 5.10).

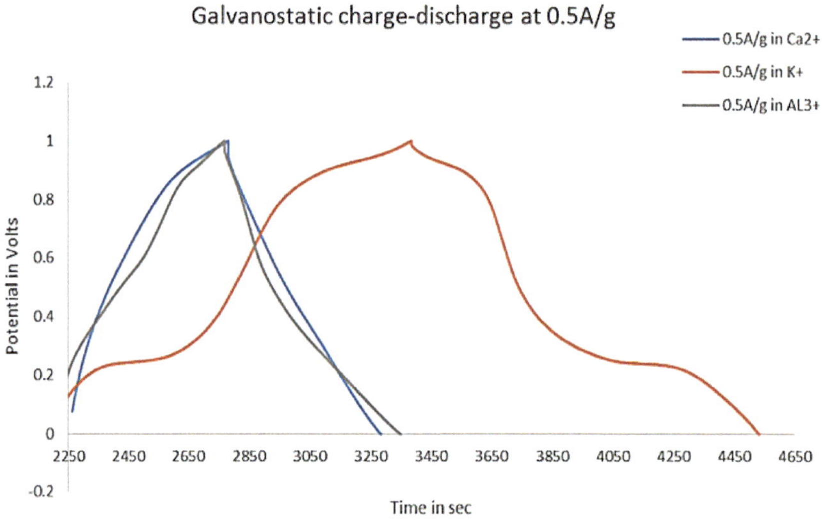

Figure 5.10 Galvanostatic constant current charge-discharge cycling of PB electrodes in aqueous solutions of several cations. In this case pseudocapacitive charge-discharge profiles are shown for Ca^{2+} and Al^{3+} ions, while K^{+} ions show a Faradaic profile with plateaus in potential [44].

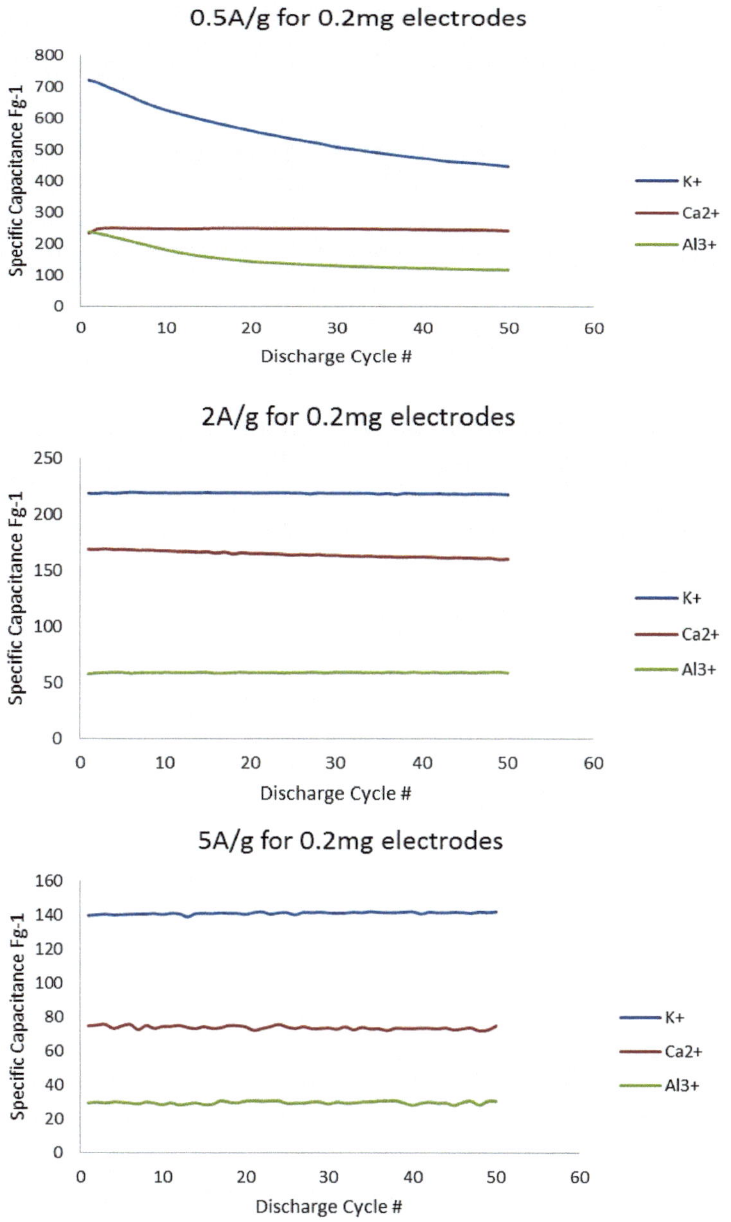

Figure 5.11 Specific capacitance of PB electrodes as a function of number of charge-discharge cycles for several different ions at current densities of 0.5 A/g, 2 A/g, and 5 A/g [44].

Specific capacitance for the PB electrodes varies from about 30 F/g to over 700 F/g depending on the ion used and the current for galvanostatic cycling in a 1.0 V potential range (Fig. 5.11). The largest variation was seen for K^+ ions, which have a Faradaic electrochemical cycling profile. K^+ ions also showed a drop off in specific capacitance with extended cycling. Using Ca^{2+} ions with PB electrodes, however, produced stable pseudocapacitance with values of 250 F/g at a current density of 0.5 A/g and 75 F/g at a current density of 5 A/g.

There are other recent reports of pseudocapacitive electrodes based on PB. Wang et al. [10] reported using a "super-activated carbon" electrode with PB in the pores of the material to achieve a specific capacitance of 263.7 F/g at a current density of 5 A/g using a voltage window of 0.8 V in acidic 1M KCl. In this report the porosity of activated carbon was first enhanced with a KOH hydrothermal process to achieve a porosity that included a mesoporous tunnel structure with a specific surface area of 3000 m^2/g. They then formed PB in situ in the pores by first adding $FeCl_2$ to a suspension of the material in alcohol, then adding a solution of $K_3[Fe(CN)_6]$ to produce embedded PB nanoparticles with a diameters of 50–100 nm. The electrode achieved 1500 constant current charge-discharge cycles with only a 5.2% reduction in capacity. However, both CV and charge-discharge measurements clearly showed Faradaic profiles in this work.

Another report also used a composite electrode with PB to achieve a specific capacitance value of 253 F/g at a current density of 2 A/g using a voltage window of 1.2 V in 1M KCl [56]. Cycling performance was good with only a 6% drop in capacity after 1000 galvanostatic charge-discharge cycles at 2 A/g. In this case the electrode was a composite of graphene/reduced graphene oxide/PB that was formed by starting with single layer graphene oxide nanosheets.

5.7.4 Prussian Blue Analogue Pseudocapacitors

Substitution of one of the iron atoms in PB with another transition metal creates structures with new properties called Prussian blue analogues, as mentioned previously. These materials can have quite different electrochemistry from PB, even though the lattice structure and dimensions are quite similar to PB. A number of these materials have been investigated for use in supercapacitors [7, 21, 22, 35, 49, 54, 57]. Table 5.2 summarizes the experimental conditions and results obtained for these compounds in supercapacitor applications.

Table 5.2 Comparing PB pseudocapacitor performance

Electrode composition	Ion	Specific capacitance	Capacity Retention	Ref.
Activated carbon/ PVDF/$Fe[Fe(CN)_6]$	K^+	220 F/g@2 A/g 140 F/g@5 A/g	100%/50 cycles 100%/50 cycles	[44]
Activated carbon/ PVDF/$Fe[Fe(CN)_6]$	Ca^{2+}	170 F/g@2 A/g 75 F/g@5 A/g	94%/50 cycles 97%/50 cycles	[44]
Super-activated carbon/$Fe[Fe(CN)_6]$	K^+ (1M)	264 F/g@5 A/g	95%/1500 cycles	[10]
Graphene-rGO/ $Fe[Fe(CN)_6]$	K^+ (1M)	253 F/g@2 A/g	94%/1000 cycles	[56]
Carbon black/PTFE $Fe[Fe(CN)_6]$	K^+ (1M)	400 F/g@1.4 A/g	95%/1000 cycles	[7]
Carbon black/PTFE $Ni[Fe(CN)_6]$	K^+ (1M)	450 F/g@1.4 A/g	64%/1000 cycles	[7]
Carbon black/PTFE $Co[Fe(CN)_6]$	K^+ (1M)	180 F/g@1.4 A/g	88%/1000 cycles	[7]
Carbon black/PTFE $Co[Fe(CN)_6]$	Na^+ (1M)	250 F/g@1 A/g 200 F/g@5 A/g	94%/5000 cycles	[57]
Carbon black/PTFE $Ni[Fe(CN)_6]$	Na^+ (2M)	184 F/g@1 A/g	64%/100 cycles	[54]
Carbon black/PTFE $Cu[Fe(CN)_6]$	Na^+ (2M)	200 F/g@1 A/g	25%/100 cycles	[54]
Carbon black/PTFE $Co[Fe(CN)_6]$	Na^+ (2M)	314 F/g@1 A/g	43%/100 cycles	[54]
Graphene/ $Ni[Fe(CN)_6]$	K^+ (1M)	460 F/g@1 A/g 350 F/g@10 A/g	95%/1000 cycles	[49]
Graphene/ $Cu[Fe(CN)_6]$	K^+ (1M)	590 F/g@1 A/g 550 F/g@10 A/g	96%/1000 cycles	[49]

Electrode composition	Ion	Specific capacitance	Capacity Retention	Ref.
Carbon black/PTFE $Mn[Fe(CN)_6]$	Na^+ (1M)	238 F/g@0.5A/g	82%/1000 cycles	[21]
Carbon black/PTFE $Mn[Fe(CN)_6]$	K^+ (2M)	690 F/g@1.3A/g 297 F/g@10 A/g	98%/10000 cycles	[35]
Carbon spheres/ PVDF/$Co[Fe(CN)_6]$	Na^+ (1M)	298 F/g@1A/g 213 F/g@5A/g	94%/1000 cycles	[22]

The earliest report of PBA being used for pseudocapacitor applications was published in 2008 by Chen et al. [7]. Testing was done with PB and Ni and Co hexacyanoferrates where one of the Fe ions is replaced with Ni or Co. The electrodes were prepared by coating a mixture of the active material, acetylene black (20% as a conductivity enhancer), and PTFE (5% as a binder) onto a stainless steel grid. All of these materials have a cubic lattice structure with similar lattice parameters: 10.21 Å for FeHCF, 10.23 Å for NiHCF, and 10.296 Å for CoHCF. Charge-discharge measurements carried out in 1M aqueous KNO_3 showed a pronounced battery-like plateau for FeHCF (PB), and a sloped curve with an inflection for NiHCF and CoHCF when measured in a 3-electrode cell with a platinum foil counter-electrode and a saturated calomel reference electrode. The highest pseudo-capacitance values were obtained for NiHCF (450 F/g at a constant current of 1.5 A/g), and the capacitance retention after 1000 cycles was 87%, higher than both FeHCF and CoHCF.

Other recent studies of metal hexacyanoferrate (MHCF) PBA were done in a similar manner using electrodes with either a mixture of materials with a carbon conductivity enhancer and a binder [21, 22, 35, 54, 57], or with a thin film of the MHCF deposited directly on a carbon electrode [49]. The presence or absence of the binder does not appear to make much of a difference in the capacitance observed (compare references [7] and [49]). However, the binderless electrode with graphene showed a much better capacity retention over 1000 cycles than the mixed electrode (96% vs. 64%) for NiHCF pseudocapacitors. Although many of the galvanostatic charge-discharge curves for

PBA electrodes have Faradaic electrochemical charge curves indicating battery-type behavior, for many of the electrodes the capacity retention over 1000 or more cycles can be quite high (>95%) and power levels can also be high (10 A/g).

Thus supercapacitor storage capacity can be significantly improved by adding surface coatings of PB or PBA materials that undergo fast, reversible redox reactions. Depending on the type of ion and the potential range used, the surface reaction can have either the electrochemical signature of a battery (galvanostatic charge-discharge that is nonlinear, typically with a plateau in potential) or the signature of a capacitor (linear change in current with potential). Pseudocapacitors with surface coatings based on PB and its analogues can have either type of electrochemical profile, and can add substantial storage capacity to EDLCs without much degradation of capacity over thousands of electrochemical cycles.

5.8 Summary

A wide range of applications, ranging from electrical vehicles to uninterrupted renewable (e.g., solar and wind) energy sources demand electrical storages devices that will store more energy, have long lifetime, and cost far less than current prices. Batteries and supercapacitors based on PB or PBAs potentially offer a solution for these applications. Energy storage is made possible by reversible insertion of ions into the open framework of these molecules. Moreover, these materials are easy to synthesize, are low cost and non-toxic, and robust that allow for hundreds to thousands of charge/discharge cycling. Both aqueous and nonaqueous secondary batteries have shown promise, opening up avenues for use in small and large-scale applications. In order to realize the full capabilities of PBAs, synthesizing highly crystalline materials with low defects and small to no amounts of interstitial and coordinated water is required. Additionally, studying the effect the overall crystal structure and stoichiometry have on the electrochemical performance of the materials has led to better performances.

References

1. A. Kraft, On the discovery and history of Prussian Blue, *Bull Hist Chem.*, 33 (2008) 61–67.
2. V. D. Neff, Electrochemical Oxidation and Reduction of Thin Films of Prussian Blue, *J. Electrochem. Soc.*, 125 (1978) 886–887. doi:10.1149/1.2131575.
3. J. Bartoll, The early use of Prussian blue in paintings, in: Proc. 9th Int. Conf. NDT Art, 2008. https://pdfs.semanticscholar.org/500c/970e0785ad2f55cd743590ef244f3bc995a1.pdf (accessed March 30, 2017).
4. A. Ludi, Prussian blue, an inorganic evergreen, *J. Chem. Educ.*, 58 (1981) 1013. doi:10.1021/ed058p1013.
5. N. P. C. Services, *Modern Technology of Paints*, Varnishes & Lacquers (2nd Edition) by Niir Board, 2nd ed., n.d. http://www.niir.org/books/book/modern-technology-paints-varnishes-lacquers-2nd-edition-niir-board/isbn-8178330881/zb,42,a,0,0,a/index.html (accessed March 30, 2017).
6. H. H. Kamerbeek, A. G. Rauws, M. ten Ham, A. N. P. van Heijst, Prussian Blue in Therapy of Thallotoxicosis, *Acta Med. Scand.*, 189 (1971) 321–324. doi:10.1111/j.0954-6820.1971.tb04383.x.
7. M. Altagracia-Martinez, J. Kravzov-Jinich, J. M. Martínez-Núñez, C. Ríos-Castañeda, F. López-Naranjo, Prussian blue as an antidote for radioactive thallium and cesium poisoning, *Orphan Drugs Res. Rev.* (2012) 13. doi:10.2147/ODRR.S31881.
8. F. Ricci, G. Palleschi, Sensor and biosensor preparation, optimisation and applications of Prussian Blue modified electrodes, *Biosens. Bioelectron.*, 21 (2005) 389–407. doi:10.1016/j.bios.2004.12.001.
9. R. J. Mortimer, Electrochromic Materials, *Annu. Rev. Mater. Res.*, 41 (2011) 241–268. doi:10.1146/annurev-matsci-062910-100344.
10. J. F. Keggin, F. D. Miles, Structures and formulæ of the Prussian blues and related compounds, *Nature*, 137 (1936) 577–578.
11. H. J. Buser, D. Schwarzenbach, W. Petter, A. Ludi, The crystal structure of Prussian blue: $Fe_4[Fe(CN)_6]_3 \cdot xH_2O$, *Inorg. Chem.*, 16 (1977) 2704–2710. doi:10.1021/ic50177a008.
12. F. Herren, P. Fischer, A. Ludi, W. Hälg, Neutron diffraction study of Prussian Blue, $Fe_4[Fe(CN)_6]_3 \cdot xH_2O$. Location of water molecules and long-range magnetic order, *Inorg. Chem.*, 19 (1980) 956–959.

13. A. G. Sharpe, The chemistry of cyano complexes of the transition metals, London; New York: Academic Press, 1976. http://trove.nla.gov.au/version/12758209 (accessed March 28, 2017).

14. C. A. Lundgren, R. W. Murray, Observations on the composition of Prussian blue films and their electrochemistry, *Inorg. Chem.*, 27 (1988) 933–939. doi:10.1021/ic00278a036.

15. C. D. Wessells, R. A. Huggins, Y. Cui, Copper hexacyanoferrate battery electrodes with long cycle life and high power, *Nat. Commun.*, 2 (2011) 550. doi:10.1038/ncomms1563.

16. P. Padigi, J. Thiebes, M. Swan, G. Goncher, D. Evans, R. Solanki, Prussian green: A high rate capacity cathode for potassium ion batteries, *Electrochim. Acta*, 166 (2015) 32–39. doi:10.1016/j.electacta.2015.03.084.

17. N. Kuperman, P. Padigi, G. Goncher, D. Evans, J. Thiebes, R. Solanki, High performance Prussian blue cathode for nonaqueous Ca-ion intercalation battery, *J. Power Sources*, 342 (2017) 414–418. doi:10.1016/j.jpowsour.2016.12.074.

18. J. Rodríguez-Carvajal, Recent advances in magnetic structure determination by neutron powder diffraction, *Phys. B Condens. Matter.*, 192 (1993) 55–69. doi:10.1016/0921-4526(93)90108-I.

19. R. Y. Wang, B. Shyam, K. H. Stone, J. N. Weker, M. Pasta, H.-W. Lee, M. F. Toney, Y. Cui, Reversible multivalent (monovalent, divalent, trivalent) ion insertion in open framework materials, *Adv. Energy Mater.*, 5 (2015) n/a-n/a. doi:10.1002/aenm.201401869.

20. Y. Mizuno, M. Okubo, E. Hosono, T. Kudo, H. Zhou, K. Oh-ishi, Suppressed activation energy for interfacial charge transfer of a Prussian blue analog thin film electrode with hydrated ions (Li^+, Na^+, and Mg^{2+}), *J. Phys. Chem. C*, 117 (2013) 10877–10882. doi:10.1021/jp311616s.

21. A. A. Karyakin, Prussian blue and its analogues: Electrochemistry and analytical applications, *Electroanalysis*, 13 (2001) 813–819.

22. D. Moscone, D. D'Ottavi, D. Compagnone, G. Palleschi, A. Amine, Construction and analytical characterization of Prussian blue-based carbon paste electrodes and their assembly as oxidase enzyme sensors, *Anal. Chem.*, 73 (2001) 2529–2535. doi:10.1021/ac001245x.

23. N. Imanishi, T. Morikawa, J. Kondo, R. Yamane, Y. Takeda, O. Yamamoto, H. Sakaebe, M. Tabuchi, Lithium intercalation behavior of iron cyanometallates, *J. Power Sources*, 81–82 (1999) 530–534. doi:10.1016/S0378-7753(98)00228-6.

24. Y. Lu, L. Wang, J. Cheng, J. B. Goodenough, Prussian blue: A new framework of electrode materials for sodium batteries, *Chem. Commun.*, 48 (2012) 6544–6546. doi:10.1039/C2CC31777J.
25. M. Jayalakshmi, F. Scholz, Performance characteristics of zinc hexacyanoferrate/Prussian blue and copper hexacyanoferrate/ Prussian blue solid state secondary cells, *J. Power Sources*, 91 (2000) 217–223. doi:10.1016/S0378-7753(00)00475-4.
26. Z. Jia, J. Wang, Y. Wang, Electrochemical sodium storage of copper hexacyanoferrate with a well-defined open framework for sodium ion batteries, *RSC Adv.*, 4 (2014) 22768–22774. doi:10.1039/ C4RA02559H.
27. M. Pasta, C. D. Wessells, N. Liu, J. Nelson, M. T. McDowell, R. A. Huggins, M. F. Toney, Y. Cui, Full open-framework batteries for stationary energy storage, *Nat. Commun.*, 5 (2014). doi:10.1038/ncomms4007.
28. D. Asakura, M. Okubo, Y. Mizuno, T. Kudo, H. Zhou, K. Ikedo, T. Mizokawa, A. Okazawa, N. Kojima, Fabrication of a cyanide-bridged coordination polymer electrode for enhanced electrochemical ion storage ability, *J. Phys. Chem. C*, 116 (2012) 8364–8369. doi:10.1021/ jp2118949.
29. X. Wu, M. Sun, S. Guo, J. Qian, Y. Liu, Y. Cao, X. Ai, H. Yang, Vacancy-free Prussian blue nanocrystals with high capacity and superior cyclability for aqueous sodium-ion batteries, *ChemNanoMat*, 1 (2015) 188–193. doi:10.1002/cnma.201500021.
30. Y. Jiang, S. Yu, B. Wang, Y. Li, W. Sun, Y. Lu, M. Yan, B. Song, S. Dou, Prussian blue@C composite as an ultrahigh-rate and long-life sodium-ion battery cathode, *Adv. Funct. Mater.*, 26 (2016) 5315–5321. doi:10.1002/adfm.201600747.
31. D. Yang, J. Xu, X.-Z. Liao, H. Wang, Y.-S. He, Z.-F. Ma, Prussian blue without coordinated water as a superior cathode for sodium-ion batteries, *Chem Commun.*, 51 (2015) 8181–8184. doi:10.1039/ C5CC01180A.
32. X. Wu, C. Wu, C. Wei, L. Hu, J. Qian, Y. Cao, X. Ai, J. Wang, H. Yang, Highly crystallized $Na_2CoFe(CN)_6$ with suppressed lattice defects as superior cathode material for sodium-ion batteries, *ACS Appl. Mater. Interfaces*, 8 (2016) 5393–5399. doi:10.1021/acsami.5b12620.
33. Y. You, X.-L. Wu, Y.-X. Yin, Y.-G. Guo, High-quality Prussian blue crystals as superior cathode materials for room-temperature sodium-ion batteries, *Energy Environ. Sci.*, 7 (2014) 1643–1647. doi:10.1039/C3EE44004D.
34. J. Song, L. Wang, Y. Lu, J. Liu, B. Guo, P. Xiao, J.-J. Lee, X.-Q. Yang, G. Henkelman, J. B. Goodenough, Removal of interstitial H_2O in

hexacyanometallates for a superior cathode of a sodium-ion battery, *J. Am. Chem. Soc.*, 137 (2015) 2658–2664. doi:10.1021/ja512383b.

35. S. J. R. Prabakar, J. Jeong, M. Pyo, Highly crystalline Prussian blue/graphene composites for high-rate performance cathodes in Na-ion batteries, *RSC Adv.*, 5 (2015) 37545–37552. doi:10.1039/C5RA04769B.

36. D. Deng, Li-ion batteries: Basics, progress, and challenges, *Energy Sci. Eng.*, 3 (2015) 385–418. doi:10.1002/ese3.95.

37. C. D. Wessells, S. V. Peddada, R. A. Huggins, Y. Cui, Nickel hexacyanoferrate nanoparticle electrodes for aqueous sodium and potassium ion batteries, *Nano Lett.*, 11 (2011) 5421–5425. doi:10.1021/nl203193q.

38. L. Wang, J. Song, R. Qiao, L. A. Wray, M. A. Hossain, Y.-D. Chuang, W. Yang, Y. Lu, D. Evans, J.-J. Lee, S. Vail, X. Zhao, M. Nishijima, S. Kakimoto, J. B. Goodenough, Rhombohedral Prussian white as cathode for rechargeable sodium-ion batteries, *J. Am. Chem. Soc.*, 137 (2015) 2548–2554. doi:10.1021/ja510347s.

39. L. Shen, Z. Wang, L. Chen, Prussian blues as a cathode material for lithium ion batteries, *Chem. Eur. J.*, 20 (2014) 12559–12562. doi:10.1002/chem.201403061.

40. L. Zhang, L. Chen, X. Zhou, Z. Liu, Morphology-dependent electrochemical performance of zinc hexacyanoferrate cathode for zinc-ion battery, *Sci. Rep.*, 5 (2015) 18263. doi:10.1038/srep18263.

41. Y. Mizuno, M. Okubo, E. Hosono, T. Kudo, K. Oh-ishi, A. Okazawa, N. Kojima, R. Kurono, S. Nishimura, A. Yamada, Electrochemical Mg^{2+} intercalation into a bimetallic CuFe Prussian blue analog in aqueous electrolytes, *J. Mater. Chem. A*, 1 (2013) 13055–13059. doi:10.1039/C3TA13205F.

42. P. Padigi, N. Kuperman, J. Thiebes, G. Goncher, D. Evans, R. Solanki, Calcium cobalt hexacyanoferrate cathodes for rechargeable divalent ion batteries, *J. New Mater. Electrochem. Syst.*, 19 (2016) 43–115.

43. P. Padigi, G. Goncher, D. Evans, R. Solanki, Potassium barium hexacyanoferrate–A potential cathode material for rechargeable calcium ion batteries, *J. Power Sources*, 273 (2015) 460–464. doi:10.1016/j.jpowsour.2014.09.101.

44. A. González, E. Goikolea, J. A. Barrena, R. Mysyk, Review on supercapacitors: Technologies and materials, *Renew. Sustain. Energy Rev.*, 58 (2016) 1189–1206. doi:10.1016/j.rser.2015.12.249.

Chapter 6

Prussian Blue for Battery Electrodes

Masashi Okubo

Department of Chemical System Engineering,
School of Engineering, The University of Tokyo,
Hongo 7-3-1, Bunkyo-ku, Tokyo 113-8656, Japan

m-okubo@chemsys.t.u-tokyo.ac.jp

6.1 Batteries for Sustainability

Distributed electrochemical energy storage resources in power grids are crucial for sustainable clean-energy society. Their seamless integration increases our energy control and operational efficiency, allowing to balance uneven energy demands and intermittent power from renewable sources. Therefore, it is highly desirable to develop high-performance batteries, which can be deployed widely in power grids [1]. Lithium-ion batteries (LIBs) are state-of-the-art electrochemical energy storage devices, which are commercialized mainly for portable electronics, because of their relatively high gravimetric/volumetric-energy densities (200–250 Wh/kg and 500–600 Wh/L). However, these energy densities of current LIBs are not satisfactory for larger-scale applications such as electric vehicles [2].

Prussian Blue-Type Nanoparticles and Nanocomposites: Synthesis, Devices, and Applications
Edited by Yannick Guari and Joulia Larionova

ISBN 978-981-4800-05-1 (Hardcover), 978-0-429-02474-0 (eBook)
www.panstanford.com

From a theoretical point of view, gravimetric-energy density of a LIB, which operates with a reaction of Li(positive electrode) + (negative electrode) ↔ (positive electrode) + Li(negative electrode), is expressed as $|\mu_{Li}^{P} - \mu_{Li}^{N}|/(M^{P} + M^{N} + M^{cell})$ where μ_{Li}^{P} and μ_{Li}^{N} are the lithium chemical potential of positive and negative electrodes, M^{P} and M^{N} are the molar mass per charge/discharge reactions of positive and negative electrodes, and M^{cell} is the relative weight of other components (e.g., electrolytes, separators). Therefore, increasing the gravimetric-energy density of LIBs is achieved by (i) decreasing M^{P} (increasing a specific capacity of a positive electrode), (ii) decreasing M^{N} (increasing a specific capacity of a negative electrode), or (iii) increasing $|\mu_{Li}^{P} - \mu_{Li}^{N}|$ (raising a cell voltage). However, all these parameters (μ_{Li}^{P}, μ_{Li}^{N}, M^{P} and M^{N}) are inherent to electrode materials, thereby exploration for novel electrode materials is of primary importance in battery researches.

In this context, coordination compounds have attracted recent attention as potential LIB electrode materials, because of their structural flexibility, versatility, designability, and diversity. Their porous frameworks consisting of transition metals and linker ligands enable reversible solid-state redox reaction of the transition metals in association with Li^{+} (de)intercalation (Fig. 6.1). In particular, in terms of a large theoretical capacity (>150 mAh/g), Prussian blue analogues (PBAs) $A_xM_A[M_B(CN)_6]_{1-y}\cdot\square_y\cdot nH_2O$ have been studied intensively as LIB electrode materials (Section 5.3).

Another important requisite for advanced batteries is elemental abundance. Because lithium resources are globally maldistributed and hence expensive, development of batteries consisting of more abundant elements is needed. For example, owing to abundant sodium resources, sodium-ion batteries (SIBs) are one of the promising candidates for advanced batteries beyond LIBs, and much effort has been devoted to researches on SIB electrode materials [3]. Several research groups reported that PBAs operate as high-performance SIB electrode materials (Section 5.4).

Aqueous batteries are the other important target for large-scale application, because aqueous electrolytes are expected to reduce a total battery cost drastically. Moreover, aqueous electrolytes generally have high ionic conductivity, which enables battery operation at extremely high-charge/discharge rates. Such high (dis)charge-rate operation is essentially required for regulating intermittent power from renewable sources (Section 5.5) [2].

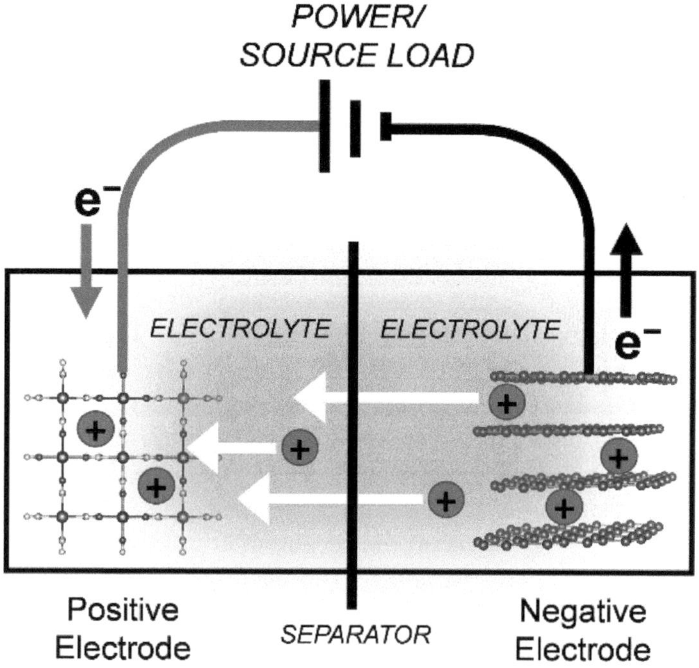

Figure 6.1 Schematic illustration of discharging a battery consisting of a Prussian blue analogue positive electrode and a graphite negative electrode. Electron and ion flows reverse upon charge.

In this chapter, after describing fundamental solid-state electrochemistry of PBAs (Section 5.2), the recent progress in their applications to various batteries is summarized.

6.2 Solid-State Electrochemistry of Prussian Blue Analogues

Prussian blue (PB) and its analogues have a perovskite related structure, where transition metals are bridged by cyanide. The formula is $A_xM_A[M_B(CN)_6]_{1-y}\cdot\square_y\cdot nH_2O$ (A: alkali meta cation,

M_A, M_B: transition metal, □: $[M_B(CN)_6]$ vacancy), where, without the vacancy, M_A is octahedrally coordinated by six nitrogen atoms while M_B is octahedrally coordinated by six carbon atoms. As shown in Fig. 6.1, a three-dimensional open channel exists in PBAs, in which A^+ can diffuse if A^+ is smaller than a bottleneck of the open channel (*ca.* 3 Å). Furthermore, electrons can migrate through overlapping orbitals by thermally activated hopping as polaron. These two features (called mixed conductivity) enable electrochemical ion (de)intercalation in PBA, which was firstly discovered by V. Neff in 1978 [4].

In general, electrochemical reactions of PBAs are described as $(2-4y)A^+ + (2-4y)e^- + M_A[M_B(CN)_6]_{1-y}\cdot\square_y\cdot nH_2O \leftrightarrow A_{2-4y}M_A[M_B(CN)_6]_{1-y}\cdot\square_y\cdot nH_2O$ (Fig. 6.2). A specific capacity of battery electrodes and hence gravimetric-energy density of batteries increase with decreasing a PBA molar mass for one-electron redox reaction ($M^{PBA}/(2-4y)$, M^{PBA}: the molar mass of PBA). Therefore, decreasing the amount of vacancies (y) is an efficient strategy to improve the energy density of batteries consisting of PBAs.

Another important factor governing the electrochemical properties of PBAs is a redox potential of transition metals. In PBAs, Ti, Cr, Mn, Fe, and Co can occupy a carbon-coordinated M_B site. However, for example, a standard redox potential of $[Cr^{III}(CN)_6]^{3-}/[Cr^{II}(CN)_6]^{4-}$ is −1.14 V vs. standard hydrogen electrode (SHE), which are too low to be used as battery positive electrodes. On the other hand, a standard redox potential of $[Fe^{III}(CN)_6]^{3-}/[Fe^{II}(CN)_6]^{4-}$ is 0.36 V vs. SHE (3.4 V vs. Li/Li^+, 3.1 V vs. Na/Na^+), which is suitable to positive electrodes in various batteries such as LIBs or SIBs. Therefore, most studies on PBA battery electrodes have focused on hexacyanoferrate-based PBAs. On the other hand, most 4th period transition metals can occupy a nitrogen-coordinated M_A site: For a battery positive electrode purpose, standard redox potentials of V^{3+}/V^{2+} and Cr^{3+}/Cr^{2+} are too low, while those of Ni^{3+}/Ni^{2+}, Cu^{3+}/Cu^{2+}, and Zn^{3+}/Zn^{2+} are too high. Therefore, PBAs consisting of M_A = Mn, Fe, and Co and M_B = Fe are major targets for high-capacity PBA electrodes. Note that, depending on the amount of a hexacyanometallate vacancy, a coordination environment of M_A changes from M_AN_6 to $M_AN_{6-6y}(OH_2)_{6y}$, leading to a redox-potential change of M_A^{3+}/M_A^{2+}. However, the effect of the coordination environment on the redox potential is yet to be understood. Nevertheless, because transition metals in PBAs play an important role in battery application, hereafter, we denote $A_xM_A[M_B(CN)_6]_{1-y}\cdot\square_y\cdot nH_2O$ as M_AM_B-PBA for simplicity and clarity.

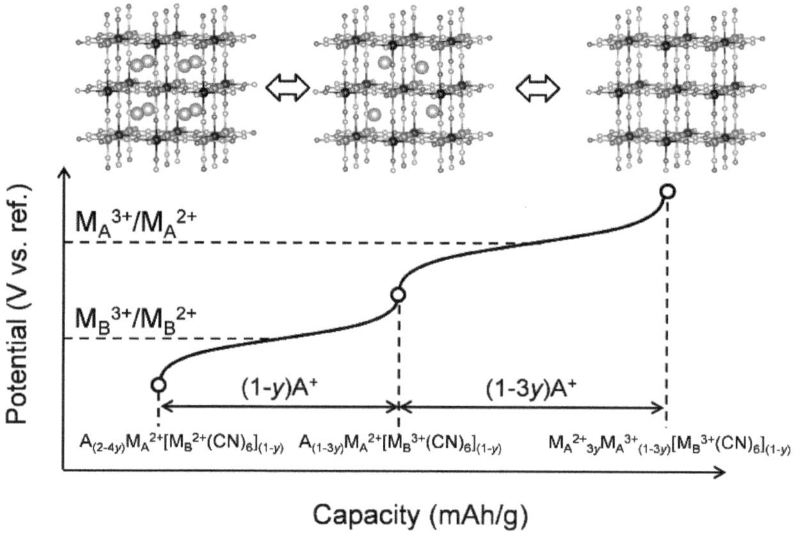

Figure 6.2 Schematic potential profile of Prussian blue analogues $A_{2-4y}M_A[M_B(CN)_6]_{1-y}\cdot\square_y\cdot nH_2O$ upon electrochemical ion (de)intercalation. The amount of an intercalated ion A^+ decreases with decreasing the amount of a $[M_B(CN)_6]$ vacancy, leading to a decrease in a specific capacity.

It is important to note that the electrode potential of PBAs is also influenced by intercalants A^+, because $\mu_A^P = \mu_{e-}^P + \mu_{A+}^P$ where μ_{e-}^P and μ_{A+}^P are electron and ion chemical potentials of PBA, respectively. In nonaqueous electrolytes, large organic electrolyte solvents cannot be co-intercalated, and desolvated A^+ is intercalated into a PBA framework. Therefore, as an ionic radius of A^+ increases (i.e., $Li^+ < Na^+ < K^+$), the ion electrochemical potential μ_{A+}^P decreases due to reduced coulombic repulsion, leading to a higher electrode potential (E_A) (Fig. 6.3) [5]. In aqueous electrolytes, hydrated A^+ is intercalated into the PBA framework. Although $\mu^{P}_{Li+} > \mu^{P}_{Na+} > \mu^{P}_{K+}$ ($E_{Li} < E_{Na} < E_{K}$) still applies, the potential difference is not so large as that in nonaqueous electrolytes, because a hydration shell screens coulombic repulsion (Fig. 6.3) [5].

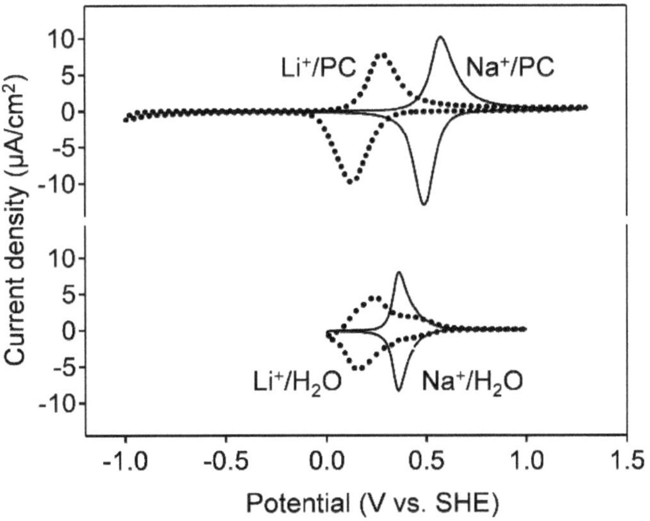

Figure 6.3 Cyclic voltammetry curves of a NiFe-PBA film electrode with various electrolytes [5].

For high charge/discharge-rate operation, poor ion diffusivity and poor electronic conductivity of PBA frameworks are main obstacles, because a resulting large resistance within the electrode substantially degrades battery efficiency. One of the effective ways to reduce these resistances is particle-size reduction, where fast ion diffusion/electron conduction occurs. Furthermore, a large surface area of the small PBA particles reduces the areal current density. Indeed, a particle size reduction of NiFe-PBA from 295 to 140 nm suppresses a polarization significantly in charge-discharge potential profiles (Fig. 6.4) [6]. Alternatively, or in parallel, nano- or mesostructuring of PBAs could also enhance a high charge/discharge-rate capability by efficient charge transport [7].

To summarize, the electrode performance of PBAs is essentially governed by various factors such as the constituent transition metals, the amount of vacancies, the intercalants, the particle size, and the electrolytes. In this context, a wide range of approaches has been exploited to improve the PBA electrode performance.

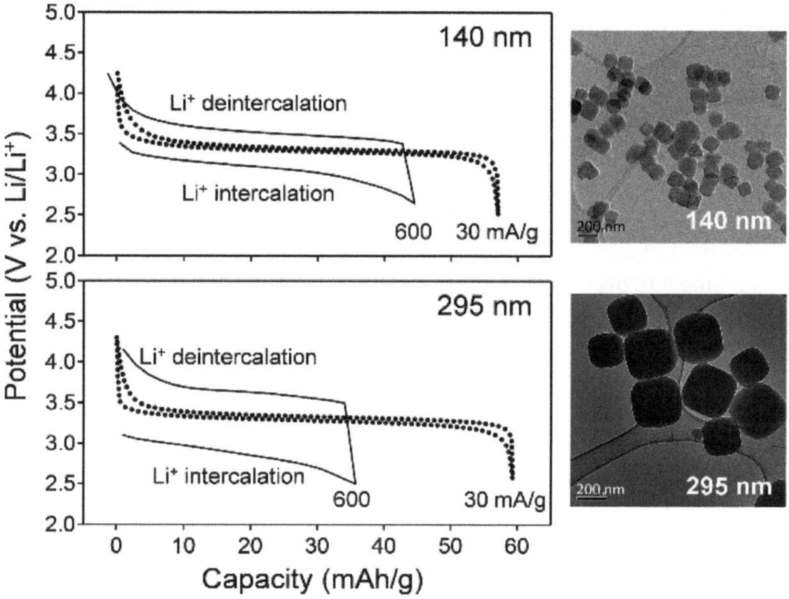

Figure 6.4 Particle size effect on high charge/discharge-rate capability of NiFe-PBA electrodes. Internal resistances within the large-particle electrode induce a large polarization, which leads to a low efficiency and a small capacity [6].

6.3 Application to Lithium-Ion Batteries

The first application of PBAs to LIB electrodes was conducted by Imanishi and coworkers in 1999 [8, 9]. They evaluated the electrode performance of M_AFe-PBA (M_A = V, Mn, Fe, Co, Ni, and Cu), and found that FeFe- and CuFe-PBAs deliver large capacities of 110 and 140 mAh/g at averaged voltages approximately of 2.9 and 3.1 V vs. Li/Li^+, owing to two-electron redox contributions from both M_A (Fe^{3+}/Fe^{2+} or Cu^{2+}/Cu^+) and M_B (Fe^{3+}/Fe^{2+}) sites. However, the maximal capacity is smaller than that of conventional LIB positive electrode materials such as $LiCoO_2$ (150 mAh/g) and $LiFePO_4$ (170 mAh/g), and the cycle stability was poor, especially for the large-capacity PBAs. In 2010, Okubo and coworkers re-examined the performance of MnFe-, NiFe-, and CuFe-PBA electrodes, and demonstrated that MnFe-PBA ($K_{0.1}Mn[Fe(CN)_6]_{0.7}\cdot\square_{0.3}\cdot 4.3H_2O$) gives a reversible capacity of *ca.* 60 mAh/g over 100 charge/discharge cycles (Fig. 6.5) [10–12]. Importantly, electrochemical impedance spectroscopy revealed that instability of the large-capacity two-electron redox PBA electrodes such as CuFe-PBA arises from surface degradation rather than from bulk degradation [11]. Indeed, a core-shell PBA (two-electron redox CuFe-PBA core with one-electron redox NiFe-PBA shell) provides a large capacity from the two-electron redox CuFe-PBA over 50 cycles, which is attributable to a surface protection effect of the stable one-electron redox NiFe-PBA shell [13, 14].

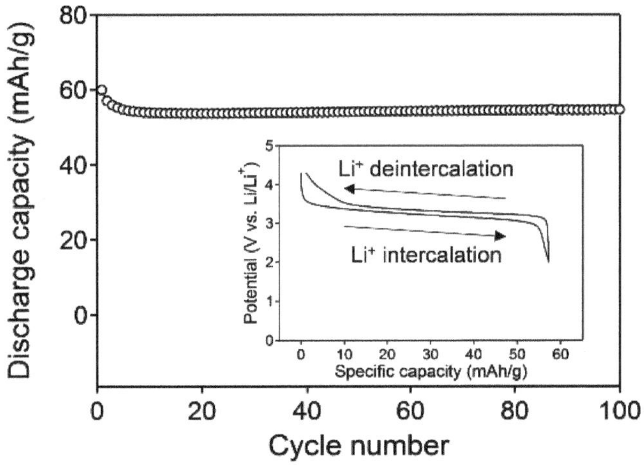

Figure 6.5 Cycle stability of MnFe-PBA in 1 M $LiClO_4$/PC. An inset shows a charge/discharge curve at 20 mA/g.

Okubo and coworkers also pointed to an importance of reducing the amount of hexacyanometallate vacancies for high-capacity PBAs, and showed that vacancy-less MnMn-PBA ($Mn[Mn(CN)_6]_{0.93}\cdot\square_{0.07}\cdot 0.65H_2O$), where all Mn ions are redox-active (Mn^{3+}/Mn^{2+}), gives a large capacity of 200 mAh/g at an averaged potential of 3.1 V vs. Li/Li^+ [15]. Wu and coworkers hydrothermally synthesized vacancy-less FeFe-PBA ($Fe[Fe(CN)_6]_{0.94}\cdot\square_{0.06}\cdot 1.6H_2O$) that provides a large capacity of 160 mAh/g (Fig. 6.6) [16].

Despite much effort devoted to application of PBAs to LIB electrodes, one of main obstacles to practical applications has been a difficulty to fabricate a lithiated PBA with a chemical composition of $Li_2M_A[M_B(CN)_6]$.

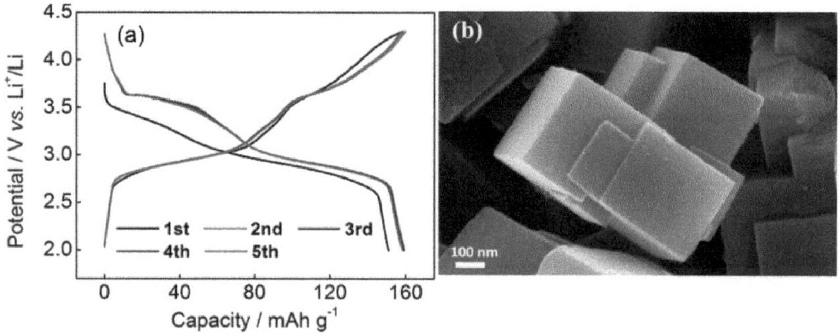

Figure 6.6 (a) Charge-discharge curves at 24 mA/g in 1 M LiPF6/EC-DEC, (b) SEM image for vacancy-less FeFe-PBA [16].

If a PBA were to be used in a practical LIB with a conventional carbonaceous negative electrode, the PBA positive electrode is delithiated upon an initial charge process, which requires a lithiated PBA as a pristine state. However, Li^+ is too small to occupy porous space of PBA frameworks, which hinders fabrication of highly lithiated PBAs. As explained in the next section, highly sodiated PBAs can be synthesized more easily, and application of PBAs to SIB electrodes has become a more active research area.

6.4 Application to Sodium-Ion Batteries

Application of PBAs to SIB electrodes was proposed by Goodenough and coworkers in 2012 [17]. They evaluated a SIB electrode performance of M_AFe-PBA (M_A = Mn, Fe, Co, Ni, Cu and Zn), delivering a specific capacity of 95 mAh/g at an averaged potential of 3.3 V vs. Na/Na^+ for M_A = Fe due to a two-electron redox reaction. Although the PBAs in the first report by Goodenough et al. contain a large amount of hexacyanoferrate vacancies that decreases a capacity, the same group synthesized highly sodiated vacancy-less MnFe- and FeFe-PBAs ($Na_{1.72}Mn[Fe(CN)_6]_{0.99}\cdot\square_{0.01}\cdot nH_2O$ and $Na_{1.92}Fe[Fe(CN)_6]\cdot 0.08H_2O$) [18–20].

The charge/discharge curves in Fig. 6.7 indicate that highly sodiated vacancy-less PBAs exhibit two-electron redox reactions of both M_A (Fe^{3+}/Fe^{2+} or Cu^{2+}/Cu^{+}) and M_B (Fe^{3+}/Fe^{2+}) sites, leading to specific capacities of 135 mAh/g and 150 mAh/g at 3.3 and 3.2 V vs. Na/Na^{+}, respectively. It is important to note that these performances are almost comparable to those of oxide (e.g., $NaFe_{0.5}Co_{0.5}O_2$, 150 mAh/g, 3.2 V vs. Na/Na^{+}) [21] and polyanionic (e.g., $Na_2Fe_2(SO_4)_3$, 100 mAh/g, 3.8 V vs. Na/Na^{+}) [22] electrode materials.

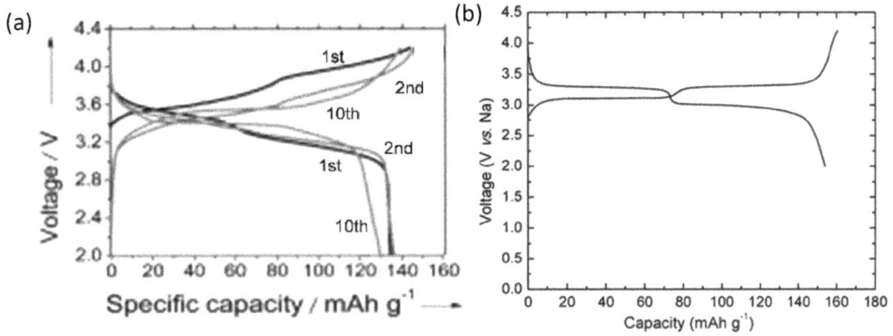

Figure 6.7 Charge-discharge curves for highly sodiated vacancy-less (a) MnFe-PBA, and (b) FeFe-PBA in 1 M $NaPF_6$/EC-DEC [18, 19].

Cui and coworkers reported a SIB electrode performance of a highly sodiated vacancy-less MnMn-PBA ($Na_2Mn[Mn(CN)_6]$) [23]. In their report, $Na_2Mn[Mn(CN)_6]$ delivers a surprisingly large specific capacity of 210 mAh/g, which were claimed to correspond to a three-electron redox reaction of $Na_3Mn^{II}[Mn^{I}(CN)_6] \leftrightarrow Mn^{III}[Mn^{III}(CN)^6]$. Although neither an electronic state nor a detailed structure of the most reduced form was examined, excess sodiation of PBAs exceeding a conventional composition limit has a potential to further extend the solid-state electrochemistry of PBAs.

From a practical application viewpoint, one of the major concerns of PBA battery electrodes is a small volumetric capacity, and hence small volumetric energy density. Although a specific capacity discussed above is important for battery electrodes to fabricate light batteries, a volumetric capacity is another important criterion of battery electrodes to fabricate small batteries: A density of PBAs is much smaller than that of other electrode materials such as oxides because of their porous frameworks. However, SIBs are expected to be distributed as low-cost batteries in power grids, and a size of SIBs should not be a crucial issue that inhibits their wide deployment. Therefore, PBAs are one of the most promising SIB positive electrode materials, and further studies including electrode optimization, full-cell optimization, long cycle tests, and safety tests are necessary.

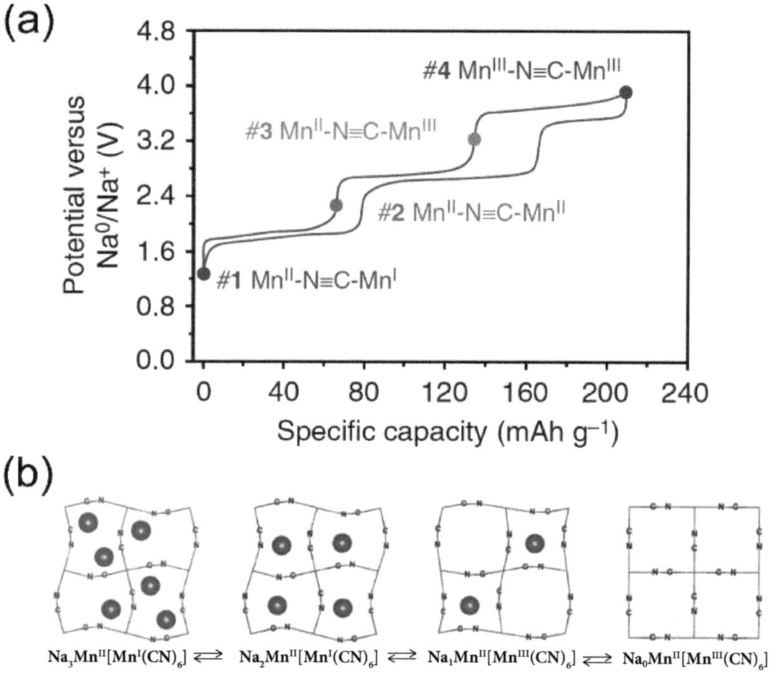

Figure 6.8 (a) Charge-discharge curves, and (b) proposed structural evolution upon (de)sodiation of a highly sodiated vacancy-less MnMnPBA [23].

6.5 Application to Aqueous Batteries

Aqueous batteries consisting of PBA electrodes have been studied [24–26] since the discovery of the solid-state electrochemistry of PBAs by Neff in 1978. However, in 2011, Cui and coworkers re-raised research interests in their use as grid-scale energy storage devices [27–29]. They demonstrated that, owing to rapid diffusion of hydrated ions in PBA frameworks as well as high ionic conductivity of the aqueous electrolytes, PBA electrodes retain 66% of the 1C capacity even at a very high charge/discharge rate (41.7 C for NiFe-PBA and 83 C for CuFe-PBA) and the long cycle life over 5000 cycles (Fig. 6.9a) [27].

Furthermore, Cui et al. fabricated an aqueous battery full cell consisting of a CuFe-PBA ($Na_{0.036}K_{0.40}Cu[Fe(CN)_6]_{0.60}\cdot\square_{0.4}\cdot3.17H_2O$) positive electrode and a chemically reduced MnMn-PBA ($Na_{0.80}K_{0.06}Mn[Mn(CN)_6]_{0.83}\cdot\square_{0.17}\cdot2.17H_2O$) negative electrode, and demonstrated that a full cell operates at 0.95 V even at a 6 min (10C) cycling rate (Fig. 6.9b) [29]. The round-trip energy efficiency reaches 95.4% at a 10C rate and 84.2% at a 50C rate. All these features are attractive for grid-scale low cost batteries for integration of renewable energy.

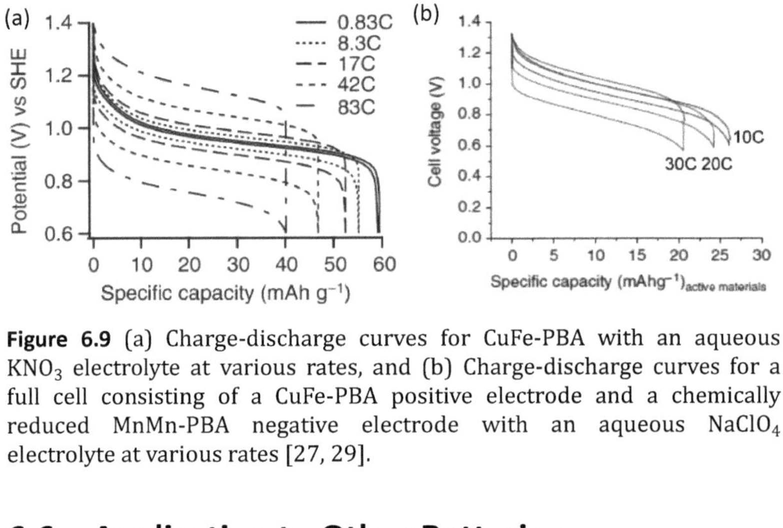

Figure 6.9 (a) Charge-discharge curves for CuFe-PBA with an aqueous KNO_3 electrolyte at various rates, and (b) Charge-discharge curves for a full cell consisting of a CuFe-PBA positive electrode and a chemically reduced MnMn-PBA negative electrode with an aqueous $NaClO_4$ electrolyte at various rates [27, 29].

6.6 Application to Other Batteries

PBAs have also been applied to other batteries such as magnesium-ion batteries, calcium-ion batteries, or aluminum-ion batteries. For example, CuFe-PBA ($K_{0.1}Cu[Fe(CN)_6]_{0.7}\cdot\square_{0.3}\cdot3.6H_2O$) in an aqueous $Mg(NO_3)_2$ electrolyte exhibits reversible (de)intercalation of $0.3Mg^{2+}$ per the formula unit, which is potentially used as electrode reaction of aqueous magnesium-ion batteries [5, 30]. MnFe-PBA in a nonaqueous Ca(trifluoromethylsulfonylimide)$_2$

electrolyte exhibits reversible Ca^{2+} (de)intercalation for nonaqueous calcium-ion batteries [31]. Reversible Al^{3+} (de)intercalation from a nonaqueous Al(fluoromethanesulfonate)$_3$ electrolyte was also demonstrated for CuFe-PBA for nonaqueous aluminum-ion batteries [32].

Nonaqueous potassium-ion batteries are emerging applications of PBA electrodes: Large K^+ can be (de)intercalated reversibly into open PBA frameworks. While FeFe-PBA ($K_{0.22}Fe[Fe(CN)_6]_{0.805}\cdot\square_{0.195}\cdot 4.01H_2O$) delivers a specific capacity of 70 mAh/g approximately at 3.2 V vs. K/K^+ [33], vacancy-less MnFe- and FeFe-PBA ($K_{1.75}Mn[Fe(CN)_6]_{0.93}\cdot\square_{0.07}\cdot 0.16H_2O$ and $K_{1.64}Fe[Fe(CN)_6]_{0.89}\cdot\square_{0.11}\cdot 0.15H_2O$) achieve large specific capacities of *ca.* 130 mAh/g from two-electron redox reactions (Fig. 6.10) [34]. Furthermore, a nonaqueous potassium-ion battery full cell consisting of the MnFe-PBA positive electrode and a graphite negative electrode results in an energy density of 385 Wh/(kg-MnFe-PBA).

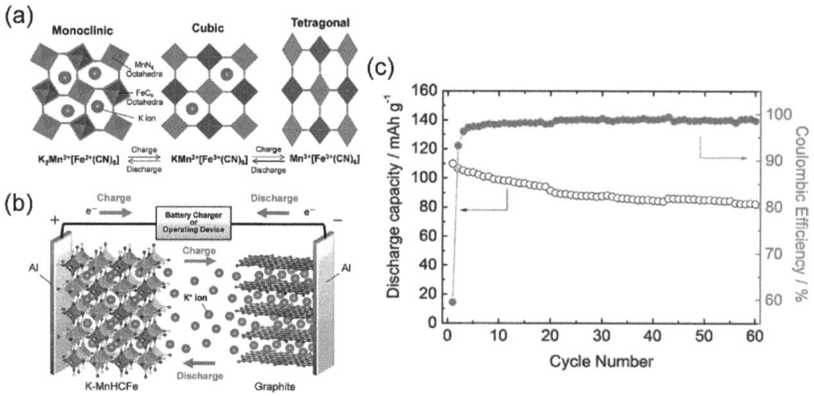

Figure 6.10 Application to potassium-ion batteries. (a) Structural change of MnFe-PBA upon K^+ (de)intercalation. (b) Schematic illustration of a potassium-ion battery consisting of a MnFe-PBA positive electrode and a graphite negative electrode. (c) Cycle stability and coulombic efficiency of a potassium-ion battery full cell [34].

Although all the battery systems here are at the early stage of development, and have several severe obstacles to commercialization (e.g., capacity, rate, voltage, stability, safety, cost), PBA open frameworks offer a versatile platform for intercalation electrode reactions in batteries.

6.7 Conclusion and Outlooks

After the commercialization of LIBs by SONY in 1991, battery researchers have intensively improved their performance for two decades. At present, the available energy density of current LIBs consisting of oxide positive electrodes and carbonaceous negative electrodes approaches a theoretical limit. Therefore, of primary importance is exploration for novel electrode materials for LIBs and other batteries. As discussed in this chapter, although there are several severe technical concerns in application of PBA electrodes to LIBs, their application to SIBs has considerable advantages to other positive electrode materials in terms of energy density, cycle stability, cost, and elemental abundance. Furthermore, preliminary studies on PBA applications to other advanced batteries, especially aqueous batteries and nonaqueous potassium-ion batteries, have already succeeded in demonstration of their performance far superior to other positive electrode materials, even in full cell configurations. All these results put forward an important prospect that PBAs are promising battery electrode materials for advanced batteries in sustainable clean society. For commercialization of batteries consisting of PBAs, it is highly desirable to conduct more collaborative R&D from both fundamental and industrial viewpoints.

References

1. Van Noorden R. (2014). A better battery, *Nature*, **507**(7490), 26–28.
2. Dunn B., Kamath H., Tarascon J. M. (2011). Electrical energy storage for the grid: A battery of choices, *Science*, **334**(6058), 928–935.
3. Yabuuchi N., Kubota K., Dahbi M., Komaba S. (2014). Research development of sodium-ion batteries, *Chem. Rev.*, **114**(23), 11636–11682.
4. Neff V. D. (1978). Electrochemical oxidation and reduction of thin-films of Prussian blue, *J. Electrochem. Soc.*, **125**(6), 886–887.
5. Mizuno Y., Okubo M., Hosono E., Kudo T., Zhou H. S., Oh-ishi K. (2013). Suppressed activation energy for interfacial charge transfer of a Prussian blue analog thin film electrode with hydrated ions (Li^+, Na^+, and Mg^{2+}), *J. Phys. Chem. C*, **117**, 10877–10882.

6. Li C. H., Nanba Y., Asakura D., Okubo M., Talham D. R. (2014). Li-ion and Na-ion insertion into size-controlled nickel hexacyanoferrate nanoparticles, *RSC Adv.*, **4**, 24955–24961.
7. Jiang Y. Z., Yu S. L., Wang B. Q., Li Y., Sun W. P., Lu Y. H., Yan M., Song B., Dou S. X. (2016). Prussian blue@C composite as an ultrahigh-rate and long-life sodium-ion battery cathode, *Adv. Funct. Mater.*, **26**(29), 5315–5321.
8. Imanishi N., Morikawa T., Kondo J., Takeda Y., Yamamoto O., Kinugasa N., Yamagishi T. (1999). Lithium intercalation behavior into iron cyanide complex as positive electrode of lithium secondary battery, *J. Power Sources*, **79**(2), 215–219.
9. Imanishi N., Morikawa T., Kondo J., Yamane R., Takeda Y., Yamamoto O., Sakaebe H., Tabuchi M. (1999). Lithium intercalation behavior of iron cyanometallates, *J. Power Sources*, **81**, 530–534.
10. Okubo M., Asakura D., Mizuno Y., Kim J. D., Mizokawa T., Kudo T., Honma I. (2010). Switching redox-active sites by valence tautomerism in Prussian blue analogues $A_xMn_y[Fe(CN)_6]\cdot nH_2O$ (A: K, Rb): Robust framework for reversible Li storage, *J. Phys. Chem. Lett.*, **1**(14), 2063–2071.
11. Okubo M., Asakura D., Mizuno Y., Kudo T., Zhou H. S., Okazawa A., Kojima N., Ikedo K., Mizokawa T., Honma I. (2011). Ion-induced transformation of magnetism in a bimetallic CuFe Prussian blue analogue, *Angew. Chem. Int. Ed.*, **50**(28), 6269–6273.
12. Mizuno Y., Okubo M., Asakura D., Saito T., Hosono E., Saito Y., Oh-ishi K., Kudo T., Zhou H. S. (2012). Impedance spectroscopic study on interfacial ion transfers in cyanide-bridged coordination polymer electrode with organic electrolyte, *Electrochem. Acta*, **63**, 139–145.
13. Asakura D., Li C. H., Mizuno Y., Okubo M., Zhou H. S., Talham D. R. (2013). Bimetallic cyanide-bridged coordination polymers as lithium ion cathode materials: Core@shell nanoparticles with enhanced cyclability, *J. Am. Chem. Soc.*, **135**(7), 2793–2799.
14. Li C. H., Peprah M. K., Asakura D., Meisel M. W., Okubo M., Talham D. R. (2015). Stepwise reduction of electrochemically lithiated core-shell heterostructures based on the Prussian blue analogue coordination polymers $K_{0.1}Cu[Fe(CN)_6]_{0.7}\cdot 3.5H_2O$ and $K_{0.1}Ni[Fe(CN)_6]_{0.7}\cdot 4.4H_2O$, *Chem. Mater.*, **27**(5), 1524–1530.
15. Asakura D., Okubo M., Mizuno Y., Kudo T., Zhou H. S., Ikedo K., Mizokawa T., Okazawa A., Kojima N. (2012). Fabrication of a cyanide-bridged coordination polymer electrode for enhanced electrochemical ion storage ability, *J. Phys. Chem. C*, **116**(15), 8364–8369.

16. Wu X., Shao M., Wu C., Qian J., Cao Y., Ai X., Yang H. (2017). Low defect $FeFe(CN)_6$ framework as stable host material for high performance Li-ion batteries, *ACS Appl. Mater. Interfaces*, **8**, 23706–23712.

17. Lu Y. H., Wang L., Cheng J. G., Goodenough J. B. (2012). Prussian blue: A new framework of electrode materials for sodium batteries, *Chem. Commun.*, **48**(52), 6544–6546.

18. Wang L., Lu Y. H., Liu J., Xu M. W., Cheng J. G., Zhang D. W., Goodenough J. B. (2013). A superior low-cost cathode for a Na-ion battery, *Angew. Chem. Int. Ed.*, **52**(7), 1964–1967.

19. Wang L., Song J., Qiao R. M., Wray L. A., Hossain M. A., Chuang Y. D., Yang W. L., Lu Y. H., Evans D., Lee J. J., Vail S., Zhao X., Nishijima M., Kakimoto S., Goodenough J. B. (2015). Rhombohedral Prussian white as cathode for rechargeable sodium-ion batteries, *J. Am. Chem. Soc.*, **137**(7), 2548–2554.

20. Song J., Wang L., Lu Y. H., Liu J., Guo B. K., Xiao P. H., Lee J. J., Yang X. Q., Henkelman G., Goodenough J. B. (2015). Removal of interstitial H_2O in hexacyanometallates for a superior cathode of a sodium-ion battery, *J. Am. Chem. Soc.*, **137**(7), 2658–2664.

21. Yoshida H., Yabuuchi N., Komaba S. (2013). $NaFe_{0.5}Co_{0.5}O_2$ as high energy and power positive electrode for Na-ion batteries, *Electrochem. Commun.*, **34**, 60–63.

22. Barpanda P., Oyama G., Nishimura S., Chung S. C., Yamada A. (2014). A 3.8 V earth abundant sodium battery electrode, *Nat. Commun.*, **5**, 4358.

23. Lee H. W., Wang R. Y., Pasta M., Lee S. W., Liu N., Cui Y. (2014). Manganese hexacyanomanganate open framework as a high-capacity positive electrode material for sodium-ion batteries, *Nat. Commun.*, **5**, 5280.

24. Neff V. D. (1985). Some performance-characteristics of a Prussian blue battery, *J. Electrochem. Soc.*, **132**(6), 1382–1384.

25. Honda K., Hayashi H. (1987). Prussian blue containing Nafion composite film as rechargeable battery, *J. Electrochem. Soc.*, **134**(6), 1330–1334.

26. Kaneko M., Okada T. (1988). A secondary battery composed of multilayer Prussian blue and its reaction characteristics, *J. Electroanal. Chem.*, **255**(1–2), 45–52.

27. Wessells C. D., Huggins R. A., Cui Y. (2011). Copper hexacyanoferrate battery electrodes with long cycle life and high power, *Nat. Commun.*, **2**, 550.

28. Pasta M., Wessells C. D., Huggins R. A., Cui Y. (2012). A high-rate and long cycle life aqueous electrolyte battery for grid-scale energy storage, *Nat. Commun.*, **3**, 1149.

29. Pasta M., Wessells C. D., Liu N., Nelson J., McDowell M. T., Huggins R. A., Toney M. F., Cui Y. (2014). Full open-framework batteries for stationary energy storage, *Nat. Commun.*, **5**, 3007.

30. Mizuno Y., Okubo M., Hosono E., Kudo T., Oh-ishi K., Okazawa A., Kojima N., Kurono R., Nishimura S., Yamada A. (2013). Electrochemical Mg^{2+} intercalation into a bimetallic CuFe Prussian blue analog in aqueous electrolytes, *J. Mater. Chem. A*, **1**(42), 13055–13059.

31. Shiga T., Kondo H., Kato Y., Inoue M. (2015). Insertion of calcium ion into Prussian blue analogues in nonaqueous solutions and its application to a rechargeable battery with dual carriers, *J. Phys. Chem. C*, **119**(50), 27946–27953.

32. Liu S., Pan G. L., Li G. R., Gao X. P. (2015). Copper hexacyanoferrate nanoparticles as cathode material for aqueous Al-ion batteries, *J. Mater. Chem. A*, **3**(3), 959–962.

33. Zhang C. L., Xu Y., Zhou M., Liang L. Y., Dong H. S., Wu M. H., Yang Y., Lei Y. (2017). Potassium Prussian blue nanoparticles: A low-cost cathode material for potassium-ion batteries, *Adv. Funct. Mater.*, **27**(4), 1604307.

34. Bie X. F., Kubota K., Hosaka T., Chihara K., Komaba S. (2017). A novel K-ion battery: Hexacyanoferrate(II)/graphite cell, *J. Mater. Chem. A*, **5**(9), 4325–4330.

Chapter 7

Prussian Blue Nanoparticles' Synthesis and Their Inks

Manabu Ishizaki and Masato Kurihara

Faculty of Science, Yamagata University,
1-4-12 Kojirakawa-machi, Yamagata, Japan

kurihara@sci.kj.yamagata-u.ac.jp

7.1 Introduction

Nanoscience has rapidly developed as a new field of the scientific and industrial research since the early 1980s. Nanoparticles (NPs) and nanostructure show interesting properties, which is different of bulk crystal [1, 2]. NPs of inorganic materials have been actively investigated because the nano-sized materials lead to new and/or unique functions, called nano-size effects, for example, band-gap tuning and light-emittance of semiconductor NPs (quantum dots) [3], plasmon resonance of metal NPs [4, 5], magnetic properties of NPs [6–8]. It is essential for the syntheses of NPs to explore suitable conditions inhibiting crystal growth and aggregation. Many effective methods for the syntheses of NPs have been reported: For instance, a reversed micellar method [9–14] an organic-polymer protection method [15, 16],

Prussian Blue-Type Nanoparticles and Nanocomposites: Synthesis, Devices, and Applications
Edited by Yannick Guari and Joulia Larionova

ISBN 978-981-4800-05-1 (Hardcover), 978-0-429-02474-0 (eBook)
www.panstanford.com

aluminum- and silica-template methods [17, 18] have been commonly adopted.

The development of nanotechnology has contributed to the rapid progress of electronic devices and their miniaturization; therefore, fabrication technologies of thin films and fine patterns consisting of functional materials are indispensable. The dry processes such as sputtering and vapor deposition are advantageous techniques for nanoscale-controlled thicknesses and widths of thin films and/or fine patterns. Highly performing and reliable electronic devices have been industrially produced because the dry-processed functional materials with high crystallinity and purity are suitable for various applications [19–22]. However, the dry processes are an opposite trend in green technologies with respect to saving energies and resources. Since the advent of printed electronics, solution-processable technologies have triggered current exhaustive and competitive applications in wide fields to conduct low-consuming societies. A variety of NPs has been explored as promising "building blocks" of solution-processable films and fine patterns for low-cost, flexible, large-area, and lightweight electronic devices. Especially, low-temperature sinterable materials are a hot topic in which metal NPs of Au, Ag, and Cu have been investigated to conduct conductive wiring for printed electronics devices [23–26]. So far, dispersion solutions of NPs (NP inks) have been prepared on the basis of scientific interests; nevertheless, NP inks are evolved into their more suitable preparation methods for the industrial applications i.e., into their convenient, low-cost, scalability, high-concentration, and waste-free productions.

Prussian blue (PB, $Fe_x[Fe(CN)_6]_y$) was discovered 300 years ago, and it has been used as a blue pigment which is the inexpensive alternative of native pigment [27]. PB is composed of a Fe^{II}-CN-Fe^{III} 3D network structure [28]. In the latter part of the 20th century, electro-deposited PB films [29–32] were invented and its reversible redox function was disclosed; thereby industrial applications using PB have been widely investigated, such as electrochromic device, electrocatalysis, secondary batteries, magnets, and others [32–35]. In hybrid functions with magnetic properties, metal displaced PB analogs (PBAs) ($M_x[M'(CN)]_y$, M, M'=Fe, Ni, Cr, Co, etc.), showed a light-induced spin-transition phenomenon, and acted as promising NP magnets [36–38]. To elucidate a relationship between the magnetic properties and the NP sizes, PB and PBA NPs were synthesized using various methods.

In 2007, it was revealed that the historically so-called "insoluble PB," prepared by a usual batch method, was an aggregated form of NPs with dimensions of ~10 nm [39] from the consistency of the transmission electron microscope (TEM) images and single-crystalline sizes from the broadened signals of the powder X-ray diffractions (XRD). However, since the advent of nanotechnology many researchers would image that the aggregated forms of the insoluble PB are bulk-sized crystals because of the insolubility. Consequently, the insoluble nature of PB and PBA NPs due to their aggregation is a longtime dilemma toward systematic evaluation and wide-field applications.

In this capture, we introduce how to prepare individually isolated PB NPs in solutions, NP inks, via typical methodologies, such as surface-protection and template syntheses. We note that preparation methods of NP inks support the recent developments of PB and PBA-based electronic film devices by solution-processed techniques such as Langmuir–Blodgett (LB) [40, 41] and layer-by-layer (LBL) methods [33, 42–44], and the printed-electronics research requires industrially available PB and PBA NP inks form their mass production methods improved in scalability.

7.2 Synthesis Methods

7.2.1 Reversed Micellar Method

Amphipathic molecules (surfactants) composed of hydrophilic and hydrophobic functional groups form nanoscale water droplets as a hydrophilic nano-reactor dispersed in organic solvents, called "reversed micelles or water-in-oil microemulsions" (Fig. 7.1) [9–14]. The nano-reactors are surrounded by self-assembled surfactant molecules and each nano-reactor is isolated in organic solvents. Water-soluble salts trapped in the nano-reactors are combined via their spontaneously continued collision and fusion among the micelles, and nanocrystals are generated and grow in controlled sizes and morphologies. The nano-reactor dimension is one of the important factors to control particle sizes of NP crystals. It is possible to change the nano-reactor dimensions by controlling the water-surfactant molar ratios (w_s = $[H_2O]/[surfactant]$) [9–14]. The nano-reactor dimensions and the obtained NP sizes are increased with increasing of w_s. The nano-reactor sizes are also influenced by molar ratios of water and organic solvents (w_o = $[H_2O]/[organic\ solvent]$) and the molecular structures of surfactants [9, 13]. The reversed micellar method is one of the conventional technologies for synthesis NPs; various functional NP materials of metals (Au, Ag, Cu, etc.) and metal chalcogenides (CdS, ZnS, etc.) as semiconductors, were successfully fabricated using similar methods [9–14].

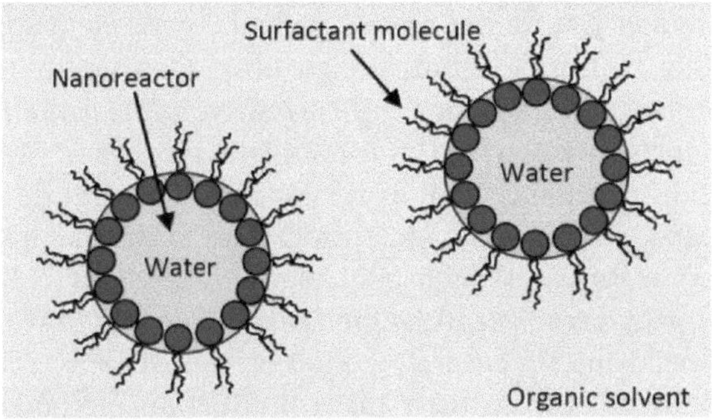

Figure 7.1 Image of reversed micelles in an organic solvent.

Mann et al. have pioneered in developing a synthetic technology for size- and morphology-controlled PB NPs using the reversed micellar method [45]. An aqueous mixture solution of $(NH_4)_3[Fe(C_2O_4)_3]$ and $(NH_4)_3[Fe(CN)_6]$ was added to an isooctane solution of sodium bis(2-ethylhexyl)sulfosuccinate (AOT) at w_{AOT} = 15 (w_{AOT} = $[H_2O]/[IO]$) in the dark. The as-prepared reversed micellar solution was exposure to daylight to slowly reduce Fe^{III} to Fe^{II} via the photo-decomposition of $C_2O_4^{2-}$, and changed into a transparent blue solution, suggesting that PB NPs were formed from the reaction between Fe^{II} and $[Fe(CN)_6]^{3-}$. No precipitation of PB was observed from the reversed micellar solution. From the TEM images of the reversed micellar solution after two days, PB NPs were a regular cubic morphology. An averaged particle size and the size-distribution deviation (σ) of the PB NPs were 16 nm and 2.7 nm, respectively. After four days, the PB nanocubes formed the superlattice structures in the TEM images up to 200 nm, where the gaps among the individual PB nanocubes were 2 to 3 nm, suggesting the coexisting bilayer of AOT molecules attached on their surfaces [46]. The particle sizes of PB nanocubes were decreased depending on decreasing the water ratios (w_{AOT}). By decreasing the water ratios (w_{AOT}), collision frequency of water drops was decreased in the reversed micellar solution, and the crystal growth was inhibited. Without AOT molecules, PB crystals showed a discrete sub-microcube (100–200 nm) with a broad size distribution. These results suggested that the limited reaction spaces in the nano-reactors were an important factor for the size- and morphology-controlled synthesis of PB NPs.

After the report of well-shaped and size-controlled PB nanocubes by a reversed micelle technique, PBA NPs were intensively synthesized by displacement of Fe with Co (Fig. 7.2) [47–49], Cr (Fig. 7.3) [48–50], Ni[50] and others (Fig. 7.3) [51, 52]. Physicochemical properties and their nano-size dependency of PB and PBA NPs were disclosed. Especially, in order to develop NP-based magnets PBA NP powders isolated from the reversed micelle solutions [47–51] were focused on the magnetization characteristics. Additionally, a reverse micellar method is effective in morphology control of the NPs—not only nanocubes and nanospheres but also nanorods (Fig. 7.4) [51] and nanorings [52].

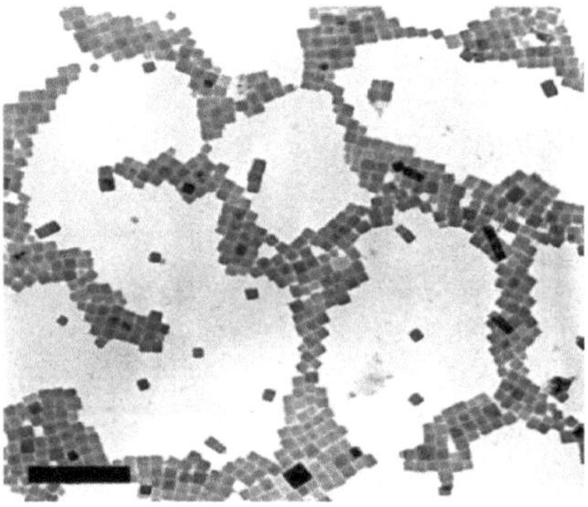

Figure 7.2 TEM image of cobalt hexacyanoferrate (Co-PBA) nanocubes and their superlattice assembly structures synthesized in $Co(AOT)_2$ water-in-oil microemulsions (reversed micelles). Scale bar is 200 nm. Reproduced from Mann et al. [47] with the permission of the American Chemical Society.

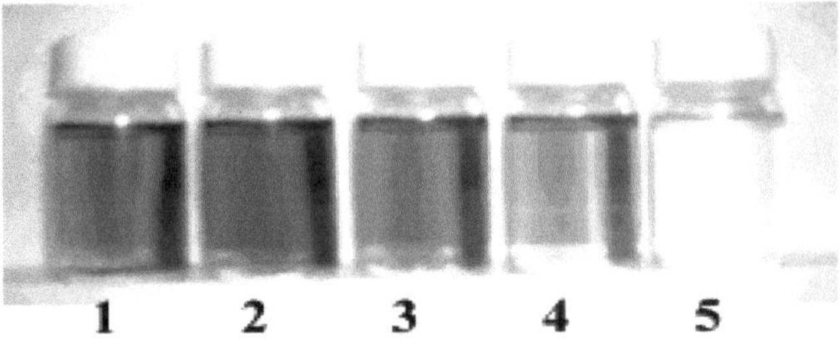

Figure 7.3 Photographs of THF dispersion solutions of $[Cr(CN)_6]$-hybrid Co-PBA NPs, $K_xCo_{x'}[Fe(CN)_6]_y[Cr(CN)_6]_{y'}$. Reproduced from Yamada, Miyake et al. [48] with the permission of the American Chemical Society.

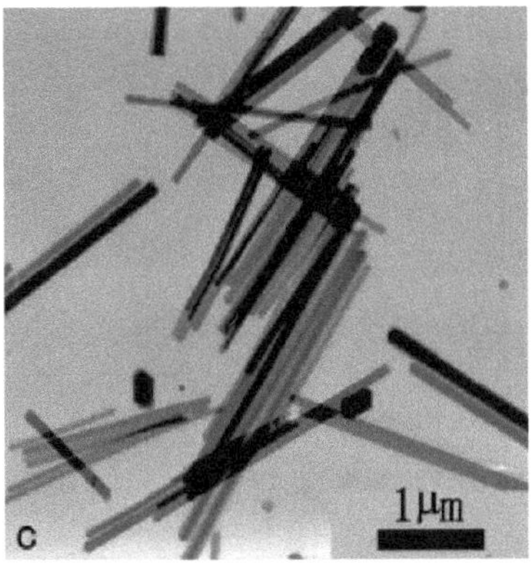

Figure 7.4 TEM image of nanorods of $Sm[Fe(CN)_6]\cdot 4H_2O$. Reproduced from Qi, Gao et al. [51] with the permission of the Royal Society of Chemistry.

The morphology- and size-controlled syntheses of PB and PBA NPs using reversed micelle techniques had an impact on many researchers because of (1) isolation of the NPs, (2) their stable dispersion in solutions, and (3) systematic measurements of their properties. However, the reversed micelle production of the NPs lacks scalability and cost performance in the industrial processes because the essentially dilute reaction conditions consume huge amounts of organic solvents as wastes.

7.2.2 Organic Polymer-Protection Method

In size-controlled syntheses of NPs, growth-inhibition processes from nucleation to large particles along with their aggregation are a crucial factor. In a variety of methods for the growth inhibition via controlling of reaction fields (spaces), collision frequencies among NPs, etc., organic polymers have been preferably adopted as surface-protecting agents of NPs to effectively inhibit their particulate growth. Various functional groups of the organic polymers endow the size- and morphology-controllable surface-protecting abilities of NPs [15, 16]. The surface-protected NPs by the organic polymers are well dispersed into solvents; therefore, the crystal growth by collision among NPs is effectively inhibited. The stable dispersion of NPs due to the affinity between the organic polymer and solvents have led to systematically explore the specific photochemical properties of NPs: The plasmonic Au, Ag, and their alloy NPs and the semiconductor NPs have been widely investigated with respect to the relationship between sizes and morphologies [16, 53, 54].

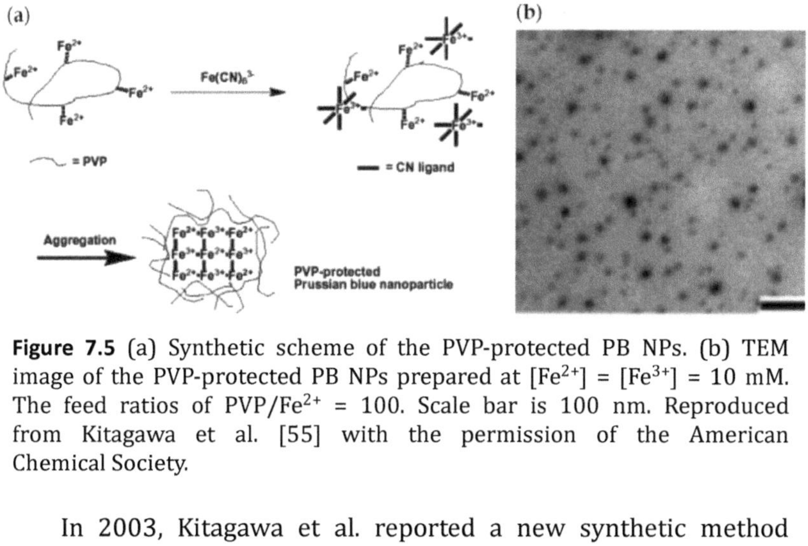

Figure 7.5 (a) Synthetic scheme of the PVP-protected PB NPs. (b) TEM image of the PVP-protected PB NPs prepared at [Fe^{2+}] = [Fe^{3+}] = 10 mM. The feed ratios of PVP/Fe^{2+} = 100. Scale bar is 100 nm. Reproduced from Kitagawa et al. [55] with the permission of the American Chemical Society.

In 2003, Kitagawa et al. reported a new synthetic method of PB NPs using a water-soluble organic polymer as a surfactant [55, 56]. The PB NPs were successfully prepared in aqueous mixtures of poly(vinylpyrrolidone) (PVP, K-30; average M_w = 40,000) and Fe^{2+} and $[Fe(CN)_6]^{3-}$ (Fig. 7.5a). The powdery samples of the PB NPs were centrifuged from the aqueous mixture by the addition of acetone. In a typical case, the TEM image showed that the well-dispersed NPs cannot retain a regular cubic shape, i.e., nanospheres of an average diameter of 27 nm with a broad particle-size distribution (Fig. 7.5b). In the IR spectrum of the PB NPs, a peak for C=O stretching at 1656 cm^{-1} together with a shoulder absorption at around 1600 cm^{-1} is assigned to the PVP amide groups. The appearance of the shoulder absorption suggests that the amide groups of PVP partially bind to the Fe ions of the PB NPs. These results indicate that the amide groups of PVP weakly coordinated to Fe ions during the nucleation and growth processes of the PB NPs. In addition, a cationic organic polymer was utilized to prepare PB NPs by a similar method. In the presence of poly(diallyldimethylammoni um chloride) (PDDA, average M_w = 100,000–200,000) instead of PVP, PDDA-protected PB NPs were obtained: The average particle sizes were within the range between 5 and 8 nm, smaller than those of the PVP-protected PB NPs. In the case of PVP, the aqueous reaction mixture immediately changed into dark blue. On the other hand, in the case of PDDA, similar color change slowly occurred to complete the formation of the PB NPs, suggesting that the cationic PDDAs and $[Fe(CN)_6]^{3-}$ were connected by the electrostatic interaction, so the reaction rate of nucleation of PB was decreased. The historically so-called insoluble PB was prepared as a precipitate from an aqueous mixture of Fe^{2+} and $[Fe(CN)_6]^{3-}$ without the polymer surfactants (PVP). It was defined that the insoluble PB was bulk particles because the TEM particle sizes were apparently over 500 nm. However, the XRD signals of the insoluble PB were significantly broadened and their half-height widths were close to that of the PVP-protected PB NPs (average diameter of 27 nm). As the results, Sherrer's single crystalline particle sizes were estimated to be ~8 nm from the half-height widths in both cases of the insoluble PB and the PVP-protected PB NPs. The inconsistency between the TEM image and Sherrer's particle sizes was ascribable to the lattice defects due to the vacant sites of $[Fe(CN)_6]^{4-}$ [28]. Consequently, the size-dependent magnetic properties of the PVP- and PDDA-protected PB NPs were first time disclosed, compared that of the insoluble PB as bulk crystals. In 2006, Hu et al. reported on a sonochemical synthesis of regular shaped PB nanocubes [57]. The nanocubes grew up to submicron sizes of ~250 nm. Sharp XRD signals appeared in the morphology-controlled submicron-sized PB cubes, that was consistent with the TEM-image sizes; nevertheless, the PB cubes bore the lattice defects due to the vacant sites of $[Fe(CN)_6]^{4-}$ in their formula. In 2007, we revealed that the insoluble PB was apparently bulk-sized particles, but each particle was an aggregated form of NPs with dimensions of ~10 nm [39] from the elaborated TEM observation. As the result, the issue on the inconsistency between the TEM image and the XRD Sherrer's particle sizes would be settled.

7.2.3 Ionic Liquid Method

Ionic liquids (ILs) are melted salts consisted of only ions under 100°C. Room-temperature ILs, of a liquid form at room temperature, have been focused on as a unique material, such as high fluidity, high thermal stability, high ionic conductivity and others. The first report of ILs is early 19th century, and the ILs have been actively investigated to apply electrolytes, catalysts, etc., since the late 19th century [58, 59]. Especially, ILs have attracted attention as alternative reaction media of the common organic solvents because of their irreplaceable properties [60]. In recent years, ILs have been employed as not only greener reaction media but also surfactants for preparing metal and metal-oxide NPs [61–66].

Larionova et al. reported a new synthetic method of PB and PBAs NPs with adjustable sizes using ILs [67]. A representative IL, 1-R-3-methyl-imidazolium tetrafluoroborate $[RMIM][BF_4]$ (R=1-butyl (BMIM) or 2-decyl (DMIM)) acts as both of a surface-stabilizing agent of NPs and a reaction medium. A precursor source of $[Fe(CN)]^{3-}$, $[BMIM]_3[Fe(CN)_6]$ was synthesized: Aqueous solutions of $K_3[Fe(CN)_6]$ and [BMIM] were mixed and stirred at room temperature; after filtration, $[BMIM]_3[Fe(CN)_6]$ was obtained as a yellow powder from the filtrate by an addition of ethanol. In a typical synthesis, various metal salts (M = Fe, Cu, Ni, and Co) were dissolved to $[RMIM][BF_4]$, and the as-synthesized $[BMIM]_3[Fe(CN)_6]$ was added to the solution of the $[RMIM][BF_4]$. After the stirring, PB and PBAs NPs were prepared as their stable dispersion solutions of ILs.

The as-synthesized PB and PBAs NPs were well dispersed in the ILs, and their particle sizes were approximately 2–3 nm with narrow size distributions, according to the TEM images. The PB NPs in $[BMIM][BF_4]$ were slightly smaller than those in [DMIM][BF_4], because the interface tension of [DMIM][BF_4] was lower than that of [BMIM][BF_4] [68]. These results suggest that the *n*-alkyl-chain lengths of imidazolium cations ([RMIM]$^+$) affect the particle-size distributions of the NPs. Moreover, the reaction temperatures influence particle sizes of the PB NPs. In the ILs, the nucleation rate is high due to the low interface tension and weak Ostwald ripening; therefore, it is a tendency that much smaller NPs are easily formed [69]. Additionally, super-lattice and/or highly ordered structure consisted of the PB and PBA NPs were formed in the ILs (Fig. 7.6) [70, 71]. Magnetic properties of the PB and the PBAs NPs were precisely investigated based on their controlled various particle sizes and self-organized solid forms.

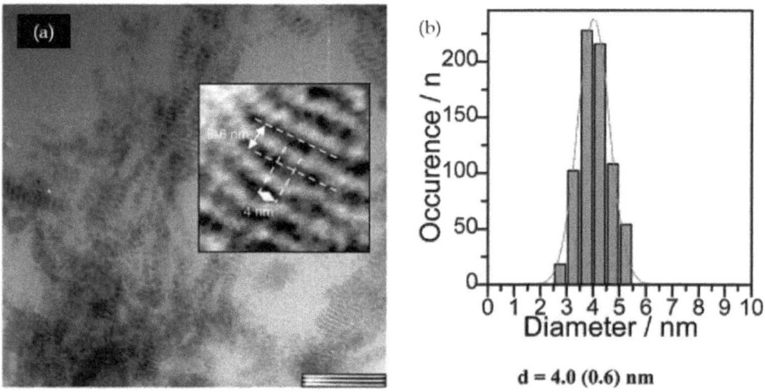

Figure 7.6 (a) TEM image of PBA NPs of $Mn_{1.5}[Cr(CN)_6]$ prepared in the IL (inset: magnification ×10) and (b) its corresponding size distribution. Scale bar is 100 nm. Reproduced from Larionova, Guari et al. [70] with the permission of the American Chemical Society.

7.2.4 Template Methods

7.2.4.1 Porous alumina template

Anodic aluminum oxides (AAO) have nanopores and/or nanochannels, which can be controlled in sizes and morphologies by the anodic oxidation conditions [17]. The AAOs were expected to be utilized to a fine-pattern printed method for replacement of a lithographic method. This method using the controlled templates of nanopores and nanochannels has been widely available for fabrication of metal, metal oxides by not only dry processes but also solution processes. As one of the advantages of this method, the AAO can be employed as an electrode; therefore, various NPs and nanowires have been prepared by an electrodeposition method [17].

Xue and Chen et al. reported the synthesis of nanowire arrays of PB into a porous AAO template by means of the electrodeposition technique [72, 73]: PB was electrically deposited on the AAO electrode in an aqueous mixture solution of $FeCl_3$ and $K_3[Fe(CN)_6]$. The mean length and diameter of the as-prepared PB nanowires are 4 μm and 55 nm, respectively, according to the pore channels of the AAO.

Figure 7.7 Cross-sectional SEM images of the electrochemically deposited PB NPs using a porous AAO template under 1, 3, 5, and 10 deposition cycles (a–d), respectively. (e) SEM image of the PB nanotubes with an outer diameter of ~300 nm, prepared under 10 deposition cycles. Reproduced from Johansson et al. [74] with the permission of the American Chemical Society.

The synthesis of PB nanowires was reported by a combination of layer-by-layer method with the AAO [74]. The AAO was immersed in an aqueous solution of $K_4[Fe(CN)_6]$. After washing with a buffer solution, the obtained $[Fe(CN]_6]$-AAO was immersed in an aqueous solution of $FeCl_3$ and washed with a buffer solution. This deposition procedure was performed several times. From the cross-section scanning electron microscope (SEM) images, PB NPs were formed in the pore channels of the AAO, and the particle size was increased with the repeated deposition cycle numbers (Fig. 7.7a–d). In the 10 deposition cycles, the PB particles were aggregated (connected) into each other. PB nanowires were isolated by etching of the AAO. The diameter and length of PB wire were ~300 nm and 1 μm, respectively, with hollows along the center of the tubes (Fig. 7.7e). The diameter of PB nanotube was accorded with the pore diameters of the AAO. This method has been adopted to synthesize nanodisks, nanorods, and nanotubes of PBAs [75].

7.2.4.2 Mesoporous silica template

Silica is one of the most common supporting materials for the syntheses of functional materials, and its property- and morphology-reforming are readily performed by surface modifications due to functional groups [76] and controlled mesoporous structures [77, 78]. Therefore, silica-supporting (-template) methods are a suitable solution-based procedures containing sol-gel and layer-by-layer techniques [79]. Additionally, silica is a biocompatible low-toxicity material for applications as bio-proves, drug delivery systems (DDS), etc. [80]. From these advantages, various functional NPs have been synthesized using the silica template method [18, 79–81].

PB NPs were synthesized using the mesoporous silica templates (Fig. 7.8) [82]. A mesoporous silica (SBA-15) with a pore size of 0.75 nm was prepared using triblock (P123) copolymer as a surfactant [77, 78], and its pores were functionalized by using $(CH_3O)_3Si(CH_2)_2C_5H_4N$ [76]. The as-functionalized mesoporous silica powder was added to a methanol solution of $FeCl_2$, and stirred overnight at room temperature. After filtration, the powder was carefully washed several times with methanol and dried under a reduced pressure at room temperature. The Fe^{II}-modified silica powder on the pyridine moieties was added to a methanol solution of $[N(C_4H_9)_4]_3[Fe(CN)_6]$ [83], and was stirred for 48 h. After filtration, the resulting powder was carefully washed with methanol and dried under a reduced pressure to obtain $Fe^{2+}/[Fe(CN)_6]^{3-}$, which was defined as one growth cycle for generation of the PB NPs. Three growth cycles were performed. The average pore size and pore area of mesoporous silica estimated from nitrogen adsorption were decreased depending on increasing in the growth cycles. After a hydrogen fluoride (HF) treatment for removal of the mesoporous silica, PB NPs were observed with a mean particle size of 5 nm (Fig. 7.9), which was slightly smaller than the pore size (7.5 nm) of the mesoporous silica.

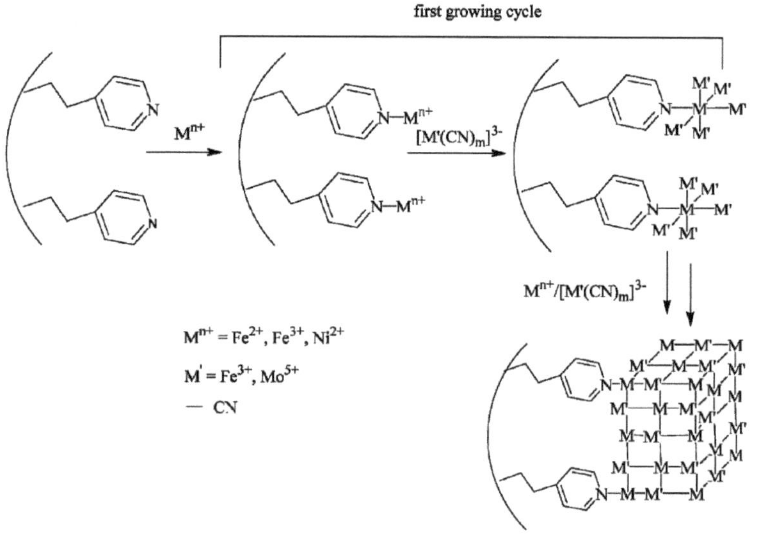

Figure 7.8 Schematic scheme of the intra-pore growth of the cyano-bridged molecule-based magnetic PB NPs. Reproduced from Guari, Larionova et al. [82] with the permission of the Royal Society of Chemistry.

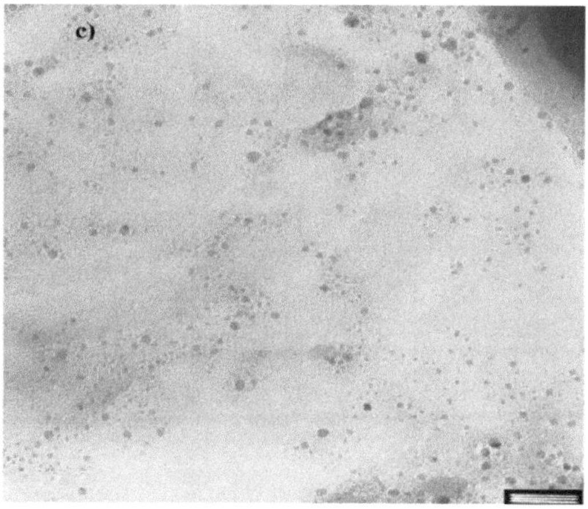

Figure 7.9 TEM image of the PB NPs obtained from the PB-mesoporous silica composite after removal of the silica matrix by a HF treatment. Scale bar is 50 nm. Reproduced from Guari, Larionova et al. [82] with the permission of the Royal Society of Chemistry.

7.2.4.3 Bio-template (apoferritin)

Template syntheses of NPs have been also investigated using a bio-templates such as proteins and cells [84, 85]. Some bio-templates have nano-sized cavities whose sizes are changed by external stimuli such as pH, photo, and others. Additionally, inner of cavities is functionalized by active moieties of amino acids, and several metal ions are trapped. Ferritin (Apoferritin) is a well-known storage protein of iron ions and composed of 24 subunits that assemble to form a vacancy of 12 nm [85]. The vacant cavity of apoferritin is available to store and transport not only iron ions but also the other metal ions along with organic molecules; therefore, apoferritin has been applied for syntheses of various NPs, such as Pd, Au, organic material, and others [85–88]. And then, the bio-template NPs have been applied to the drug-delivery systems (DDSs) [89]. Domínguez-Vera et al. proposed a new chemical method for the preparation of PB NPs based on the apoferritin cavity (Fig. 7.10) [90, 91]. The apoferritin protein was loaded with $[Fe(CN)_6]^{3-}$ ions in an aqueous solution of HCl (pH 2) and isolated. And then, the pH of the loaded apoferritin solution was slowly raised to 8.5 by an addition of a NaOH solution (0.1 M). Finally, an aqueous solution

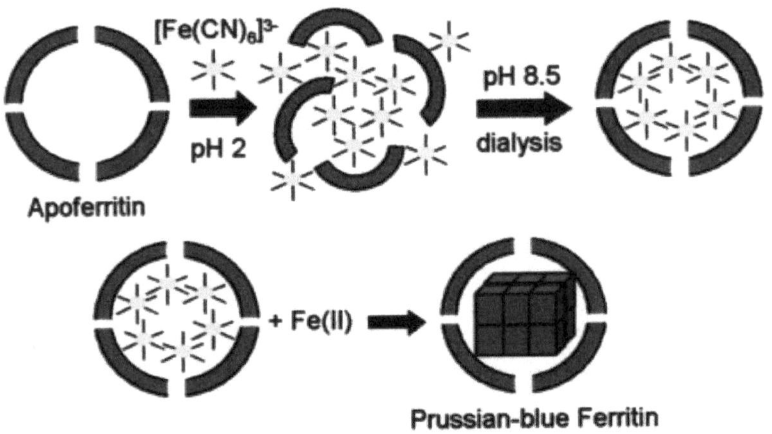

Figure 7.10 Synthetic scheme of the pH-induced dissociation-reformation process of apoferritin in the presence of hexacyanoferrate(III) and the subsequent reaction with iron(II) giving rise to PB NPs within the apoferritin cavity. Reproduced from Domínguez-Vera et al. [90] with the permission of the American Chemical Society.

of Fe(II) was added to the pH-controlled $[Fe(CN)_6]^{3-}$-loaded apoferritin dispersion solution. The color of the solution was immediately changed to blue, suggesting that PB was synthesized. From the TEM images, an average size of PB NPs was approximately 5 nm.

7.2.4.4 Bio-template (chitosan)

Chitosan has a similar structure of cellulose focused as a functional material (Fig. 7.11) [92, 93]. And chitosan is synthesized from chitin, which is easily obtained from crab or shrimp shells: Chitosan is obtained by displacement $NHCOCH_3$ of chitin to NH_3 in an alkali-solution treatment. Chitin and its derivatives are used as a surfactant for syntheses of nano-/micro-particles and DDSs [92, 93]. Guari et al. successfully synthesized PB and PBA NPs using chitosan beads [94], focused on the advantages: (i) the functional amino groups, (ii) the porous structure, (iii) the high water solubility, and (iv) the biocompatibility. The chitosan beads were prepared as the previous report [95]. The chitosan beads were immersed in a methanol solution of M^{2+} (M^{2+} = Ni^{2+}, Cu^{2+} and others) and $(N(C_4H_9)_4)_3[Fe(CN)_6]$ [83]. At each immersion process, the chitosan beads were carefully washed with methanol and dried under reduced pressure. The obtained PBA/chitosan beads were dispersed into water (pH = 4–5). The aqueous dispersion solutions were remarkably stable for several months (Fig. 7.12). The TEM images of the as-prepared PBA/chitosan beads showed the homogeneously isolated PBA NPs, whose mean particle size was 2.8 nm in the case of Ni-PBA (Fig. 7.13).

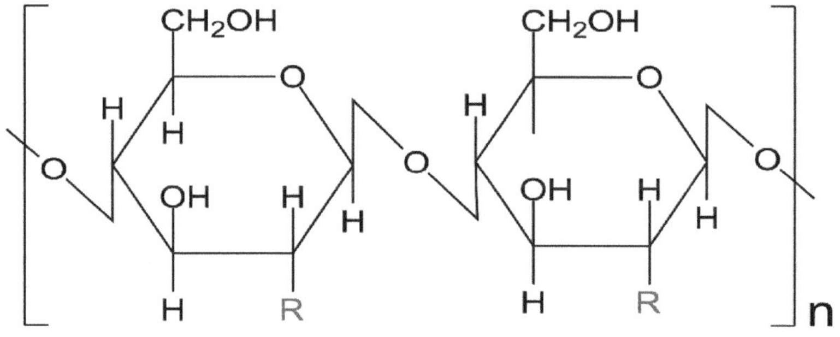

Figure 7.11 Structures of cellulose (R = OH), chitin (R = $NHCOCH_3$) and chitosan (R = NH_2).

Figure 7.12 Photographs of the nanocomposite beads and aqueous solutions consisting of PB/PBAs and chitosan. Reproduced from Guari, Larionova et al. [94] with the permission of the Royal Society of Chemistry.

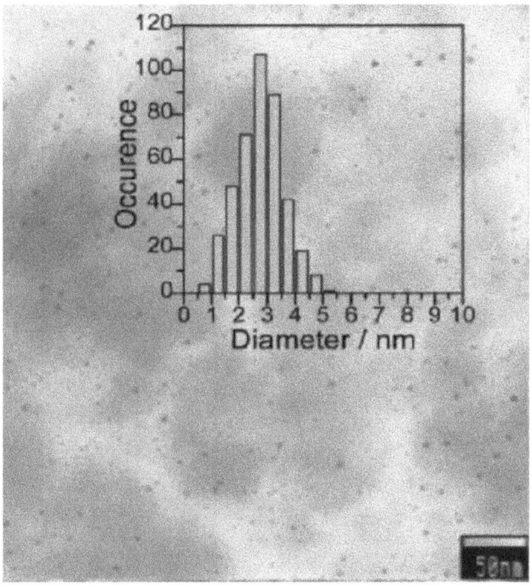

Figure 7.13 TEM image of the nanocomposite Ni-PBA/chitosan. The insert shows a size-distribution histogram of the Ni-PBA NPs. Reproduced from Guari, Larionova et al. [94] with the permission of the Royal Society of Chemistry.

7.3 PB Inks for Industrial Applications

7.3.1 Solvent-Extraction Method

Prior to the rapid progress of nanotechnology and nanoscience, in 1990, Toshima et al. reported a preparation method of PB dispersion solutions, and electrochemical properties of PB thin films by a cast method [96, 97]. A suspension solution of PB was prepared by mixing of aqueous solutions of $K_4[Fe(CN)_6]$ and $FeCl_3 \cdot 6H_2O$ at the same molar ratios, and the concentration of the suspension was adjusted to 20 mM or less. A chloroform solution containing an excess amount of cationic surfactants, such as hexadecyltrimethylammonium chloride (HTAC), was added to the PB suspension solution, and stirred to extract PB particles from the water to the chloroform phase. The PB particles were effectively extracted using the surfactants bearing longer alkyl chains than C_{14}. The HTAC-modified PB particles (PB-HTAC) were spread on an ITO glass substrate using their dispersion solutions. The electrochemical properties of the as-prepared PB-HTAC film were compared to those of the electrodeposited PB film. The electrodeposited film showed an electrochromic phenomenon by changing the applied potentials, while the PB-HTAC film did not show the electrochromism because HTAC as a surfactant inhibited electron/ion transfer among the PB particles. To improve the conductivity of the film, lithium perchlorate $LiClO_4$ in acetone was added to the extracted PB-HTAC chloroform solution, and the mixed solution was spread on ITO glass substrates (PB-HTAC-$LiClO_4$). The PB-HTAC-$LiClO_4$ film exhibited a similar electrochromic phenomenon to the case of the electrodeposited PB film. This was the first report related to preparation of the stable PB dispersion solutions and electrochemical properties due to its simple cast films.

7.3.2 Water-Soluble PB

The advent of nanotechnology and its rapid progress in the 21st century have triggered to develop the various synthetic strategies for PB and PBA NPs and their stable dispersion solutions (inks). In low-temperature solution-processable technologies for thin-film electronic devices, the precursor solutions have been indispensable to readily fabricate large-area thin films with controlled nano-scale thicknesses in industries [98–104]. For example, in an epochal discovery of perovskite solar cells, highly performing $CH_3NH_3PbI_3$ crystal thin films were successfully prepared by simple solution-based procedures [104]. These convenient processes have stimulated exhaustive research in a large number of fields. Toward realization of the printed electronics industries, dispersion solutions of surface-modified NPs have been favorably employed as a promising precursor to produce solution-processed, lightweight, flexible, thin-film electronic devices [23–26]. In the industrial applications, the above-mentioned production methods of PB and PBA NPs (Section 7.2) would lack scalability and cost-performance because of huge volumes of the reaction solvents as wastes and complicated reaction procedures.

In general, PB is well known as an insoluble blue pigment; nevertheless, the presence of a soluble form of PB has been recognized [28, 30, 31, 105]. The insoluble and soluble PB particles have been distinguished as the formula of $Fe_4[Fe(CN)_6]_3$ and $AFe[Fe(CN)_6]$ (A = K, Na and others), respectively. The soluble PB particles are allowed to release its counter cations such as K^+ and Na^+ in water; therefore, the negatively charged particles of $Fe[Fe(CN)_6]^-$ can be dispersed into water by hydrated structures and electrostatic repulsion between them; however, the dispersible property strongly depends on the particle sizes. In 2004, Hammond et al. successfully synthesized water-soluble PB NPs without any surfactants insulating among the NPs in their solution-processed thin films [33]. The water-soluble PB NPs were synthesized under excess amount conditions of $[Fe(CN)_6]^{3-}$ acting as an ionic surfactant to disperse the PB NPs into water as well as inhibition of their crystal growth. An aqueous solution (35 mL) of $FeCl_2$ (0.01 M) was dropwise added to an aqueous solution (35 mL) of $K_3[Fe(CN)_6]$ (0.05 M) and KCl (0.05 M). After completion of the addition, the mixture solution was vigorously stirred for 1 min. The excess $K_3[Fe(CN)_6]$ was removed by a permeation process using a Millipore membrane (for 48 h and the elution of at least 1000 mL), and pH of the as-obtained PB NP solution was adjusted to 4. From the TEM images, an average particle size of the PB NPs was 4–5 nm. Using the PB NP ink and linear poly(ethylene imine) (LPEI), layer-by-layer (LBL) films composed of PB/LPEI were fabricated, and the electrochemical properties and electrochromic phenomena of the hybrid films were explored as a smart window.

7.3.3 Surface Modification Methods of the Historically Insoluble PB

In 2007, Kurihara et al. invented one of the most convenient, scalability, environmentally friendly, low-cost methods to produce dense PB and PBA NP inks suitable for industrial applications [39]. It was revealed that the historical insoluble PB, i.e., an insoluble precipitate from high-concentration aqueous mixture conditions of Fe^{2+} and $[Fe(CN)_6]^{4-}$, was an aggregated form of PB NPs with diameters of 10–20 nm or less based on the consistency between the TEM image and the XRD Sherrer's particle sizes. These findings conducted to develop a direct transformation method from the insoluble PB to PB NP inks using various solvents by a surface-modification technology using active surface-unsaturated coordination sites of each PB NP as open frameworks of continuous 3-D coordination bonds exposed on the surface (Fig. 7.14).

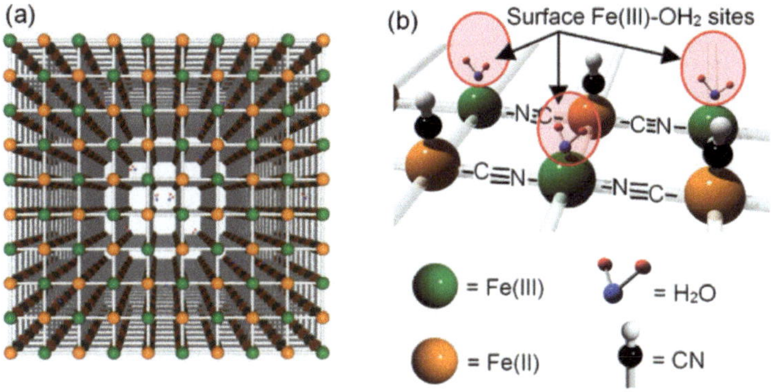

Figure 7.14 (a) 3-D Fe(II)–CN–Fe(III) bonding network structure and (b) surface reaction sites with Fe(III)–OH_2 of PB NPs. The PB NPs are drawn as an ideal cubic shape (nanocube) surrounded by six {100} surfaces, which consist of a 10 × 10 arrangement of irons with a 0.5 nm × 2 dimension. The 10-nm nanocubes are a three-dimensional 21 × 21 × 21 structure whose surface Fe(III)–OH_2 sites occupy 15%, numerically, of all Fe ions. Reproduced from Kurihara et al. [106] with the permission of the Royal Society of Chemistry.

The insoluble PB was prepared by a conventional batch method: An aqueous solution (30 mL) of $Fe(NO_3)_3 \cdot 9H_2O$ (16.2 g,

40.1 mmol) was added to an aqueous solution (60 mL) of $Na_4[Fe(CN)_6]\cdot10H_2O$ (14.5 g, 30.0 mmol). The high-concentration aqueous mixture was vigorously stirred for 5 min. The generated insoluble blue pigment was centrifuged, washed with water and methanol, and then dried under a reduced pressure. The as-isolated insoluble PB (1.00 g, 0.885 mmol as a formula, $Fe_4[Fe(CN)_6]\cdot15H_2O$) was stirred with toluene (15 mL) and water (2 mL), and then oleylamine (0.650 g, 2.43 mmol) was added. The blue suspension was further stirred for 1 h at room temperature. PB NPs were eluted to the toluene layer during stirring. The separated deep blue toluene layer was dehydrated using anhydrous Na_2SO_4 and filtered through a celite layer. Finally, a stable toluene-dispersion solution of PB NPs was obtained (Fig. 7.15a). After evaporation of the toluene-dispersion solution, the surface modified PB NPs with oleylamine molecules as a solid residue can be well re-dispersed in the usual organic solvents.

In the same report [39], water-soluble PB NPs were also directly transformed from the insoluble PB. The insoluble PB solid (0.40 g, 0.35 mmol) was suspended in water (8 mL). $Na_4[Fe(CN)_6]\cdot10H_2O$ (180 mg, 0.37 mmol) of 15 mol% of total Fe of PB was added to the suspension and stirred at room temperature for several hours. During the reaction period, the suspension completely changed into a transparent deep blue solution of the PB NPs, a PB NP ink, without any PB solids. In the PB NP ink, the PB NPs were almost independently dispersed into water via their negative zeta potentials, suggesting that they were surface modified with $[Fe(CN)_6]^{4-}$ on the active surface-unsaturated coordination sites (Fig. 7.15b). The Ni and Co-displaced PBA NPs could be dispersed into solvents by similar experimental procedures. It is generally difficult to stably disperse NPs into environmentally friendly alcohols having hydrophilic-and-hydrophobic boundary properties. The alcohol-dispersed PB NPs were successfully developed through the hybrid surface modification using $[Fe(CN)_6]^{4-}$ and various chain-length alkylamines in controlling electrostatic repulsion among each PB NP due to the negative zeta potentials and the affinity of the alkylamine to the solvents (Fig. 7.15c) [106]. The as-prepared water and alcohol inks of the surface-modified PB NPs were well spread on

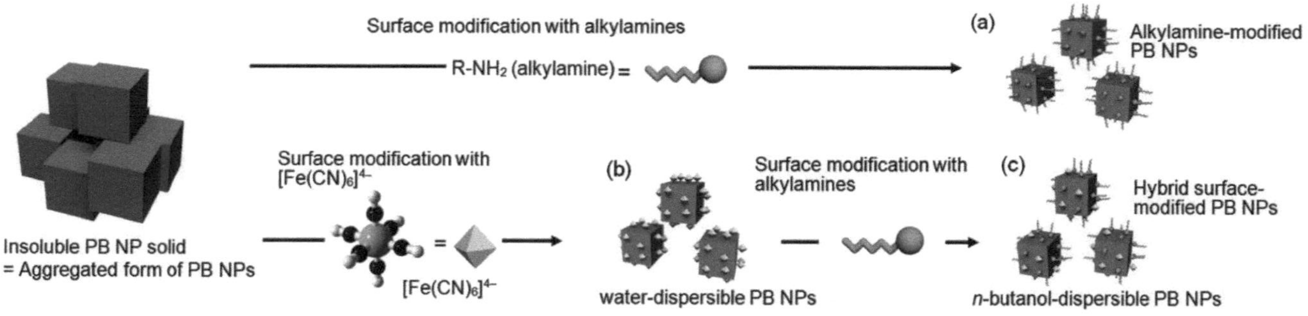

Figure 7.15 Synthetic strategy of PB NPs dispersible into greener solvents, water and alcohols, from the insoluble PB (aggregated form of PB NPs) via reactions of the H_2O ligand of the surface Fe(III) sites (Fig. 7.14b): (a) surface modification with alkylamines, and (b) hybrid surface modification with $[Fe(CN)_6]^{4-}$ and (c) alkylamines. The insoluble PB NPs are transformed into *n*-butanol-dispersible PB NPs by the hybrid surface modification with 5% of $[Fe(CN)_6]^{4-}$ and 10% of shorter-chain alkylamines, numerically, of all Fe ions of PB. Reproduced from Kurihara et al. [106] with the permission of the Royal Society of Chemistry.

various substrates using wet-processes, such as a spin-coating method. By means of a low-temperature heating process of the spin-coating films at less than 120°C, discontinuous coordination networks among the PB and PBA NPs were successfully restored to function as electrochromic (Fig. 7.16) [106, 107], electron-rectifier [108], and grain-boundary free super proton conductor films [109]. The simple, low-cost, and scalability synthetic method of PB and PBA NPs has been currently utilized in various fields of applications [110–112].

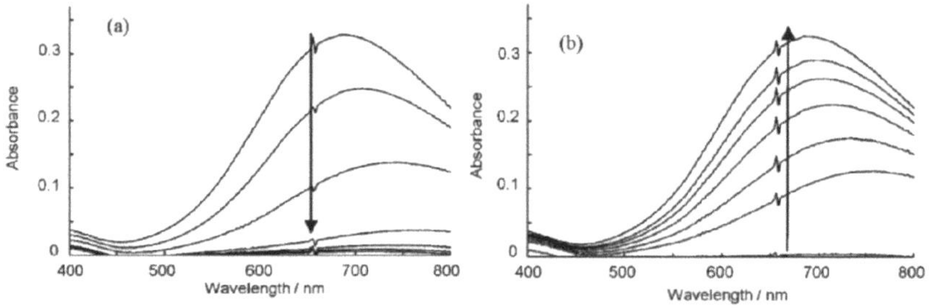

Figure 7.16 Time-course change in the UV-Vis absorption spectra, i.e., an electrochromic phenomenon, of the spin-coated and low-temperature heated PB NP thin film at the applied potentials of −0.5 V (a) and 0.5 V (b) vs. Ag/AgCl. Reproduced from Kurihara et al. [106] with the permission of the Royal Society of Chemistry.

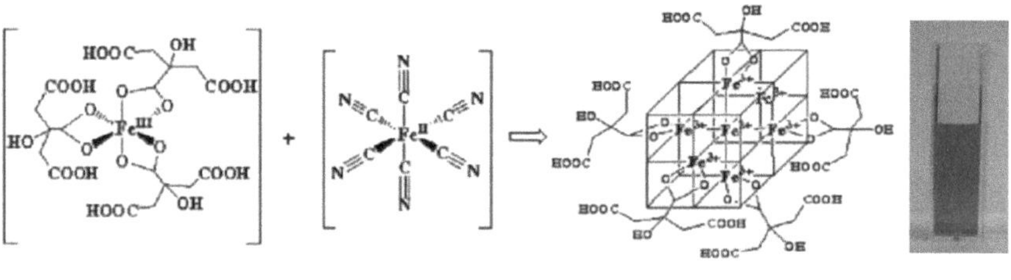

Figure 7.17 Synthetic scheme of the citrate-modified PB NPs. Reproduced from Huang et al. [120] with the permission of the Royal Society of Chemistry.

7.3.4 Citrate-Modified Method

Citric acid is a common reagent for the syntheses of metal NPs [113–115], acting as a reducing agent of metal ions and a surface-stabilizer for metal NPs. Citric acid and its compounds have been applied to food and pharmaceuticals because of low toxicity [116, 117]. The citrate-modified metal NPs usually are stably dispersed into water by their negative surface charges [118]. PB and PBAs have been explored in a range of bio-applications. Currently, the non-toxic water dispersible NPs are required: Since 2010, the water ink of the citrate-modified PB NPs has been continuously reported (Fig. 7.17), and been availed as such bio-probes and magnetic resonance imaging agents (MRI) [119, 120]. In a typical synthesis, an aqueous solution (20 mL) of $FeCl_3$ (1.0 mM) containing citric acid (0.5 mmol) was added to an equimolar aqueous solution of $K_4[Fe(CN)_6]$ containing citric acid (0.5 mmol) with vigorous stirring at room temperature, and the solution color was immediately changed to blue. After adding acetone to the blue solution, the blue pigment was isolated by centrifugation. The XRD patterns of the blue pigment showed the cubic face-centered structure of PB. A TEM average particle size was 13 ± 5 nm. The citrate-modified PB NPs can be re-dissolved into water [121], and have been widely used such as photo-device consisting of a PB thin film [122] and a bio-probe using the dye-connected PB NPs [119, 120].

7.4 Summary

Prussian blue (PB = iron hexacyanoferrate) has a three-century-old history as a blue pigment. Currently, it is defined as the first artificial porous coordination polymer. The multiple functionalities originated from the extended π-electron coordination bonding networks and porous spaces with PB and its analogs (PBAs) have renewed wide-fields of scientific and industrial interests. Toward the rapid progress in miniaturization of lightweight and/or flexible electronic devices, the discovery of novel nanoparticle (NP)-based functions is indispensable for actual applications, especially in printed electronics. In this chapter, a trend of a variety of synthetic strategies of PB and PBA NPs along with their rapid progress in the 21st century has been summarized. The related nano-chemistry has presented industrially available PB and PBA NP inks with respect to scalability, low-cost, easy-procedure, and green productions saving energies, resources, and solvents. The 3-D coordination bonding networks of PB and PBA NPs are discontinuous on their surfaces which operate as unsaturated coordination active sites. These findings of the surface active sites in coordination polymer NPs lead to the surface-modification and surface-architecture technologies. Although the surface modification to improve dispersible natures of NPs into solvents generally has been obstacle to communicate among the NPs through electrons/holes or ions in the solution-processed film devices composed of the NPs, recently NP-based film technologies using PB and PBAs is overcoming the longtime dilemma [106–108].

References

1. Ozin G. A. (1992). Nanochemistry: Synthesis in diminishing dimensions, *Adv. Mater.,* **4**(10), 612–649.
2. Xia Y., Gates B., Yin Y., Lu Y. (2000). Monodispersed colloidal spheres: Old materials with new applications, *Adv. Mater.,* **12**(10), 693–713.
3. Alivisatos A. P. (1996). Semiconductor clusters, nanocrystals, and quantum dots, *Science,* **271**(5251), 993–937.
4. Ghosh S. K., Pal T. (2007). Interpartidcle coupling effect on the surface plasmon resonance of gold nanoparticles: From theory to application, *Chem. Rev.,* **207**(11), 4797–4862.
5. Comin A., Manna L. (2014). New materials for tunable plasmonic colloidal nanocrystal, *Chem. Soc. Rev.,* **43**(11), 3957–3975.
6. Park S.-J., Kim S., Lee S., Khim Z. G., Char K., Hyeon T. (2000). Synthesis and magnetic studies of uniform Iron, *J. Am. Chem. Soc.,* **122**(35), 8581–8582.
7. Sun S., Murray C. B., Weller D., Folks L., Moser A. (2000). Monodisperse FePt nanoparticles and ferromagnetic FePt nanocrystal superlattices, *Science,* **287**(5460), 99, 1989–1992.
8. Sun S., Murray C. B. (1999). Synthesis of monodisperse cobalt nanocrystals and their assembly into magnetic superlattices (invited), *J. Appl. Phys.,* **85**(8), 4325–4330.
9. Pileni M. P. (1993). Reverse micelles as microreactors, *J. Phys. Chem.,* **97**(27), 6961–6973.
10. Petit C., Pileni M. P. (1988). Synthesis of cadmium sulfide in situ in reverse micelles and in hydrocarbon gels, *J. Phys. Chem.,* **92**(8), 2282–2286.

11. Petit C., Lixon P., Pileni M. P. (1993). In situ synthesis of silver nanocluster in AOT reverse micelles, *J. Phys. Chem.*, **97**(49), 12974–12983.
12. Taleb A., Petit C., Pileni M. P. (1997). Synthesis of highly monodisperse silver nanoparticles from AOT reverse micelles: A way to 2D and 3D self-organization, *Chem. Mater.*, **9**(4), 950–959.
13. Usković V., Drofenik M. (2005). Synthesis of materials within reverse micelles, *Surf. Rev. lett.*, **12**(2), 239–277.
14. Lianos P., Thomas J. K. (1986). Cadmium sulfide of small dimensions produced in inverted micelles, *Chem. Phys. Lett.*, **125**(3), 299–302.
15. Mayer A. B. R., Mark J. E. (1996). Polymer-protected, colloidal platinum nanoparticles, *Polym. Bull.*, **37**(5), 683–690.
16. Mayer A., Antonietti M. (1998). Investigation of polymer-protected noble metal nanoparticles by transmission electron microscopy: Control of particle morphology and shape, *Colloid. Polym. Sci.*, **276**(9), 769–779.
17. Chik H., Xu J. M. (2004). Nanometric superlattices: Non-lithographic fabrication, materials, and prospectives, *Mater. Sci. Eng. R*, **43**(4), 103–138.
18. Vondrova M., Klimczuk T., Miller V. L., Kirby B. W., Yao N., Cava R. J., Bocarsly A. B. (2005). Supported superparamagnetic Pd/Co alloy nanoparticles prepared from a Silica/Cyanogel co-gel, *Chem. Mater.*, **17**(25), 6216–6218.
19. Nomura K., Ohta H., Takagi A., Kamiya T., Hirano M., Hosono H. (1993). Room-temperature fabrication of transparent flexible thin-film transistors using amorphous oxide semiconductors, *Nature*, **432**(7016), 488–492.
20. Shrotriya V., Li G., Yao Y., Chu C.-W., Yang Y. (2006). Transition metal oxides as the buffer layer for polymer photovoltaic cells, *Appl. Phys. Lett.*, **88**(7), 073508.
21. Fortunato E., Barquinha P., Martins R. (2012). Oxide semiconductor thin-film transistors: A review of recent advances, *Adv. Mater.*, **24**(22), 2945–2986.
22. Pergolesi D., Fabbri E., D'Epifanio A., Di Bartolomeo E., Tebano A., Sanna S., Licoccia S., Balestrino G., Traversa E. (2010). High proton conduction in grain-boundary-free yttrium-doped barium zirconate films grown by pulsed laser deposition, *Nat. mater.*, **9**(10), 846–852.

23. Kamyshny A., Steinke J., Magdassi S. (2011). Metal-based inkjet inks for printed electronics, *Open Appl. Phys. J.*, **4**(1), 19–36.

24. Perelaer J., Smith P. J., Mager D., Soltman D., Volkman S. K., Subramanian V., Korvink J. G., Schubert U. S. (2010). Printed electronics: The challenges involved in printing devices, interconnects, and contacts based on inorganic materials, *J. Mater. Chem.*, **20**(39), 8446–8453.

25. Anto B. T., Sivaramakrishnan S., Chua L.-L., Ho P. K. H. (2010). Hydrophilic sparse ionic monolayer-protected metal nanoparticles: Highly concentrated nano-Au and nano-Ag "inks" that can be sintered to near-bulk conductivity at 150°C, *Adv. Funct. Mater.*, **20**(2), 296–303.

26. Yamada T., Fukuhara K., Matsuoka K., Minemawari H., Tsutsumi J., Fukuda N., Aoshima K., Arai S., Makita Y., Kubo H., Enomoto T., Togashi T., Kurihara M., Hasegawa T. (2016). Nanoparticle chemisorption printing technique for conductive silver patterning with submicron resolution, *Nat. Commum.*, **7**, 11402.

27. Kraft A. (2008). On the discovery and history of Prussian blue, *Bull. Hist. Chem.*, **33**(2), 61–67.

28. Buser H. J., Schwarzenbach D., Petter W., Ludi A. (1977). The crystal structure of Prussian blue: $Fe_4[Fe(CN)_6]_3 \cdot nH_2O$, *Inorg. Chem.*, **16**(11), 2704–2710.

29. Neff V. D. (1979). Electrochemical oxidation and reduction of thin films of Prussian blue, *J. Electrochem. Soc.*, **125**(6), 886–887.

30. Itaya K., Ataka T., Toshima S. (1982). Spectroelectrochemistry and electrochemical preparation method of Prussian blue modified electrodes, *J. Am. Chem. Soc.*, **104**(18), 4767–4772.

31. Itaya K., Uchida I., Neff V. D. (1986). Electrochemistry of polynuclear transition metal cyanides: Prussian blue and its analogues, *Acc. Chem. Res.*, **19**(6), 162–168.

32. Itaya K., Shibayama K. (1982). Prussian blue-modified electrodes: An application for a stable electrochromic display device, *J. Appl. Phys.*, **53**(1), 804–805.

33. DeLongchamp D. M., Hammond P. T. (2004). High-contrast electrochromism and controllable dissolution of assembled Prussian blue/polymer nanocomposites, *Adv. Funct. Mater.*, **14**(3), 224–232.

34. Kong B., Selomulya C., Zheng G., Zhao D. (2015). New faces of porous Prussian blue: Interfacial assembly of integrated hetero-

structures for sensing applications, *Chem. Soc. Rev.,* **44**(22), 7997–8018.

35. Wessells C. D., Huggins R. A., Cui Y. (2011), Copper hexacyanoferrate battery electrodes with long cycle life and high power, *Nat. Commun.,* **2**, 550.

36. Sato O., Einaga Y., Fujishima A., Hashimoto K. (1999). Photoinduced long-range magnetic ordering of a cobalt-Iron cyanide, *Inorg. Chem.*, **38**(20), 4405–4412.

37. Sato O. (2004). Photoinduced magnetization in molecular compounds, *J. Photoch. Photobio. C,* **5**(3), 203–223.

38. Verdaguer M., Bleuzen A., Train C., Garde R., Fabrizi de Biani F., Desplanches C. (1999). Room-temperature molecule-based magnets, *Phil. Trans. R. Soc. Lond. A,* **357**(1762), 2959–2976.

39. Gotoh A., Uchida H., Ishizaki M., Satoh T., Kaga S., Okamoto S., Ohta M., Sakamoto M., Kawamoto T., Tanaka H., Tokumoto M., Hara S., Shiozaki H., Yamada M., Miyake M., Kurihara M. (2007). Simple synthesis of three primary colour nanoparticle inks of Prussian blue and its analogues, *Nanotechnology,* **18**(34), 345609.

40. Ravaine S., Lafuente C., Mingotaud C. (1998). Electrochemistry of Langmuir–Blodgett films based on Prussian blue, *Langmuir,* **14**(22), 6347–6349.

41. Ohnuki H., Saiki T., Kusakari A., Endo H., Ichihara M., Izumi M. (2007). Incorporation of glucose oxidase into Langmuir–Blodgett films based on Prussian blue applied to amperometric glucose biosensor, *Langmuir,* **23**(8), 4675–4681.

42. Bharathi S., Nogami M., Ikeda S. (2001). Layer by layer self-assembly of thin films of metal hexacyanoferrate multilayers, *Langmuir,* **17**(24), 7468–7471.

43. Pyrasch M., Tieke B. (2001). Electro- and photoresponsive films of Prussian blue prepared upon multiple sequential adsorption, *Langmuir,* **17**(24), 7706–7709.

44. Bera A., Dey S., Pal A. J. (2013). Magnetic moment assisted layer-by-layer film formation of a Prussian blue analog, *Langmuir,* **29**(7), 2159–2165.

45. Vaucher S., Li M., Mann S. (2000). Synthesis of Prussian blue nanoparticles and nanocrystal superlattice in reverse microemulsion, *Angew. Chem. Int. Ed.,* **39**(10), 1783–1796.

46. Li M., Schnablegger H., Mann S. (1999). Coupled synthesis and self-assembly of nanoparticles to give structures with controlled organization, *Nature,* **402**(6760), 393–395.

47. Vaucher S., Fielden J., Li M., Dujardin E., Mann S. (2002). Molecule-based magnetic nanoparticles: Synthesis of cobalt hexacyanoferrate, cobalt pentacyanonitrosylferrate, and chromium hexacyanochromate coordination polymers in water-in-oil microemulsions, *Nano lett.*, **2**(3), 225–229.

48. Yamada M., Arai M., Kurihara M., Sakamoto M., Miyake M. (2004). Synthesis and isolation of cobalt hexacyanoferrate/chromate metal coordination nanopolymers stabilized by alkylamino ligand with metal elemental control, *J. Am. Chem. Soc.*, **126**(31), 9482–9483.

49. Arai M., Miyake M., Yamada M. (2008). Metal(II) hexacyanochromate(III) MCr (M = Co, Cu, Fe) corrdination nanoparticles stabilized by alkyl surface corrdination ligand: Downsizing effect on their crystal structure and magnetic properties, *J. Phys. Chem. C*, **112**(6), 1953–1962.

50. Catala L., Gacoin T., Boilot J.-P., Pivière É., Paulsen C., Lhotel E., Mallah T. (2003). Cyanide-bridged Cr^{III}-Ni^{II} superparamagnetic nanoparticels, *Adv. Mater.*, **15**(10) 826–829.

51. Sun H.-L., Shi H., Zhao F., Qi L., Gao S. (2005). Shape-dependent magnetic properties of low-dimensional nanoscale Prussian blue (PB) analogue $SmFe(CN)_6 \cdot 4H_2O$, *Chem. Commun.*, (34), 4339–4341.

52. Kondo N., Kurihara M., Yamada M., Miyake M., Nishijima M., Ohsuna T., Mizukami F., Sakamoto M. (2005). Hexagonal ring submicro- and nanocrystals of a La–hexacyanoferrate coordination polymer, *Chem. Lett.*, **34**(4), 590–591.

53. Zhou Y., Wang C. Y., Zhu Y. R., Chen Z. Y. (1999). A novel ultraviolet irradiation technique for shape-controlled synthesis of gold nanoparticles at room temperature, *Chem. Mater.*, **11**(10), 2310–2312.

54. Kuo P.-L., Chen W.-F. (2003). Formation of silver nanoparticles under structured amino groups in pseudo-dendritic Poly(allylamine) derivatives, *J. Phys. Chem. B*, **107**(41), 11267–11272.

55. Uemura T., Kitagawa S. (2003). Prussian blue nanoparticles protected by Poly(vinylpyrrolidone), *J. Am. Chem. Soc.*, **125**(26), 7814–7815.

56. Uemura T., Ohba M., Kitagawa S. (2004). Size and surface effects of Prussian blue nanoparticles protected by organic polymers, *Inorg. Chem.*, **43**(23), 7339–7345.

57. Wu X., Cao M., Hu C., He X. (2006). Sonochemical synthesis of Prussian blue nanocubes from a single-source precursor, *Cryst. Growth Des.*, **6**(1), 26–28.

58. Wasserscheid P., Keim W. (2000). Ionic liquids-new "solutions" for transition metal catalysis, *Angew. Chem. Int. Ed.*, **39**(21), 3772–3789.

59. Welton T. (1999). Room-temperature ionic liquids. Solvents for synthesis and catalysis, *Chem. Rev.*, **99**(8), 2071–2083.

60. Wasserscheid P., van Hal R., Bosmann A. (2002). 1-*n*-Butyl-3-methylimidazolium ([bmim]) octylsulfate—an even 'greener' ionic liquid, *Green Chem.*, **4**(4), 400–404.

61. Wei G.-T., Yang Z., Lee C.-Y., Yang H.-Y., Chris Wang C. R. (2004). Aqueous-organic phase transfer of gold nanoparticles and gold nanorods using an ionic liquid, *J. Am. Chem. Soc.*, **126**(16), 5036–5037.

62. Itoh H., Naka K., Chujo Y. (2004). Synthesis of gold nanoparticles modified with ionic liquid based on the imidazolium cation, *J. Am. Chem. Soc.*, **126**(10), 3026–3027.

63. Scheeren C. W., Machado G., Dupont J., Fichtner P. F. P., Texeira S. R. (2003). Nanoscale Pt(0) particles prepared in imidazolium room temperature ionic liquids: Synthesis from an organometallic precursor, characterization, and catalytic properties in hydrogenation reactions, *Inorg. Chem.*, **42**(15), 4738–4742.

64. Huang J., Jiang T., Han B., Gao H., Chang Y., Zhao G., Wu W. (2003). Hydrogenation of olefins using ligand-stabilized palladium nanoparticles in an ionic liquid, *Chem. Commun.*, (14), 1654–1655.

65. Zhu Y.-J., Wang W.-W., Qi R.-J., Hu X.-L. (2004). Microwave-assisted synthesis of single-crystalline tellurium nanorods and nanowires in ionic liquids, *Angew. Chem. Int. Ed.*, **43**(11), 1410–1414.

66. Nakashima T., Kimizuka N. (2003). Interfacial synthesis of hollow TiO_2 microspheres in ionic liquids, *J. Am. Chem. Soc.*, **125**(21), 6386–6387.

67. Clavel G., Larionova J., Guari Y., Guérin C. H. (2006). Synthesis of cyano-bridged magnetic nanoparticles using room-temperature ionic liquids, *Chem. Eur. J.*, **12**(14), 3798–3804.

68. Law G., Watson P. R. (2001). Surface tension measurements of *n*-alkylimidazolium ionic liquids, *Langmuir*, **17**(20), 6138–6141.

69. Antonietti M., Kuang D., Smarsly B., Zhou Y. (2004). Ionic liquids for the convenient synthesis of functional nanoparticles and other inorganic nanostructures, *Angew. Chem. Int. Ed.*, **43**(38), 4988–4992.

70. Larionova J., Guari Y., Blanc C., Dieudonné P., Tokarev A., Guérin C. H. (2009). Toward organization of cyano-bridged coordination polymer nanoparticles within an ionic liquid crystal, *Langmuir*, **25**(2), 1138–1147.

71. Larionova J., Guari Y., Tokarev A., Chelebaeva E., Luna C., Sangregorio C., Caneschi A., Guérin C. H. (2008). Coordination polymer nano-objects into ionic liquids: Nanoparticles and superstructures, *Inorg. Chim. Acta.*, **361**(14), 3988–3996.

72. Zhou P., Xue D., Luo H., Chen X. (2002). Fabrication, structure, and magnetic properties of highly ordered Prussian blue nanowire arrays, *Nano Lett.*, **2**(8), 845–847.

73. Luo H., Chen X., Zhou P., Shi H., Xue D. (2004). Prussian blue nanowires fabricated by electrodeposition in porous anodic aluminum oxide, *J. Electrochem. Soc.*, **151**(9), C567–C570.

74. Johansson A., Widenkvist E., Lu J., Boman M., Jansson U. (2005). Fabrication of high-aspect-ratio Prussian blue nanotubes using a porous alumina template, *Nano Lett.*, **5**(8), 1603–1606.

75. Sabzia R. E., Kantb K., Losicb D. (2010). Electrochemical synthesis of nickel hexacyanoferrate nanoarrays with dots, rods and nanotubes morphology using a porous alumina template, *Electrochimi. Acta*, **55**(5), 1829–1835.

76. Cauvel A., Renard G., Brunel D. (1997). Monoglyceride synthesis by heterogeneous catalysis using MCM-41 type silicas functionalized with amino groups, *J. Org. Chem.*, **62**(3), 749–751.

77. Zhao D., Huo Q., Feng J., Chmelka B. F., Stucky G. D. (1998). Nonionic triblock and star diblock copolymer and oligomeric surfactant syntheses of highly ordered, hydrothermally stable, mesoporous silica structures, *J. Am. Chem. Soc.*, **120**(24), 6024–6036.

78. Zhao D., Feng J., Huo Q., Melosh N., Fredrickson G. H., Chmelka B. F., Stucky G. D. (1998). Triblock copolymer syntheses of mesoporous silica with periodic 50 to 300 angstrom pores, *Science*, **279**(5350), 548–552.

79. Wang Y., Angelatos A. S., Caruso F. (2008). Template synthesis of nanostructured materials via layer-by-layer assembly, *Chem. Mater.*, **20**(3), 848–858.

80. LEE J. E., Lee N., Kim T., Kim J., Hyeon T. (2011). Multifunctional mesoporous silica nanocomposite nanoparticles for theranostic applications, *Acc. Chem. Res.*, **44**(10), 893–902.

81. Song H., Rioux R. M., Hoefelmeyer J. D., Komor R., Niesz K., Grass M., Yang P., Somorjai G. A. (2006). Hydrothermal growth of mesoporous SBA-15 silica in the presence of PVP-stabilized Pt nanoparticles: Synthesis, characterization, and catalytic properties, *J. Am. Chem. Soc.*, **128**(9), 3027–3037.

82. Clavel G., Guari Y., Larionova J., Guérin C. H. (2005). Formation of cyano-bridged molecule-based magnetic nanoparticles within hybrid mesoporous silica, *New J. Chem.*, **29**(2), 275–279.

83. Mascharak P. K. (1986). Convenient synthesis of *Tris*(tetra-ethylammonium) Hexacyanoferrate(III) and its use as an oxidant with tunable redox potential, *Inorg. Chem.*, **25**(3), 245–247.

84. Niemeyer C. M. (2001). Nanoparticles, proteins, and nucleic acids: Biotechnology meets materials science, *Angew. Chem. Int. Ed.*, **40**(22), 4128–4158.

85. Yamashita I., Iwahori K., Kumagai S. (2010). Ferritin in the field of nanodevices, *Biochim. Biophys. Acta*, **1800**(8), 846–857.

86. Harrison P. M., Andrews S. C., Artymuik P. J., Ford G. C., Guest J. R., Hirzmann J., Lawson D. M., Livingstone J. C., Smith J. M. A., Treffery A., Yewdall S. J. (1991). Probing structure-function relations in ferritin and bacterioferritin, *Adv. Inorg. Chem.*, **36**, 449–486.

87. Douglas T., Dickson D. P. E., Betteridge S., Charnock J., Garner C. D., Mann S. (1995). Synthesis and structure of an iron(III) sulfide-ferritin bioinorganic nanocomposite, *Science*, **269**(5220), 54–57.

88. Meldrum F. C., Heywood B. R., Mann S. (1992). Magnetoferritin: In vitro synthesis of a novel magnetic protein, *Science*, **257**(5069), 522–523.

89. MaHam A., Tang Z., Wu H., Wang J., Lin Y. (2009). Protein-based nanomedicine platforms for drug delivery, *Small*, **5**(15), 1706–1721.

90. Domínguez-Vera J. M., Colacio E. (2003). Nanoparticles of Prussian blue ferritin: A new route for obtaining nanomaterials, *Inorg. Chem.*, **42**(22), 6983–6985.

91. Gálvez N., Sánchez P., Domínguez-Vera J. M. (2005). Preparation of Cu and CuFe Prussian blue derivative nanoparticles, *Dalton Tran.*, (15), 2492–2494.

92. Ravi Kumar M. N. V., Muzzarelli R. A. A., Muzzarelli C., Sashiwa H., Domb A. (2004). Chitosan chemistry and pharmaceutical perspectivs, *Chem. Rev.*, **104**(12), 6017–6084.

93. Ravi Kumar M. N. V. (2000). A review of chitin and chitosan applications, *React. Funct. Polym.*, **46**(1), 1–27.

94. Guari Y., Larionova J., Molvinger K., Folch B., Guérin, C. H. (2006). Magnetic water-soluble cyano-bridged metal coordination nano-polymers, *Chem. Commun.*, **24**, 2613–2615.

95. Molvinger K., Quignard F., Brunel D., Boissière M., Devoisselle J.-M. (2004). Porous chitosan-silica hybrid microspheres as a potential catalyst, *Chem. Mater.*, **16**(17), 3367–3372.

96. Toshima N., Lin R.-J., Kaneko M. (1990). A newpreparation method of electrochromic Prussian blue films by casting a solution and their electrochemical properties, *Chem. Lett.*, **19**(4), 485–488.

97. Lin R.-J., Toshima N. (1991). Spectroelectrochemistry of new Prussian blue films prepared by a cast method, *Bull, Chem, Soc, Jpn.*, **64**(1), 136–148.

98. Minemawari H., Yamada T., Matsui H., Tsutsumi J., Haas S., Chiba R., Kumai R., Hasegawa T. (2011). Inkjet printing of single-crystal films, *Nature*, **475**(7356), 364–367.

99. Singh M., Haverinen H. M., Dhagat P., Jabbour G. E. (2010). Inkjet printing—process and its applications, *Adv. Mater.*, **22**(6), 673–685.

100. Matsuhisa N., Kaltenbrunner M., Yokota T., Jinno H., Kuribara K., Sekitani T., Someya T. (2015). Printable elastic conductors with a high conductivity for electronic textile applications, *Nat. Commun.*, **6**, 7461.

101. Fukuda K., Takeda Y., Yoshimura Y., Shiwaku R., Tran L. T., Sekine T., Mizukami M., Kumaki D., Tokito S. (2014). Fully-printed high-performance organic thin-film transistors and circuitry on one-micron-thick polymer films, *Nat. Commun.*, **5**, 4147.

102. Lau P. H., Takei K., Wang C., Ju Y., Kim J., Yu Z., Takahashi T., Cho G., Javey A., (2013). Fully printed, high performance carbon nanotube thin-film transistors on flexible substrates, *Nano Lett.*, **13**(8), 3864–3869.

103. Søndergaard R. R., Hösel M., Krebs F. C. (2013). Roll-to-Roll fabrication of large area functional organic materials, *J. Polym. Sci. Part B Polym. Phys.*, **51**(1), 16–34.

104. Lee M. M., Teuscher J., Miyasaka T., Murakami T. N., Snaith H. J. (2012). Efficient hybrid solar cells based on meso-superstructured organometal halide perovskites, *Science*, **338**(6107), 643–647.

105. Robin M. B. (1962). The color and electronic configurations of Prussian blue, *Inorg. Chem.* **1**(2), 337–342.

106. Ishizaki M., Kanaizuka K., Abe M., Hoshi Y., Sakamoto M., Kawamoto T., Tanaka H., Kurihara M. (2012). Preparation of electrochromic Prussian blue nanoparticles dispersible into various solvents for realization of printed electronics, *Green Chem.*, **14**(5), 1537–1544.

107. Shiozaki H., Kawamoto T., Tanaka H., Hara S., Tokumoto M., Gotoh A., Satoh T., Ishizaki M., Kurihara M., Sakamoto (2008). Electrochoromic thin film fabricated using a water-dispersible ink of Prussian blue nanoparticles, *Jpn. J. Appl. Phys.*, **47**(2), 1242–1244.

108. Ono K., Ishizaki M., Soma S., Kanaizuka K., Togashi T., Kurihara M. (2015). A low-temperature sintered heterostructure solid film of coordination polymer nanoparticles: an electron-rectifier function based on partially oxidised/reduced conductor phases of Prussian blue, *RSC Adv.*, **5**(117), 96297–96304.

109. Ono K., Ishizaki M., Kanaizuka K., Togashi T., Yamada T., Kitagawa H., Kurihara M. (2017). Grain-boundary-free super-proton conduction of a solution-processed Prussian-blue nanoparticle film, *Angew. Chem. Int. Ed.*, DOI: 10.1002/anie.201701759.

110. Ishizaki M., Ono K., Suzuki K., Naito W., Kanaizuka K., Kawamoto T., Tanaka H., Kurihara M. (2017). Fine-tunable electronic energy levels of mixed-metal Prussian-blue alloy nanoparticles, *ChemNanoMat*, DOI: 10.1002/cnma.201700084.

111. Yeh M.-H., Lin L., Yang P.-K., Wang Z. L. (2015). Motion-driven electrochromic reactions for self-powered smart window system, *ACS Nano*, **9**(5), 4757–4765.

112. Chen R., Tanaka H., Kawamoto T., Asai M., Fukushima C., Kurihara M., Watanabe M., Arisaka M., Nankawa T. (2012). Preparation of a film of copper hexacyanoferrate nanoparticles for electrochemical removal of cesium from radioactive wastewater, *Electrochem. Commun.*, **25**, 23–25.

113. Enüstün B. V., Turkevich J. (1963). Coagulation of colloidal gold, *J. Am. Chem. Soc.*, **85**(21), 3317–3328.

114. Henglein A., Giersig M. (1999). Formation of colloidal silver nanoparticles: Capping action of citrate, *J. Phys. Chem. B*, **103**(44), 9533–9539.

115. Kuo C.-H., Chiang T.-F., Chen L.-J., Huang M. H. (2004). Synthesis of highly faceted pentagonal- and hexagonal-shaped gold nanoparticles with controlled sizes by sodium dodecyl sulfate, *Langmuir*, **20**(18), 7820–7824.

116. Dhillon G. S., Brar S. K., Verma M., Tyagi R. D. (2011). Recent advances in citric acid bio-production and recovery, *Food Bioprocess Technol.*, **4**(4), 505–529.

117. Soccol C. R., Vandenberghe L. P. S., Rodrigues C., Pandey A. (2006). New perspectives for citric acid production and application, *Food Technol. Biotechnol.*, **44**(2), 141–149.

118. El Badawy A. M., Silva R. G., Morris B., Scheckel K. G., Suidan M. T., Tolaymat T. M. (2011). Surface charge-dependent toxicity of silver nanoparticles, *Environ. Sci. Technol.*, **45**(1), 283–287.

119. Shokouhimehr M., Soehnlen E. S., Khitrin A., Basu S., Huang S. D. (2010). Biocompatible Prussian blue nanoparticles: Preparation, stability, cytotoxicity, and potential use as an MRI contrast agent, *Inorg. Chem. Commun.*, **13**(1), 58–61.

120. Shokouhimehr M., Soehnlen E. S., Hao J., Grisword M., Flask C., Fan X., Pasilison, J. P., Basu S., Huang S. D. (2010). Dual purpose Prussian blue nanoparticles for cellular imaging and drug delivery: A new generation of T_1-weighted MRI contrast and small molecule delivery agents, *J. Mater. Chem.*, **20**(25), 5251–5259.

121. Zaręba J. K., Szeremeta J., Waszkielewicz M., Nyk M., Samoć M. (2016). Nonlinear-optical response of Prussian blue: Strong three-photon absorption in the IR region, *Inorg. Chem.*, **55**(19), 9501–9504.

122. Wang W., Feng J., Ye Y., Lyu F., Liu Y.-S., Guo J., Yin Y. (2017). Photocatalytic color switching of transition metal hexacyanometalate nanoparticles for high-performance light-printable rewritable paper, *Nano Lett.*, **17**(2), 755–761.

Chapter 8

Prussian Blue Nanoparticles and Nanocomposites for Cs Decontamination

Akira Takahashi, Hisashi Tanaka, and Tohru Kawamoto

Nanomaterials Research Institute, AIST,
1-1-1 Higashi, Tsukuba 305-8565, Japan

tohru.kawamoto@aist.go.jp

8.1 Introduction

The technology for extracting radioactive caesium from an aqueous solution is quite important for treatment of wastewater from damaged nuclear power plants or any radioactive isotope leakage accident. Among the radioactive isotopes in radioactive waste, because ^{137}Cs and ^{90}Sr are heat-generating nuclides, the partitioning of isotopes is expected to reduce their volume at the final disposal site. On the other hand, for radioisotope leakage accidents, ^{134}C, ^{137}Cs, ^{133}Xe, and ^{131}I, which have a lower boiling point, are emitted to the atmosphere [1, 2]. Especially, ^{134}Cs and ^{137}Cs are more severe than gaseous Xenon or ^{131}I, with short half-lives. Actually, today, three decades after the Chernobyl accident, radioactive Cs persists in the environment [3]. Regarding

Prussian Blue-Type Nanoparticles and Nanocomposites: Synthesis, Devices, and Applications
Edited by Yannick Guari and Joulia Larionova

ISBN 978-981-4800-05-1 (Hardcover), 978-0-429-02474-0 (eBook)
www.panstanford.com

environmental decontamination in the Fukushima region, radioactive Cs is specifically examined herein.

The use of adsorbents is an effective method for radioactive Cs removal. As Cs-adsorbents, various materials such as Prussian blue (PB) and its analogues (PBAs) [4], zeolite [5], crown calix arenes [6, 7], silicotitanate [8, 9], and ammonium phosphomolybdate [10] have been investigated and used. Among them, the benefits of Prussian blue-type complexes include their adsorption capacity and selectivity. Additionally, they are economically reasonable in terms of cost. In fact, they have been used as Cs-adsorbents from salty aqueous solutions such as seawater and leachate from disposal sites. Medicines, including PBA, have been commercialized for patients with internal contamination from radioactive caesium [11].

Preparation of PBA nanoparticles and composites containing them are important to reach the full potential of PBA during practical use. In general, operation procedures are limited if the adsorbent is a powder or slurry. Especially, using them after loading material into a column, after moulding such as granulation, or after immobilization onto matrixes is effective. Nevertheless, to maintain their adsorption performance even after moulding, fabrication of composites must be done with control of the structure on a nanoscale to millimetre scale.

This chapter introduces the Cs-adsorption properties of PBA along with examples of nanoparticle and composite preparation.

8.2 Cs-adsorption by Prussian Blue Analogues (PBA)

8.2.1 History

The first report describing Cs-adsorption properties was published around 1950. Initially, PBA was used at US nuclear development facilities to treat radioactive wastewater and to remove radioactive Cs. After the Chernobyl accident in 1986, Prussian blue was used mostly for the veterinary treatment of cows [12].

Figure 8.1 portrays effects of a PB-bolus on the transfer of radioactive Cs to milk. When Prussian blue is fed to cows, it adsorbs radioactive Cs from the body of the cow. The Prussian blue is not digested. Therefore, it will be discharged with faeces: The radioactive Cs contamination of milk is decreased. A treatment program was started around 1990 and was continued even after 2000.

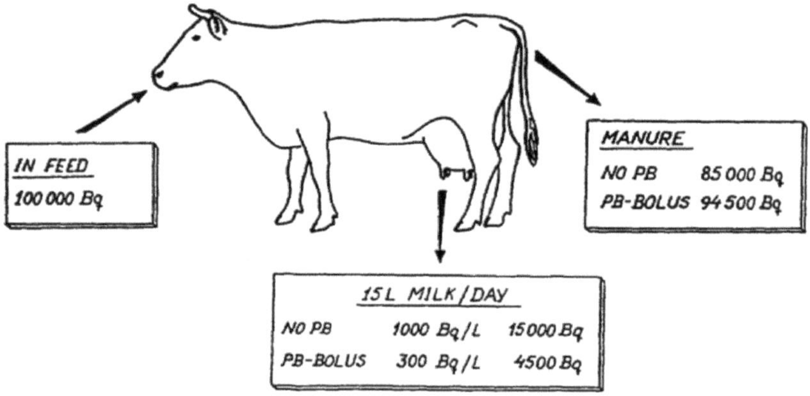

Figure 8.1 Effects of PB-bolus on the transfer of radioactive Cs (Bq/d) to milk: the radioactive Cs budget for a cow with PB boli and for an untreated cow. Constant radioactive Cs levels in milk and manure were obtained after several weeks of grazing on a contaminated pasture. Reproduced from [12] by permission of the International Atomic Energy Agency.

Regarding wastewater treatment, Fortum Keilaniemi (Fortum, Finland) has long been developing methods. Its adsorbent was used at a nuclear power plant [13]. After the Fukushima accident, further development of decontamination technology with PBA adsorbent has been undertaken in Japan to fit the circumstances. For instance, decontamination technologies have been developed with adsorbents for ash from incinerated general municipal wastes [14–17], and rapid evaluation techniques to evaluate radioactive Cs concentrations in environmental water [18, 19].

8.2.2 Advantages of PBA in Cs Adsorption

The most important benefits of the Prussian blue family as Cs-adsorbent are their large distribution coefficient K_d and their selectivity. The distribution is defined by the ratio of the Cs-concentration in the adsorbent (ρ_{Se}) to that in the solution (ρ_{Le}) in equilibrium conditions, calculated as

$$K_d = \frac{\rho_{Se}}{\rho_{Le}} = \frac{C_{L0} - C_{Le}}{C_{Le}} \cdot \frac{V}{m}, \quad (8.1)$$

where C_{L0}, C_{Le}, V, and m, respectively, represent the Cs concentration in solution at the initial condition, that in the equilibrium condition, the volume of the solution, and the mass of the adsorbent. For example, we consider a case in which the adsorbent is added to the Cs-solution with the ratio of V/m = 1,000 (mL/g): 1,000 ppm. With $K_d = 10^4$, 10^5, 10^6 (mL/g), C_{Le} is decreased to about 10%, 1%, and 0.1% of C_{L0}.

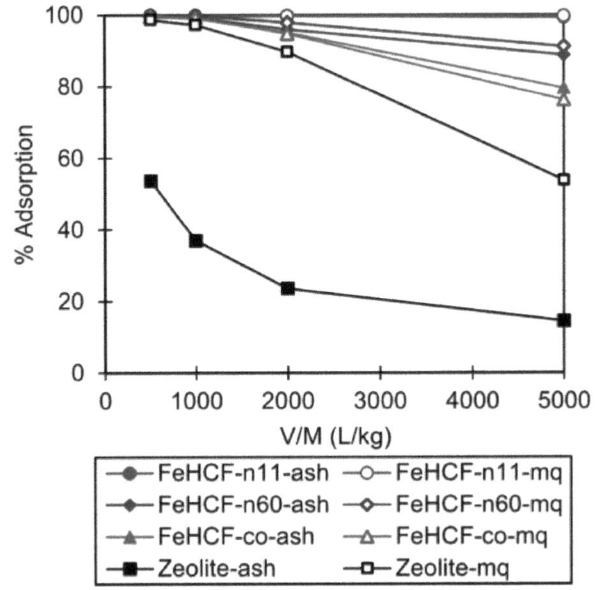

Figure 8.2 Difference in Cs adsorption behaviours of FeHCF (Prussian blue) in various forms and natural zeolite Ayashi from pure water and an ash extract solution: [Cs] = 1 mg/L 100 min mixing at 30°C, 600 rpm [15]. Published by Elsevier.

Figure 8.2 presents differences in Cs-adsorption behaviours among various adsorbents [15]. It is readily apparent that the zeolite shows less adsorption behaviour than Prussian blues, even for pure water, indicating lower Cs-capacity. In addition, for ash-extraction solutions with high concentrations of other alkali cations, the adsorption ratio is drastically decreased because of poor selectivity. Regarding Prussian blue, three types were investigated: FeHCF-n11, FeHCF-n60, and FeHCF-co, respectively, represent materials with nanoparticles of 11 μm secondary particles size, with 60 μm, and commercially available Prussian blue. Results show that all have higher adsorption performance than zeolite. Especially, FeHCF-n11 shows almost 100% adsorption, even from the ash extract solution, where K_d reaches 1.6×10^6 mL/g.

8.2.3 Mechanism of Cs Adsorption by PBA

The Cs-adsorption mechanism by Prussian blue analogue is generally regarded as ion exchange. The general formula of PBA used for the Cs adsorbent is described as $A_yM[M'(CN)_6]_{1-x}\cdot zH_2O$, where A represents a mono-cation such as Li^+, Na^+, K^+, and NH_4^+. M is a metal ion, e.g. Fe, Co, Ni, Cu, and Zn. Fe and Co are generally implemented for M′. When Cs^+ is adsorbed, the ion exchange reaction occurs as

$$A_yM[M'(CN)_6]_{1-x}\cdot zH_2O + \alpha Cs^+ \rightarrow A_{y-\alpha}Cs_\alpha[M'(CN)_6]_{1-x}\cdot zH_2O + \alpha A^+.$$

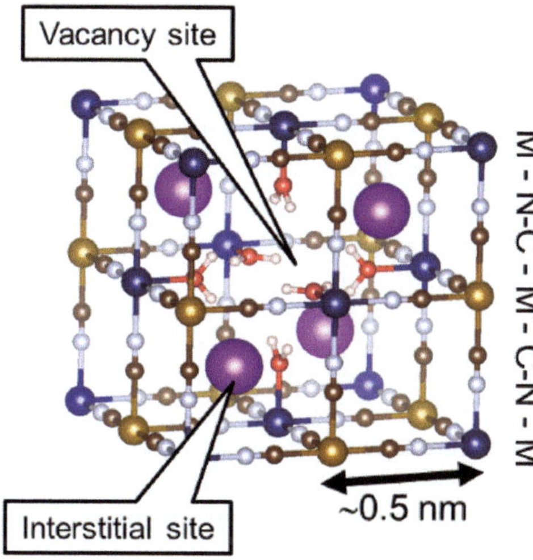

Figure 8.3 Crystal structure of Prussian blue analogues.

Prussian blue has adsorption sites of two kinds, as shown in Fig. 8.3: interstitial sites and vacancy sites. Although H_2O would adsorb into both sites, alkali cations, including Cs^+, are regarded as located at the interstitial site. The maximum adsorption capacity is determined by the number of A^+ implemented into the crystal at the synthesis with a few exceptions, e.g. in the case of Prussian blue, insoluble Prussian blue showed higher adsorption performance than soluble Prussian blue, although insoluble Prussian blue has no intrinsic mono-cation for ion exchange. In this case, the ion exchange between H^+ in water molecules and Cs^+ was proposed [20].

Although the Cs^+ adsorption mechanism remains controversial, the Prussian blue cages are regarded as adsorbing the alkali cations with a smaller hydrated radius. Cs^+ has the smallest hydrated radius among the major cations, except for H^+[21].

8.3 PBA Nanoparticles for Cs Decontamination

8.3.1 Preparation of Prussian Blue Family Nanoparticles

Prussian blue and its analogues are often synthesized as nanoparticles. In fact, one Prussian blue used for pigment has approximately 50 nm primary particle size [22]. Prussian blue analogues can be synthesized using various chemical compositions. For example, two kinds of Prussian blue are well known: soluble and insoluble. Although both are insoluble, the soluble Prussian blue is dispersible into aqueous solution, mainly used for pigment. The insoluble Prussian blue is easier to synthesize merely by mixing an aqueous solution of iron salt and that of hexacyanoferrate anion as follows:

$$Fe^{3+} + 0.75[Fe^{2+}(CN)_6]^{4-} \rightarrow Fe[Fe(CN)_6]_{0.75} \quad (8.2)$$

Among Prussian blue analogues, $Ni[Fe(CN)_6]_x$ and $Cu[Fe(CN)_6]_x$ tend to be obtained as <100 nm nanoparticles [23]. $Zn[Fe(CN)_6]_x$ is obtained as larger particles [24]. Consequently, the Prussian blue family nanoparticles can often be obtained easily. However, Cs-adsorption performance can be improved with further tuning to improve the downsizing of improving homogeneity of particle size.

8.3.2 Size Control of NPs in Micromixer Synthesis

The effect of producing smaller particles to improve particle size homogeneity was clarified for nanoparticles synthesized using micromixing [23]. Usually, Prussian blue analogues are obtained using a batch method for precipitation after mixing two aqueous solutions of M cation and one of $[M'(CN)_6]$ anion. In the batch method, the PBA crystallization starts immediately after contact of the two solutions. This contact causes fluctuation of the synthesis conditions such as the concentration of raw materials, resulting in difficulty of size control.

Micromixing, which can be used for synthesizing uniform particles from liquid phase, is categorized as a flow method with mixing of two solutions in a Y (or T) type mixer with less than 1 mm internal diameter. The reaction starts at the micromixer junction and proceeds with flow in the tube. Micromixing has three salient advantages for nanoparticle synthesis. The first is high efficiency of mixing because of the small mixing volumes of the junction and tube. The second is immediate supersaturation of the target material because of the rapid mixing speed, promoting crystal nucleation. The third is ease of adaptation to mass synthesis. The former two are suitable for nanoparticle synthesis.

Potassium copper hexacyanoferrate ($K_{0.66}Cu[Fe(CN)_6]_{0.66}$, KCuHCF) is an example for controlling the particle size with a micromixer [23]. For the study, KCuHCF was synthesized by mixing 0.6 mol/L of $CuSO_4$ aqueous solution and 0.4 mol/L of $K_4[Fe(CN)_6]$ aqueous solution using a Y type micromixer with inner diameter of 150 μm at various flow rates of 5, 10, 40, and 100 mL/min (Fig. 8.4). The KCuHCF particles synthesized using a micromixer were smaller than those synthesized using batch method, which have non-uniform shape and size. Furthermore, the particle size decreases along with increase of the flow rate as shown in Fig. 8.5.

The KCuHCF chemical compositions were evaluated. The ratios of $[Fe(CN)_6]$ to Cu were 0.63–0.68, which is similar to the mixing ratio of the raw materials: 0.67.

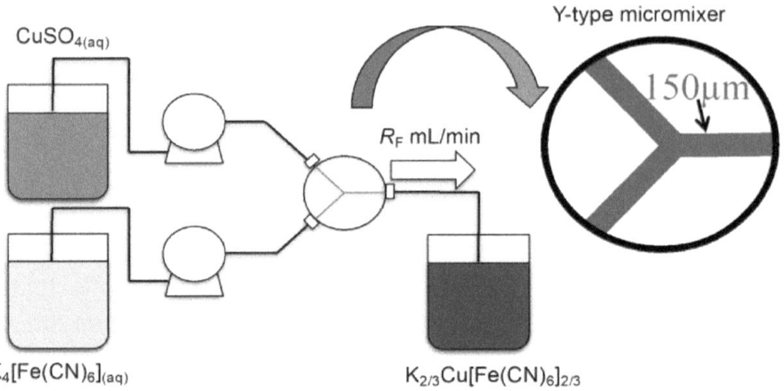

Figure 8.4 KCuHCF synthesis using a micromixer system.

Figure 8.5 FE–SEM images of KCuHCF NPs synthesized using the micromixer at different flow rates, and batch-synthesized NPs [23]. Published by The Royal Society of Chemistry.

We evaluated the capacity and speed for Cs-adsorption of KCuHCFs synthesized with batch and micromixer methods at different flow rates. KCuHCF adsorbs Cs^+ with ion exchange with K^+. Therefore, the adsorption capacity can be estimated with the amount of the implemented K^+ from the chemical composition as shown in Fig. 8.6. The caesium adsorption capacity of KCuHCF with the batch synthesis was less than the estimated value, perhaps because of shielding of some KCuHCF in large particles.

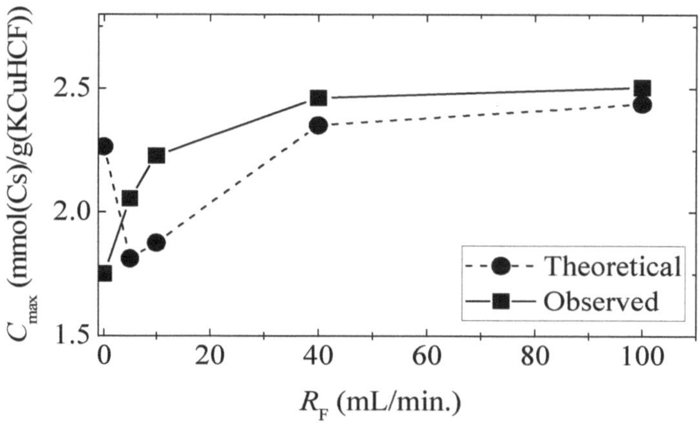

Figure 8.6 Variation of the saturated capacity of KCuHCF-NPs with the flow-rate R_F used for synthesis. R_F = 0 represents the batch-synthesized KCuHCF-NPs [23]. Published by The Royal Society of Chemistry.

This low caesium adsorption property of the batch-synthesized KCuHCF was also found in the distribution co-efficient, K_d. For each sample, K_d was evaluated under conditions with 10 mg of KCuHCF in 10 mL of 50 mg/L Cs^+ solution. The K_d of the batch-synthesized KCuHCF and that of KCuHCF synthesized using a micromixer at the flow rate of 100 mL/min were, respectively, 4.1×10^5 and 1.2×10^6 (mL/g). This low K_d of batch-synthesized KCuHCF would also result from the shielding effect. In terms of practical applications, using KCuHCF by micromixer-synthesis requires only 1/3 of the batch-synthesized KCuHCF the adsorbent amount.

As shown in Fig. 8.7, smaller particles also raised the Cs-adsorption speed. The Cs-adsorption speeds of KCuHCF synthesized using a micromixer with 100 mL/min, the fastest flow rate in the experiment, showed the highest adsorption speed among the batch-synthesized and flow-synthesized KCuHCFs. This result is expected to derive from the highest surface area because of the smallest particle size. The time dependence of caesium concentrations was evaluated with pseudo-second adsorption equation:

$$\frac{dq(t)}{dt} = k_2(q_e - q(t))^2.$$

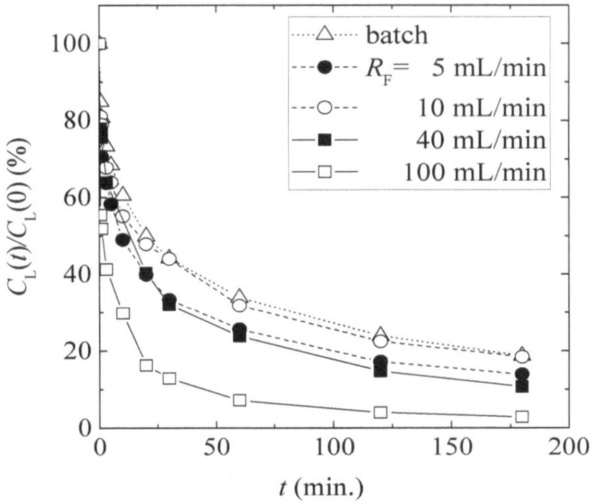

Figure 8.7 Variation of Cs-concentration in the solution, $C_L(t)/C_L(0)$, as a function of the shaking time t [23]. Published by The Royal Society of Chemistry.

The adsorption speed coefficient k_2 of the batch-synthesized KCuHCF and that synthesized using a micromixer with a 100 mL/min flow-rate were estimated, respectively, as 9.14 and 50.8 mL/g·min. KCuHCF produced using a micromixer was 7.7 times larger than the batch-synthesized one. This result shows that the caesium adsorption rate is dependent on the primary particle size and demonstrates that nanocrystallization enhances caesium adsorption.

8.4 Recycling of PBA for Cs Uptake

As described above, the highly Cs-selective adsorption of Prussian blue-type compounds (metal hexacyanoferrate, MHCF) has been known for more than half a century. Moreover, Cs^+ removal from radioactive waste liquids at nuclear power plants has been discussed [25]. To reduce secondary radioactive waste amounts, the adsorbent should preferably be used repeatedly with regeneration. Therefore, regeneration methods of the adsorbent using chemical reagents have been studied. For example, A_2^+CuHCF(II) (A = alkali metal, (II) shows the valence number of Fe in hexacyanoferrate) adsorbed Cs^+;

it becomes $Cs_2^+CuHCF(II)$ via ion exchange. The Cs^+ adsorbed $CuHCF(II)^{2-}$ was oxidized to CuHCF(III) using concentrated nitric acid (3–5 mol/L HNO_3); Cs^+ was released. Then, the CuHCF(III) was regenerated to $A_2^+CuHCF(II)$ by adding hydrazinium sulphate as a reductant [26].

On the other hand, the electrochemical properties of MHCFs have been reported for two decades. The potentiometric selectivity coefficients of electro-deposited CuHCF membranes for alkali metal ions have been reported, with especially high values for Cs^+ [27]. A schematic figure of electrochemical Cs^+ adsorption and desorption with MHCF electrode is shown in Fig. 8.8. The Cs^+ is adsorbed selectively onto the MHCF electrode from the adsorption solution by electrochemical reduction; then, the Cs^+ on the MHCF electrode is eluted into the desorption solution by electrochemical oxidation. Lilga et al. reported the electrically deposited MHCFs membrane as oxidized and reduced electrochemically. Reportedly, it is possible to adsorb and desorb Cs^+ ions by electric switching: so-called electrically switched ion exchange (ESIX) [28, 29].

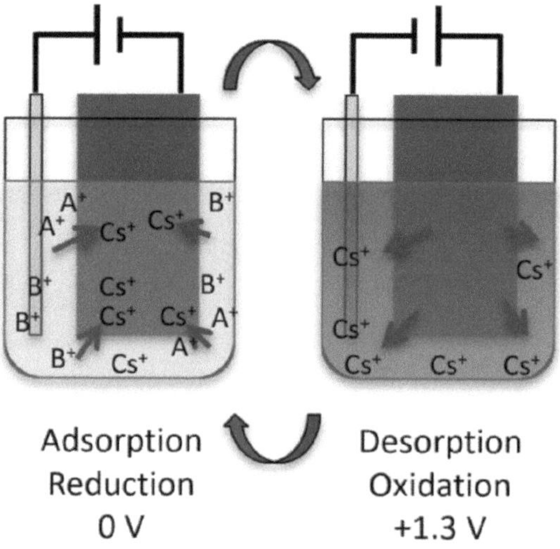

Figure 8.8 Schematic figure showing electrochemical adsorption–desorption.

The merits of electrochemical adsorption/desorption control are the following: (1) Making redox with no more chemical reagents can reduce radioactive waste; (2) Cs^+ selectivity of MHCF electrode is enhanced by restricting electric potential ranges; and (3) by adopting the column system as described below, the Cs^+ is separated by controlling the solution flow and electrode-redox; both are conducted by electric switching.

However, the electrical deposition membrane is thin for ESIX. Complicated deposition process used to fabricate high-quality membrane is also a problem. To solve the problems, the preparation of an MHCF film electrode by wet-printing, and prepared a series of water-dispersible inks of MHCF nanoparticles under the precise control of particle sizes and compositions has been reported [23, 30]. Because this water-dispersible ink is readily able to fabricate high-quality thin MHCF film on the metal electrode, a repeated recycling system of Cs^+ adsorption–desorption with this MHCF electrode was achieved.

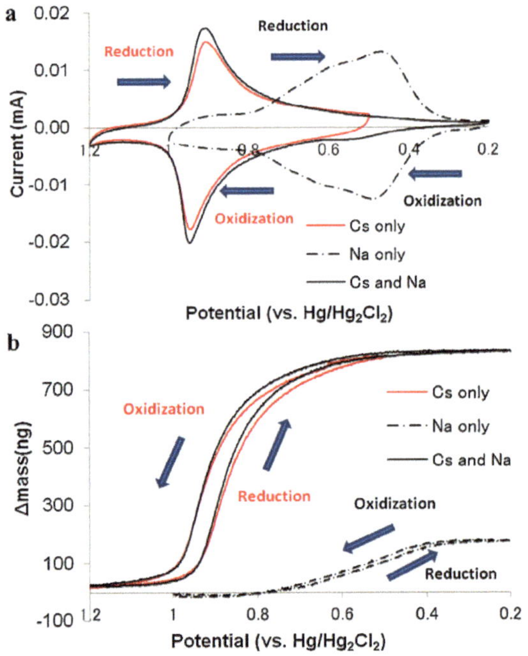

Figure 8.9 (a) Cyclic voltammograms (CV) of CuHCF films employed in a 0.1 mol/L $NaNO_3$ solution, a 0.1 mol/L $CsNO_3$ solution, and a mixture solution of 0.1 mol/L $CsNO_3$ and 0.1 mol/L $NaNO_3$. (b) Mass change in an electrochemical quartz crystal microbalance recorded concurrently with the CV of (a), with scan rate = 5 mV/s [31]. Published by Elsevier.

As an example of Cs^+ selectivity enhancement, Fig. 8.9 shows a large difference of redox potentials of CuHCF film electrode in solutions containing Cs^+ and/or Na^+ [31]. The CuHCF film electrode adsorbs and desorbs Cs^+ selectively by controlling electric potentials, even in the case of seawater. The high Cs^+ selectivity of CuHCF is also effective in a highly acidic solution such as 1M HNO_3 [32].

Furthermore, the electrochemical column system was invented by packing the CuHCF electrodes into a cylindrical housing as shown in Fig. 8.10. With control of the solution-flow and CuHCF electrode-redox, more than 97% of Cs^+ in 1 mol/L HNO_3 adsorption solution was removed by electrochemical adsorption; more than 96% of the adsorbed Cs^+ was released into a desorption solution [33].

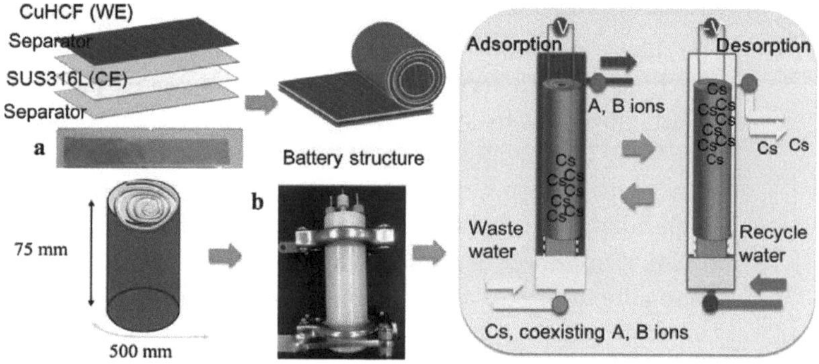

Figure 8.10 Schematic figures showing structures of electric column and Cs recovery with the column system [33]. Published by Elsevier.

8.5 PBA-NP Nanocomposite for Cs Uptake

8.5.1 Magnetic Nanocomposites

Composites consisting of Cs-adsorbing PBA and ferromagnetic materials can be collected by magnetic force after Cs-adsorption. The collection process is magnetic separation, where the adsorbent can be recovered easily using a permanent magnet or electromagnet. Although Prussian blue has magnetic properties and although it becomes a magnet at low temperatures, it is too weak to use for magnetic separation at room temperature

[34, 35]. Similarly, magnetic iron oxide itself is unsuitable for magnetically separable Cs-adsorbent, although it has Cs-adsorption properties with 68.49 mg/g and 114.9 mg/g of Cs adsorption capacity, respectively, for natural and synthesized magnetite (Fe_3O_4) [36] because their distribution coefficient K_d is less than 1,000 mL/g. Based on these results, a composite of Cs-adsorbing PBA and magnet is expected for appropriate magnetic-separable Cs-adsorbent with high-K_d and high magnetization at room temperature.

Several composites of PBA and ferrite have already been reported as magnetically separable caesium adsorbents. Three methods can be used to prepare PBA/ferrite composites: (1) attaching PBA to ferrite, (2) partial transformation from ferrite to PB, and (3) partial transformation from PB to ferrite.

(1) Attaching PBA to ferrite

The most popular method for composite preparation is attaching PBA to ferrite. PBA/ferrite composites produced by attaching PB to synthesized ferrite have been reported [37–40]. The comparison of magnetization behaviour potassium nickel hexanoferrate/magnetite composite and single-phases of them, as shown in Fig. 8.11. Reportedly, mixing graphene oxide and organic molecules to PBA/ferrite composite improves adsorption properties [41, 42].

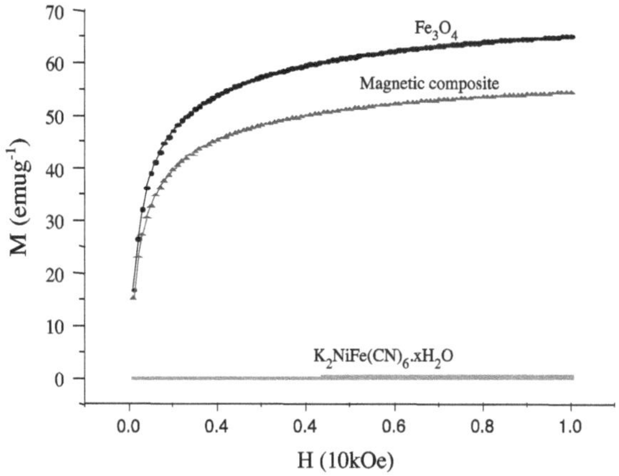

Figure 8.11 Comparison of magnetization behaviour of magnetite and potassium nickel hexacyanoferrate and magnetite [36]. Published by Elsevier.

(2) Partial transformation from ferrite to PB

A nanocomposite of PB/ferrite can be expected if a part of the ferrite particle is transformed into PB. Actually, a PB/ferrite composite synthesized with partial PB transformation from ferrite using iron cation eluted from ferrite has been reported [43], as shown in Fig. 8.12. The synthesis scheme is simple. When ferrite is put in an acidic aqueous solution containing $K_4[Fe(CN)_6]$ anion, iron cations elute to the solution and react with $[Fe(CN)_6]^{4-}$ cation into PB. The reaction proceeds on the ferrite surface. Therefore, synthesized PB attaches to the ferrite surface. However, the caesium cation capacity of this composite is only 16.2 mg g^{-1} because of the low amount of PB in the composite.

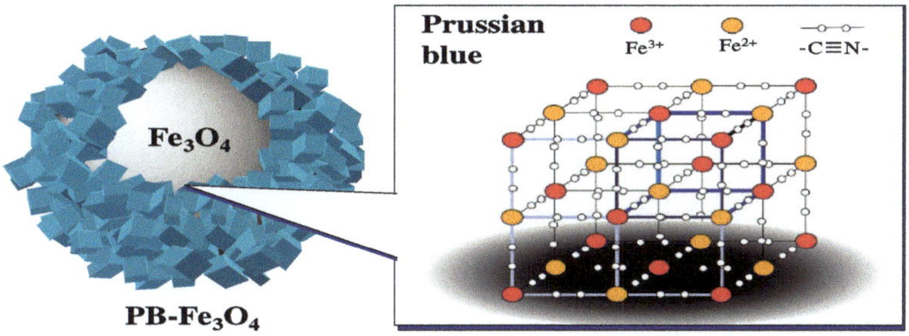

Figure 8.12 Schematic representation of PB-Fe_3O and the PB lattice structure [42].

(3) Partial transformation from PB to ferrite

PB is converted into ferrite by heating in air. Therefore, with appropriate control of heating conditions such as temperature and time, it is possible that only some PB powder is converted into ferrite [44]. In this process, the magnetically separable PB/ferrite composite is synthesized with heating alone, as shown in Fig. 8.13. Furthermore, enhancement of the caesium adsorption capacity beyond that of raw PB was reported. This enhancement is attributable to the increase of $[Fe(CN)_6]$ defect in PB with the heating process. As described above, PB having a higher amount of $[Fe(CN)_6]$ defects reportedly shows higher adsorption capacity [45].

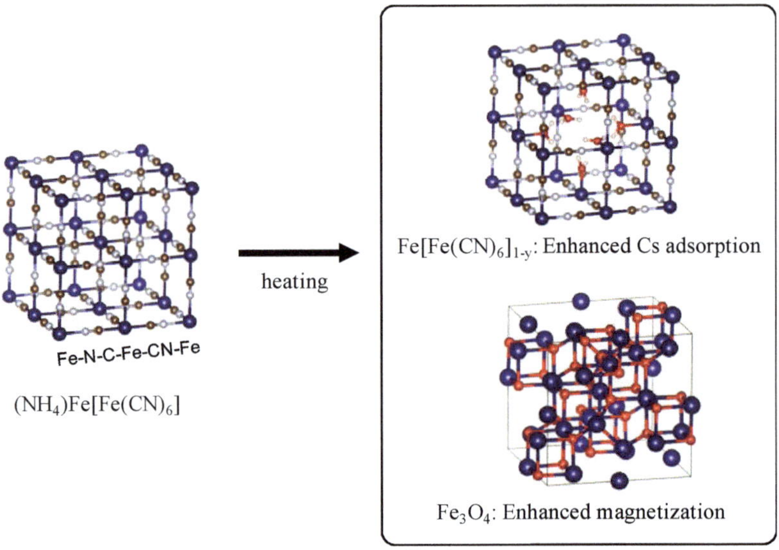

Figure 8.13 Schematic view of the transformation of PB in a PB/ferrite nanocomposite by partial PB to ferrite transformation by heating [46]. Published by The Chemical Society of Japan.

8.5.2 Granules

Although the increase of specific surface area by the nanoparticulation of adsorbent improves a material's adsorption properties, such as the adsorption rate or adsorption capacity, it remains difficult to handle, even presenting safety hazards from inhaling nanoparticles. For easier and safer practical use of fine powders, improvement of the form, such as granulation is probably necessary.

Therefore, many researchers have presented various methods to immobilize nanoparticles on filter matrixes with binders such as organic or inorganic polymers. For example, a substituted imidazole-based MOF was supported on millimetre-sized alumina beads for catalytic application [47], a Prussian blue pigment was impregnated on nonwoven organic fabric for rapid radioactive Cs^+ adsorption from water [48], and a resin containing adsorbent was coated onto the base bead material [49].

For high adsorption capacity, metal alginates have been used frequently [50] because alginate forms porous beads (microcapsule beads) containing adsorbents at high concentrations. Regarding Cs^+ adsorption with calcium alginate beads, for example, the composite adsorbent of potassium tetraphenylborate showed high equilibrium-adsorption-capacity of Cs^+ (176 mg/g) after 24 h [51]. Reportedly, the structure of alginate beads with Prussian blue and CNT has porous channels formed by calcium alginate [52].

From the perspective of porous materials, the micropore size (<2 nm) determines the adsorbent basic functions: the molecular/ionic size of materials to adsorb. For adsorbent use in water, the more important pore size is the mesopores (2–50 nm) and macropores (>50 nm), which are formed by the agglomerated secondary particles and the matrix binder supporting adsorbent because these pores maintain the liquid–gas media permeability [53].

Regarding alginate microcapsule beads, although mesopores and macropores exist immediately after preparation (wet state), the pores shrink because of heat-drying. Figure 8.14 shows the CuHCF microcapsule beads of wet, freeze-dried, and heat-dried conditions. The freeze-dried microcapsule beads appeared to have less shrinking than wet ones, compared to the heat-dried ones [53].

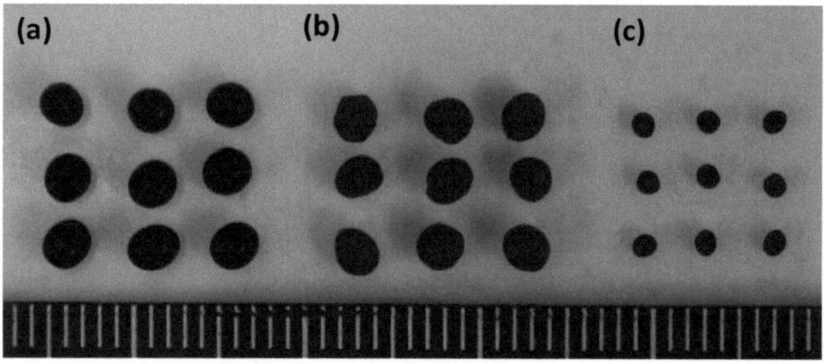

Figure 8.14 Photograph of wet CuHCF microcapsule beads (a) and dried CuHCF microcapsule beads, both freeze-dried (b) and heat-dried (c) [53]. Published by The Royal Society of Chemistry.

According to FE-SEM analysis, the inner structure of freeze-dried microcapsule beads retained micrometre-sized vacancies formed by alginate polymer cells. Reportedly, the cross-section images of freeze-dried FeHCF-alginate microcapsule beads on SEM show onion-like concentric ring structures of alginate cells [52]. Surface images of other alginate beads on SEM have been reported [50], where the freeze-dried surface was smooth and swollen state, whereas the surface of beads exposed to air exhibited a shrunken and folded structure. Regarding the abundance distribution of adsorbent nanoparticles, e.g. CuHCF-alginate beads, the aggregated nanoparticles of CuHCF were dispersed on the cell walls. However, because the alginate polymer cells of the heat-dried ones were shrunk and packed densely, little space exists for water flow, as shown in Fig. 8.15 [53].

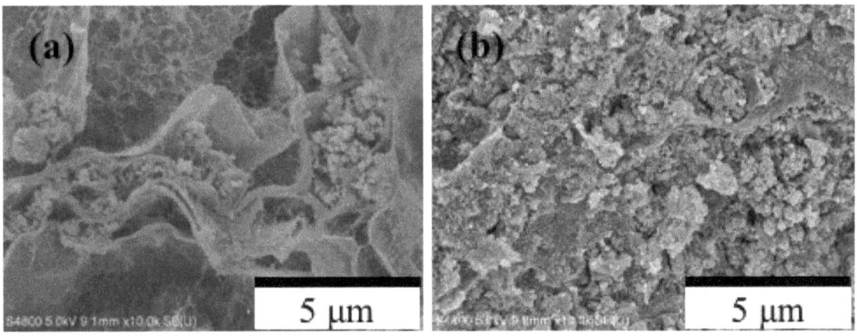

Figure 8.15 Cross-sectional FE-SEM images of freeze dried (a) and heat dried (b) microcapsule beads [53]. Published by The Royal Society of Chemistry.

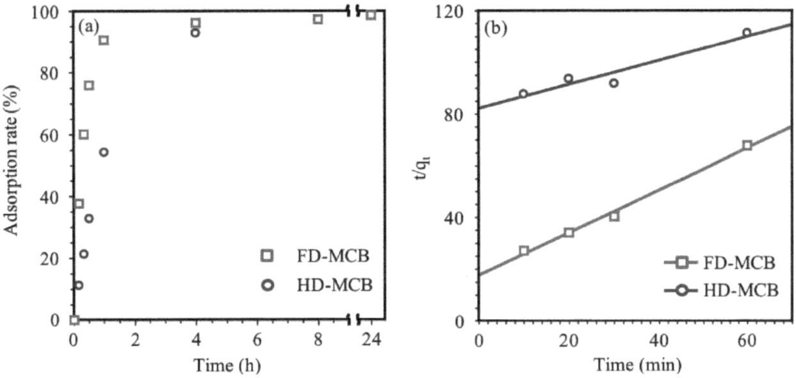

Figure 8.16 Adsorption rate (%) (a) and the pseudo-second-order kinetic model (b) for the Cs^+ ion sorption onto FD-MCB and HD-MCB [53]. Published by The Royal Society of Chemistry.

For the evaluation of the adsorption rate of the heat-dried and freeze-dried microcapsule beads, Cs^+ ion adsorption experiments were conducted by shaking for 24 h [53]. Although both MCBs adsorb more than 98% of Cs^+ ion after 24 h, marked differences were observed between the adsorption rate within 1 h, as exhibited in Fig. 8.16. A pseudo-second-order kinetic model was applied. Values of the adsorption rate constant of freeze-dried microcapsule beads were improved to about 15 times higher than those of heat-dried ones.

8.5.3 Immobilization onto Matrix

Similar to granules, immobilization onto organic matrixes is effective to prepare adsorbents for actual use. An organic matrix, non-woven fibre, or other textile is often used to fabricate the filter-type adsorbent or column packings. Advantages of an adsorbent with an organic matrix are a high adsorption rate and a large surface area. The capacity per unit of adsorbent weight becomes generally small compared with the granules. This is true because a large amount of the organic matrix is necessary for comparison with the amount of binder in the granules.

Utilization of the PBA nanoparticles is effective for fabricating an adsorbent with an organic matrix because homogeneous attachment onto the matrix is possible. For example, using water-dispersible Prussian blue nanoparticles, quite homogeneous immobilization onto cotton fibre has been reported [54].

To attach the PBA-NP, two methods have been reported. One is using a binder. Usually organic binders are used for strong immobilization onto the matrix. The other method is control of the hydrophilicity of the nanoparticles. The surface charge of the PBA nanoparticles depends on the structure of the surface termination. PBA is an ionic crystal. Therefore, the surface termination of $[Fe(CN)_6]^{\alpha-}$ anion is associated with negative surface charge. In contrast, metal cation termination causes the opposite. The surface charge can also be controlled by post-treatment after nanoparticle preparation [55]. Addition of $[Fe(CN)6]^{\alpha-}$ solution makes the nanoparticles negatively charged, and dipping into metal cation solution has opposite effect. Similarly, by dipping the PBA-NP-attached cotton fibre into a salty solution, the strength of the immobilization can be increased [54].

In addition, many reports have described composites containing granules with PBA-NP and other matrices such as porous silica [56], graphene oxide [41, 57–59], polymers [60], and walnut-shell [61] with an evaluation of Cs-adsorption performance. As explained above, an important point of the Cs-adsorbent with PBA-NP is its capacity and adsorption speed. Although the intrinsic character for Cs-adsorption is determined by PBA-NP, these two characters depend on the structure of the nanocomposite. To raise the capacity, the ratio of PBA-NP contents must be increased. For rapid adsorption, surface area enhancement is crucially important. For nanocomposite preparation, clarifying priorities is extremely important.

8.6 Summary

The Cs-adsorption performance is excellent in comparison with those of other Cs-adsorbents such as zeolite or silicotitanate. By producing smaller nanoparticles, not only the adsorption speed but also the capacity can be improved because the functional characteristics for Cs-adsorption will be enhanced.

Fabrication of nanocomposites is generally applied to improve their utility in actual use. In view of real-world applications, granulation or immobilization onto matrices is effective. For the preparation of nanocomposites, adsorption speed and capacity are the main factors used to evaluate the Cs-adsorption performance.

References

1. S. Hirao, H. Yamazawa, T. Nagae, Estimation of release rate of iodine-131 and cesium-137 from the Fukushima Daiichi nuclear power plant, *J. Nucl. Sci. Technol.* 50 (2013) 139–147. doi:10.1080/00223131.2013.757454.
2. G. Brumfiel, Fallout forensics hike radiation toll, *Nature.* 478 (2011) 435–436. doi:10.1038/478435a.
3. Environmental consequences of the Chernobyl accident and their remediation: Twenty years of experience. Report of the Chernobyl Forum Expert group "Environment," 2006. doi:10.1093/rpd/ncl163.

4. A. P. Haas, A review of information on ferrocyanide solids for removal of cesium from solutions, *Sep. Sci. Technol.* 28 (1993) 2479–2506. doi:10.1080/01496399308017493.
5. A. M. El-Kamash, Evaluation of zeolite A for the sorptive removal of Cs^{+} and Sr^{2+} ions from aqueous solutions using batch and fixed bed column operations, *J. Hazard. Mater.* 151 (2008) 432–445. doi:10.1016/j.jhazmat.2007.06.009.
6. A. Cadogan, D. Diamond, M. R. Smyth, G. Svehla, M. A. McKervey, E. M. Seward, et al., Caesium-selective poly(vinyl chloride) membrane electrodes based on calix[6]arene esters, *Analyst.* 115 (1990) 1207. doi:10.1039/an9901501207.
7. A. Casnati, A. Pochini, R. Ungaro, F. Ugozzoli, F. Arnaud, S. Fanni, et al., Synthesis, complexation, and membrane transport studies of 1,3-alternate calix[4]arene-crown-6 conformers: A new class of cesium selective ionophores, *J. Am. Chem. Soc.* 117 (1995) 2767–2777. doi:10.1021/ja00115a012.
8. A. J. Celestian, J. D. Kubicki, J. Hanson, A. Clearfield, J. B. Parise, The mechanism responsible for extraordinary Cs ion selectivity in crystalline silicotitanate, *J. Am. Chem. Soc.* 130 (2008) 11689–11694. doi:10.1021/ja801134a.
9. R. G. Anthony, R. G. Dosch, D. Gu, C. V. Philip, Use of silicotitanates for removing cesium and strontium from defense waste, *Ind. Eng. Chem. Res.* 33 (1994) 2702–2705. doi:10.1021/ie00035a020.
10. J. Krtil, V. Kouřím, Exchange properties of ammonium salts of 12-heteropolyacids. Sorption of caesium on ammonium phosphotungstate and phosphomolybdate, *J. Inorg. Nucl. Chem.* 12 (1960) 367–369. doi:10.1016/0022-1902(60)80390-9.
11. FDA, RADIOGARDASE—ferric hexacyanoferrate(ii) capsule HEYL Chem.-pharm. Fabrik GmbH & Co. KG, (2008). http://www.accessdata.fda.gov/drugsatfda_docs/label/2008/021626s007lbl.pdf (accessed March 20, 2017).
12. International Atomic Energy Agency (IAEA), The Use of Prussian Blue to Reduce Radiocaesium Contamination of Milk and Meat Produced on Territories Affected by the Chernobyl Accident, IAEA TECDOC No. 926, IAEA, Vienna (1997).
13. E. Tusa, Successful cesium removal campaign at the Loviisa NPP, Finland, in: *WM2011 Conf.*, Phoenix, 2011: p. 11002. http://www.wmsym.org/archives/2011/papers/11002.pdf (accessed March 26, 2017).

14. NIES, Removal of radioactive Cs by washing from incineration ash (in Japanese) (2014). http://www.nies.go.jp/shinsai/techrepo_r4_140414_8-10.pdf.

15. D. Parajuli, A. Kitajima, A. Takahashi, H. Tanaka, H. Ogawa, Y. Hakuta, et al., Application of Prussian blue nanoparticles for the radioactive Cs decontamination in Fukushima region, *J. Environ. Radioact.* 151 (2016) 233–237. doi:10.1016/j.jenvrad.2015.10.014.

16. Y. Nishizaki, H. Miyamae, S. Ichikawa, K. Izumiya, K. Takano, N. Kumgai, et al., New technologies for decontamination of radioactive substances scattered by nuclear accident, *Metall. Mater.*, 58 (2013) 283. http://www.imim.pl/files/archiwum/Vol1_2013/45.pdf (accessed May 14, 2016).

17. T. Kawamoto, H. Tanaka, Y. Hakuta, A. Takahashi, D. Parajuli, K. Minami, et al., Radioactive cesium decontamination technology for ash, *Synth. English Ed.* 9 (2016) 139–154. doi:10.5571/syntheng.9.3_139.

18. T. Yasutaka, H. Tsuji, Y. Kondo, Y. Suzuki, A. Takahashi, T. Kawamoto, Rapid quantification of radiocesium dissolved in water by using nonwoven fabric cartridge filters impregnated with potassium zinc ferrocyanide, *J. Nucl. Sci. Technol.* 52 (2015) 792–800. doi:10.1080/00223131.2015.1013071.

19. T. Yasutaka, S. Miyazu, Y. Kondo, H. Tsuji, K. Arita, S. Hayashi, et al., Development of a copper-substituted, Prussian blue-impregnated, nonwoven cartridge filter to rapidly measure radiocesium concentration in seawater, *J. Nucl. Sci. Technol.* (2016) 1–8. doi:10.1080/00223131.2015.1135302.

20. M. Ishizaki, S. Akiba, A. Ohtani, Y. Hoshi, K. Ono, M. Matsuba, et al., Proton-exchange mechanism of specific Cs^+ adsorption via lattice defect sites of Prussian blue filled with coordination and crystallization water molecules, *Dalt. Trans.* 42 (2013) 16049. doi:10.1039/c3dt51637g.

21. E. R. Nightingale, Phenomenological theory of ion solvation. effective radii of hydrated ions, *J. Phys. Chem.* 63 (1959) 1381–1387. doi:10.1021/j150579a011.

22. Vossen blue for the dyeing of fungicides, Degussa-AG, 1986. https://books.google.co.jp/books?id=0AMYHQAACAAJ (accessed March 26, 2017).

23. A. Takahashi, N. Minami, H. Tanaka, K. Sue, K. Minami, D. Parajuli, et al., Efficient synthesis of size-controlled open-framework

nanoparticles fabricated with a micro-mixer: route to the improvement of Cs adsorption performance, *Green Chem.* 17 (2015) 4228–4233. doi:10.1039/C5GC00757G.

24. K. Moo Lee, H. Tanaka, K. Ho Kim, M. Kawamura, Y. Abe, T. Kawamoto, et al., Improvement of redox reactions by miniaturizing nanoparticles of zinc Prussian blue analog, *Appl. Phys. Lett.* 102 (2013) 8–11. doi:10.1063/1.4800443.
25. G. B. Barton, J. L. Hepworth, E. D. McClanahan, R. L. Moore, H. H. Van Tuyl, Chemical processing wastes. Recovering fission products, *Ind. Eng. Chem.* 50 (1958) 212–216. doi:10.1021/ie50578a039.
26. K. Tanihara, A new process using a redox type of ion exchanger for cesium separation, *J. Radioanal. Nucl. Chem. Artic.* 185 (1994) 57–67. doi:10.1007/BF02042952.
27. Y. Tani, H. Eun, Y. Umezawa, A cation selective electrode based on copper(II) and nickel(II) hexacyanoferrates: dual response mechanisms, selective uptake or adsorption of analyte cations, *Electrochim. Acta.* 43 (1998) 3431–3441. doi:10.1016/S0013-4686(98)00089-9.
28. M. A. Lilga, R. J. Orth, J. P. H. Sukamto, S. M. Haight, D. T. Schwartz, Metal ion separations using electrically switched ion exchange, *Sep. Purif. Technol.* 11 (1997) 147–158. doi:10.1016/S1383-5866(97)00017-8.
29. M. Lilga, Cesium separation using electrically switched ion exchange, *Sep. Purif. Technol.* 24 (2001) 451–466. doi:10.1016/S1383-5866(01)00145-9.
30. A. Gotoh, H. Uchida, M. Ishizaki, T. Satoh, S. Kaga, S. Okamoto, et al., Simple synthesis of three primary colour nanoparticle inks of Prussian blue and its analogues, *Nanotechnology.* 18 (2007) 345609. doi:10.1088/0957-4484/18/34/345609.
31. R. Chen, H. Tanaka, T. Kawamoto, M. Asai, C. Fukushima, H. Na, et al., Selective removal of cesium ions from wastewater using copper hexacyanoferrate nanofilms in an electrochemical system, *Electrochim. Acta.* 87 (2013) 119–125. doi:10.1016/j.electacta.2012.08.124.
32. R. Chen, H. Tanaka, T. Kawamoto, M. Asai, C. Fukushima, M. Kurihara, et al., Preparation of a film of copper hexacyanoferrate nanoparticles for electrochemical removal of cesium from radioactive wastewater, *Electrochem. Commun.* 25 (2012) 23–25. doi:10.1016/j.elecom.2012.09.012.

33. R. Chen, H. Tanaka, T. Kawamoto, J. Wang, Y. Zhang, Battery-type column for caesium ions separation using electroactive film of copper hexacyanoferrate nanoparticles, *Sep. Purif. Technol.* 173 (2017) 44–48. doi:10.1016/j.seppur.2016.09.013.

34. R. E. Wilde, S. N. Ghosh, B. J. Marshall, Prussian blues, *Inorg. Chem.* 9 (1970) 2512–2516. doi:10.1021/ic50093a027.

35. F. Herren, P. Fischer, A. Ludi, W. Haelg, Neutron diffraction study of Prussian Blue, $Fe_4[Fe(CN)_6]_3 \cdot xH_2O$. Location of water molecules and long-range magnetic order, *Inorg. Chem.* 19 (1980) 956–959. doi:10.1021/ic50206a032.

36. R. R. Sheha, E. Metwally, Equilibrium isotherm modeling of cesium adsorption onto magnetic materials, *J. Hazard. Mater.* 143 (2007) 354–361. doi:10.1016/j.jhazmat.2006.09.041.

37. R. D. Ambashta, P. K. Wattal, S. Singh, D. Bahadur, Nano-aggregates of hexacyanoferrate (II)-loaded magnetite for removal of cesium from radioactive wastes, *J. Magn. Magn. Mater.* 267 (2003) 335–340. doi:10.1016/S0304-8853(03)00401-3.

38. R. R. Sheha, Synthesis and characterization of magnetic hexacyanoferrate (II) polymeric nanocomposite for separation of cesium from radioactive waste solutions, *J. Colloid Interface Sci.* 388 (2012) 21–30. doi:10.1016/j.jcis.2012.08.042.

39. Y. Namiki, T. Namiki, Y. Ishii, S. Koido, Y. Nagase, A. Tsubota, et al., Inorganic-organic magnetic nanocomposites for use in preventive medicine: A rapid and reliable elimination system for cesium, *Pharm. Res.* 29 (2012) 1404–1418. doi:10.1007/s11095-011-0628-x.

40. J. Jang, D. S. Lee, Magnetic Prussian blue nanocomposites for effective cesium removal from aqueous solution, *Ind. Eng. Chem. Res.* 55 (2016) 3852–3860. doi:10.1021/acs.iecr.6b00112.

41. H. Yang, H. Li, J. Zhai, L. Sun, Y. Zhao, H. Yu, Magnetic Prussian blue/graphene oxide nanocomposites caged in calcium alginate microbeads for elimination of cesium ions from water and soil, *Chem. Eng. J.* 246 (2014) 10–19. doi:10.1016/j.cej.2014.02.060.

42. S.-C. Jang, S.-B. Hong, H.-M. Yang, K.-W. Lee, J.-K. Moon, B.-K. Seo, et al., Removal of radioactive cesium using prussian blue magnetic nanoparticles, *Nanomaterials.* 4 (2014) 894–901. doi:10.3390/nano4040894.

43. T. Sasaki, S. Tanaka, Magnetic separation of cesium ion using Prussian blue modified magnetite, *Chem. Lett.* 41 (2012) 32–34. doi:10.1246/cl.2012.32.

44. R. Zboril, L. Machala, M. Mashlan, V. Sharma, Iron(III) Oxide nanoparticles in the thermally induced oxidative decomposition of Prussian blue, $Fe_4[Fe(CN)_6]_3$, *Cryst. Growth Des.* 4 (2004) 1317–1325. doi:10.1021/cg049748+.

45. M. Ishizaki, S. Akiba, A. Ohtani, Y. Hoshi, K. Ono, M. Matsuba, et al., Proton-exchange mechanism of specific Cs^+ adsorption via lattice defect sites of Prussian blue filled with coordination and crystallization water molecules., *Dalton Trans.* 42 (2013) 16049–55. doi:10.1039/c3dt51637g.

46. A. Takahashi, H. Tanaka, N. Minami, M. Kurihara, T. Kawamoto, Simultaneous enhancement of Cs-adsorption and magnetic properties of Prussian blue by thermal partial oxidation, *Bull. Chem. Soc. Jpn.* 88 (2015) 69–73. doi:10.1246/bcsj.20140229.

47. S. Aguado, J. Canivet, D. Farrusseng, Facile shaping of an imidazolate-based MOF on ceramic beads for adsorption and catalytic applications, *Chem. Commun.* 46 (2010) 7999. doi:10.1039/c0cc02045a.

48. T. Yasutaka, T. Kawamoto, Y. Kawabe, T. Sato, M. Sato, Y. Suzuki, et al., Rapid measurement of radiocesium in water using a Prussian blue impregnated nonwoven fabric, http://dx.doi.org/10.1080/00223131.2013.797936. (2013).

49. C. Dwivedi, A. Kumar, J. K. Ajish, K. K. Singh, M. Kumar, P. K. Wattal, et al., Resorcinol-formaldehyde coated XAD resin beads for removal of cesium ions from radioactive waste: Synthesis, sorption and kinetic studies, *RSC Adv.* 2 (2012) 5557. doi:10.1039/c2ra20054f.

50. G. Fundueanu, C. Nastruzzi, A. Carpov, J. Desbrieres, M. Rinaudo, Physico-chemical characterization of Ca-alginate microparticles produced with different methods, *Biomaterials.* 20 (1999) 1427–1435. doi:10.1016/S0142-9612(99)00050-2.

51. T. Guo, Y. Hu, X. Gao, X. Ye, H. Liu, Z. Wu, Competitive adsorption of Li, Na, K, Rb and Cs ions onto calcium alginate–potassium tetraphenylborate composite adsorbent, *RSC Adv.* 4 (2014) 24067. doi:10.1039/c4ra02606c.

52. A. K. Vipin, B. Hu, B. Fugetsu, Prussian blue caged in alginate/calcium beads as adsorbents for removal of cesium ions from contaminated water, *J. Hazard. Mater.* 258–259 (2013) 93–101. doi:10.1016/j.jhazmat.2013.04.024.

53. K.-M. Lee, T. Kawamoto, K. Minami, A. Takahashi, D. Parajuli, G. Kido, et al., Improved adsorption properties of granulated copper

hexacyanoferrate with multi-scale porous networks, *RSC Adv.* 6 (2016) 16234–16238. doi:10.1039/C5RA25388H.

54. A. Kitajima, H. Tanaka, N. Minami, K. Yoshino, T. Kawamoto, Efficient cesium adsorbent using Prussian blue nanoparticles immobilized on cotton matrices, *Chem. Lett.* 41 (2012) 1473–1474. doi:10.1246/cl.2012.1473.
55. K.-M. Lee, A. Takahashi, H. Tanaka, K. H. Kim, M. Kawamura, Y. Abe, et al., Sequential structural control of open-framework nanoparticles both in dispersion and in film for electrochemical performance tuning, *Bull. Chem. Soc. Jpn.* 88 (2015) 1561–1566. doi:10.1246/bcsj.20150199.
56. L. Lartigue, S. Oh, E. Prouzet, Y. Guari, J. Larionova, Superspin-glass behavior of $Co_3[Fe(CN)_6]_2$ Prussian blue nanoparticles confined in mesoporous silica, *Mater. Chem. Phys.* 132 (2012) 438–445. doi:10.1016/j.matchemphys.2011.11.050.
57. H. Yang, L. Sun, J. Zhai, H. Li, Y. Zhao, H. Yu, In situ controllable synthesis of magnetic Prussian blue/graphene oxide nanocomposites for removal of radioactive cesium in water, *J. Mater. Chem. A.* 2 (2014) 326–332. doi:10.1039/c3ta13548a.
58. S.-C. Jang, Y. Haldorai, G.-W. Lee, S.-K. Hwang, Y.-K. Han, C. Roh, et al., Porous three-dimensional graphene foam/Prussian blue composite for efficient removal of radioactive 137Cs, *Sci. Rep.* 5 (2015) 17510. doi:10.1038/srep17510.
59. A. A. Kadam, J. Jang, D. S. Lee, Facile synthesis of pectin-stabilized magnetic graphene oxide Prussian blue nanocomposites for selective cesium removal from aqueous solution, *Bioresour. Technol.* 216 (2016) 391–398. doi:10.1016/j.biortech.2016.05.103.
60. H. Yang, H. Li, J. Zhai, H. Yu, In situ growth of Prussian blue nanocrystal within Fe^{3+} crosslinking PAA resin for radiocesium highly efficient and rapid separation from water, *Chem. Eng. J.* 277 (2015) 40–47. doi:10.1016/j.cej.2015.04.047.
61. D. Ding, Z. Lei, Y. Yang, C. Feng, Z. Zhang, Selective removal of cesium from aqueous solutions with nickel (II) hexacyanoferrate (III) functionalized agricultural residue–walnut shell, *J. Hazard. Mater.* 270 (2014) 187–195. doi:10.1016/j.jhazmat.2014.01.056.

Chapter 9

Prussian Blue Nanoparticles for Cancer Theranostics

Sudip Mukherjee,* Bonda Rama Rao,* Rajesh Kotcherlakota,* and Chitta Ranjan Patra

Department of Applied Biology, CSIR—Indian Institute of Chemical Technology, Uppal Road, Tarnaka, Hyderabad 500007, India
Academy of Scientific and Innovative Research (AcSIR), Training and Development Complex, CSIR Campus, CSIR Road, Taramani, Chennai 600 113, India

crpatra@iict.res.in, patra.chitta@gmail.com

Since its discovery 290 years ago, Prussian blue (PB: ferric hexacyanoferrate, historically known as ancient dark blue pigment) has gained enormous attraction owing to its several exciting properties in the bulk state as well as in the nanoscale region. The porous nature and several unique properties (electro, photochemical, biochemical and magnetic properties) of PB (a mixed valance co-ordination complex) make it an appropriate candidate for several applications, including catalysis, photonics, energy storage, magneto-optical devices, and staining agent for histology study. Importantly, PB (also known as Radiogardase) was FDA approved as an antidote for radioactive cesium and thallium poisoning. Moreover, PB is highly biocompatible and biodegradable in nature, which increases the utility of PB for

*The authors contributed equally.

Prussian Blue-Type Nanoparticles and Nanocomposites: Synthesis, Devices, and Applications
Edited by Yannick Guari and Joulia Larionova

ISBN 978-981-4800-05-1 (Hardcover), 978-0-429-02474-0 (eBook)
www.panstanford.com

various biomedical applications. Recently, the nanomedicine (application of nanotechnology in biology and medicine)-based theranostics approach has gained enormous attention in healthcare owing to its vast and unusual physicochemical properties. In the past decade, several groups including ours, demonstrated various cancer theranostics applications (drug delivery, gene delivery, MRI imaging, CT-contrast imaging, photothermal therapy, etc.) of Prussian blue nanoparticles (PBNPs) and their analogue nanoparticles (PBANPs). In this chapter, we summarize the recent examples of PBNPs and PBANPs and their biomedical applications in cancer theranostics.

9.1 Introduction and Background of Prussian Blue

Prussian blue (PB: $Fe_4[Fe(CN)_6]_3 \cdot xH_2O$ termed as iron hexacyanoferrate) complex is considered as the first synthetic coordination compound, which was invented by German painter, J. J. Diesbach in 1704 [1–3]. It is also popular as iron blue, Paris blue, Milori blue, Chinese blue, pigment blue and French blue [4]. The blue color of the compound originates from the interchange of electrons between iron atoms present in coordination and ionic spheres in the complex [5, 6]. During the 18th century, PB was familiar as painter's pigment and coloring agent for textiles. PB was also used in the wine making process to remove excess iron [4]. Extensive research was carried out during the 1960s that confirm the effectiveness of PB (Radiogardase) as an antidote for the radioactive cesium and thallium [7, 8]. In 1986, the Chernobyl nuclear reactor accident led to the radioactive cesium contamination. This incident led to a renewed interest in PB and its analogues (ammonium ferric cyanoferrate: $NH_4Fe[Fe(CN)_6]_3$) for the use as an antidote for heavy metal contamination in the environment as well as in biological systems by exchanging monovalent ammonium with cesium [9, 10]. Recently, the US FDA has approved ~500 mg of PB (trade name Radiogardase) as a safe oral antidote for cesium and thallium contamination [11, 12].

In the past decade, PBNPs and PBANPs have been extensively studied for various biomedical applications, including cancer therapy, drug delivery, gene delivery, MRI imaging, CT contrast agent, photothermal therapy, biosensing, oxygen scavengers, and biological markers due to several unique magnetic and electrical properties, eco-friendliness, high biocompatibility and low cost [13–20]. Unfortunately, there are no comprehensive review articles or published book chapters that demonstrate the cancer theranostics application of PBNPs and PBANPs. Hence, in this chapter, we discuss the various cancer theranostics applications of PBNPs and PBANPs along with the future challenges that will help the upcoming researchers to gather valuable information in this area.

9.2 Chemical and Crystalline Structure of Prussian Blue

The chromic center of PB is ferric ferrocyanide ($Fe^{III}[Fe^{II}(CN)_6)$ and the Roman superscripts on the formula tell about the oxidation states of respective iron atoms. The crystalline structure of PB was first discussed by Keggin and Miles on the basis of powder diffraction patterns [21]. Later on, more precisely Ludi and coworkers from single crystals through electron and neutron diffraction measurements determined exact structure of PB [4, 22]. PB has a basic cubic structure consisting of alternating iron(II) and iron(III) located on a face-centered cubic lattice in such a way that the iron(III) ions are surrounded by octahedral nitrogen atoms, and iron(II) ions are surrounded by carbon atoms. The most intense reflections of XRD spectrum suggested that the space group is $Fm\bar{3}m$, but the actual space group (considering the weak reflections) is $Pm\bar{3}m$ [14, 23, 24]. In the insoluble form, a quarter of the $[Fe^{II}(CN)_6]^{4-}$ unit is missing from the unit cell in order to maintain composition and electro neutrality. This packing arrangement leaves some cavities in the crystal lattice that are occupied by water molecules. In addition, the coordination sphere of each Fe(III) center around the vacant site is completed by a water molecule. Therefore, the water molecules in the formula $Fe_4^{III}[Fe^{II}(CN)_6]_3\cdot nH_2O$ (n = 14–16) can be divided into two types: coordinative and zeolitic [25]. Beall et al. suggested that the Zeolitic water molecules or balancing cations occupy the interstitial sites and forms the hydrogen bond that helps to stabilize the crystal [26]. PB and PBNPs are face-centered cubic lattice and exited in such a way, that the iron (III) ions are surrounded by octahedral nitrogen atoms and iron(II) ions are surrounded by carbon atoms, respectively.

9.3 Synthesis, Characterization and Stability of PBNPs and PBANPs

The PB can be synthesized by mixing of iron (Fe^{3+}) solutions with hexacyanoferrate (Fe^{2+}) ions ($[Fe(CN)_6]^{4-}$) that results the formation of dark blue precipitate. After PB invention researchers were focused on developing various analogues of it. The transition metal cations react more effectively with anionic coordination sphere of PB that enables the formation of different analogues. Due to versatile characteristics of PBAs, they were used in various biomedical applications. The synthesis and biomedical applications of various PBNPs and PBANPs are tabulated in Table 9.1.

Table 9.1 Applications of various PBNPs and PBANPs in cancer theranostics

Complex	Shape	Capping agent	Biomedical applications
Iron	Cubic	No capping agent	Nano theranostic agent for combinatorial therapy [30]
	Cubic	CTAB	Photo thermal ablation agents for cancer therapy [13]
	Spherical	Citric acid	Effective removal of cesium in water [31]
	Spherical	Folate	Contrast agent for tumor targeted MRI [32]
	Spherical	Antibody	Targeted MRI and PTT of hepatocellular carcinoma [33]
	Spherical	Dextran	Efficient T_1 contrast agent for MRI [34]
	Rods	No capping agent	Phosphate binding agent [35]
	Cubic	PEG	MR and CT imaging agents [36]

Copper	Cubic	No capping agent	Drug delivery vehicle and fluorescent quenching [37]
	Spherical	Kenaf stem	Biosensing [38]
Gadolinium	Spherical	No capping agent	Multimodal MRI molecular imaging agents [39]
	Spherical	No capping agent	RI Contrast and hyperthermic effects [40]
	Cubic	PVP and PEG	Oral MRI imaging contrast agent for gastrointestinal tract [41]
	Spherical	PVP/Citrate	MR probes for biological markers [16]
Gallium	Spherical	PVP	Exchange and removal of iron from poisoned cells [40]
	Cubic	PVP-citrate	Bimodal MRI/PET imaging agents [42]
Bismuth	Spherical	PVP	Potential contrast agent for computed tomography [43]
Manganese	Cubic	No capping agent	Imaging of pediatric brain tumors [44]
	Spherical	Dextran	Highly efficient T1 contrast agents for MRI [34]

The PBNPs and PBANPs were characterized (morphology, size, zeta potential, crystalline structure, functional groups, chemical composition, binding energy, surface area, surface volume and magnetic moment for contrasting property) by several physicochemical techniques [27–29]. We are not discussing the synthesis procedure and characterization of PBNPs and PBANPs in detail as these are beyond the scope of our present book chapter. Size of the materials plays an important role in biomedical applications. It was identified that bigger sized nanomaterials (>100 nm) show toxicity compared to smaller sized nanomaterials (0.1–100 nm) [28]. According to previous reports, the size of the PBNPs were found to be less than 100 nm that increases the utility of these materials towards biomedical applications due to easy transmembrane permeation and excretion [45, 46]. Xiaoqing Jia et al. demonstrated the synthesis of perfluoro pentane stabilized mesoporus PB nano cubes (size less than 100 nm) and used for cancer therapy and imaging (Fig. 9.1) [29].

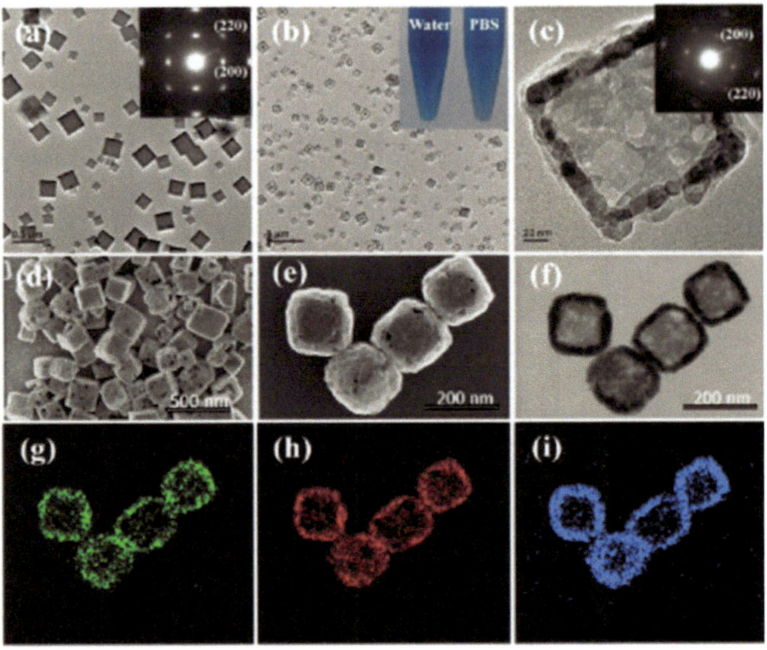

Figure 9.1 (a) TEM image of mPB; inset of (a), electron diffraction image of mPB. (b, c, f) TEM images of HPB; inset of (b), HPB suspensions in water and PBS; inset of (c), electron diffraction image of HPB. (d, e) SEM images of HPB with different magnifications. (f–h) Corresponding bright-field imaging and element mapping (g, N; h, Fe; i, C). Reprinted with permission from [29]. Copyright © 2015, American Chemical Society.

Generally, PBNPs are found to be stable attributed from the strong interaction of the ligands in the coordination sphere with central metal atom [44, 47]. Recently, our group studied the stability of copper analogue of PBNPs in various physiological buffers by different techniques that infer high colloidal stability [14]. Thermal stability of the PBNPs was also reported in earlier published literature [48]. From all these observation, PBNPs are found to be highly stable and suitable for various biomedical applications.

9.4 Prussian Blue Applications in Cancer Theranostics

Cancer is a foremost public health problem all over the world where the malignant cells grow abnormally leading death. In the United States alone, 1,688,780 new cases of cancer are expected to occur in 2018 [49]. Surgeries, chemotherapy, radiation therapies, monoclonal antibodies and stem cell transplantation are the current treatment methods applied in clinics [50]. However, the use of chemotherapeutic agents has various limitations, including side effects, nonspecific toxicity, rapid clearance and development of multidrug resistance [51]. Further, the cost of the chemotherapeutic agents is very high. In this regard, nanotechnology offers the development of different types of nanomaterials or nanoconjugates in cost effective manner that can be employed in the targeted delivery of therapeutic agents and to develop anticancer nanomaterials [52]. Nanoparticle-based delivery systems can reduce the dose of therapeutic agents along with the associated side effects. Several research groups, including ours, are working on the development of different nanoparticle-based therapeutics (using metal nanoparticles) to combat cancer. In recent years, PBNPs have also become popular in the field of cancer research because of their multifunctional activities [14, 53, 54]. In this chapter, we discuss on the recent development and applications of PBNPs for the cancer theranostics. We believe that this chapter will be useful for the future investigators in the betterment of cancer research.

9.4.1 Therapeutics

The porous nature and several unique properties (vast unusual physicochemical properties, unusual electro, photochemical, biochemical and magnetic properties, high biocompatibility, and high biodegradability) of PB, PBNPs, and PBANPs make them ideal candidates for several biomedical applications, including magneto-optical devices, staining agent for histology study, antidotes for radioactive metal removal, and cancer theranostics applications (drug delivery, gene delivery, MRI imaging, CT-contrast imaging, photothermal therapy, fluorescence imaging, etc.) [14, 53, 55–57].

The therapeutic applications of PB, PBNPs and PBANPs are summarized below under different sub headings.

9.4.1.1 Drug delivery

The distinctive physicochemical properties of metal nanoparticles showed a pronounced attraction to the researchers to develop new drug delivery carriers that can be useful in the treatment of numerous diseases including cancers [58]. Biocompatible nanoparticle-based drug delivery is one of the new strategies in modern medicine that holds the promising outcomes through nanomedicine. The efficiency and efficacy of drug delivery depends on (i) design, functionalization, surface properties, morphology and size of the vehicle (ii) capable of carrying the therapeutic agents in desired area of the body without damaging the other cells [59]. Using passive and active targeting strategies, these two criteria could be fulfilled using nanoparticle-based drug delivery. Passive targeting is mainly useful in cancer therapy as most of the tumor architecture contains the leaky vasculature, which enables the drug carriers to reach to the tumor tissue via enhanced permeability and retention (EPR) effect [60]. Active targeting, on the other hand, utilizes the targeting ligands (like small peptides, chemical agents, antibodies) which specifically binds to the target and leads to accumulation of cargo in desired site [61]. The present section describes about the drug delivery application of the PBNPs and PBANPs using either active or passive targeting strategies.

PBNPs are the new class of drug delivery vehicles with vast number of applications in biology and medicine. In recent years, researchers employed PBNPs as potent drug delivery vehicles for cancer and other diseases. For example, Lian et al. designed coordination polymer PBNPs harboring microporous framework with hollow interior and demonstrated the delivery of anticancer drug cisplatin [62]. These nanoparticles are highly biocompatible when tested in vitro and stable in different pH conditions. The results indicated that cisplatin conjugated drug delivery system exhibited the potent cytotoxic effect in T24 cells (bladder cancer cells). Considering these promising results, the authors concluded that PBNPs could be useful in future for different biomedical applications. Similarly, Li et al. designed and developed PBNPs coated with human-serum-albumin (HSA) and elucidated the delivery of doxorubicin to HeLa cervical cancer cells [55]. The system has the ability to absorb near-infrared (NIR) radiation that will be useful in photodynamic therapy for cancer. Further, the release of the drug in side cancer cells can be triggered by either pH or NIR radiation. Thus, the authors claimed that a bimodal on-demand drug release with HSA-coated PBNP system opens the door to cancer therapy in near future.

In another study, Wang et al. prepared 11-mercaptoundecanoic acid (MUA)-functionalized PBNPs loaded with STAT3 inhibitor decoy oligodeoxynucleotide (dODN) and demonstrated the anti-cancer activity in 22rv1 carcinoma cells (Fig. 9.2) [63]. The MUA on the surface of PBNPs was modified with carboxyl and thiol groups to make the 5′ NH_2-dODN-FAM and/or 5′ SH-dODN-FAM conjugation by EDC/NHS bonding, oxidative disulfide bond linkage,

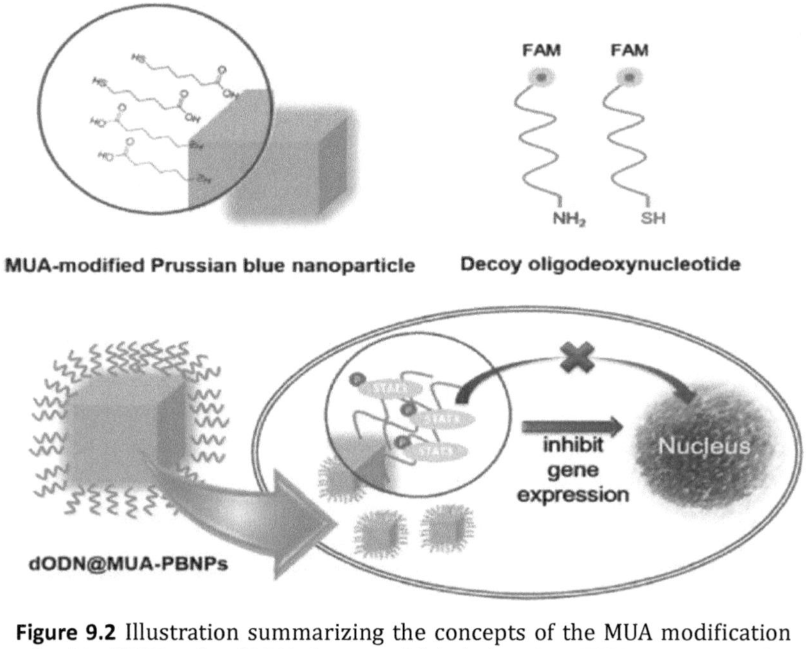

Figure 9.2 Illustration summarizing the concepts of the MUA modification of cubic PBNPs, the dODN drugs end-labeled with a FAM reporter and a linker, and the final configuration of the DNA nano drug (dODN@ MUA–PBNPs) and its potential cancer cell-killing effect. Reprinted with permission from [63]. Copyright © 2013, IOPScience (open access article).

or both. Interestingly, here the oligonucleotides were labeled with the FAM reporter molecule to detect the internalization of nanoconjugates inside cells. The results of this study demonstrated that PBNPs could successfully deliver the anticancer drug (STAT3 gene inhibitor) inside the cancer cells through the endocytosis pathway.

Our group recently fabricated the copper analogue of PB nanoparticles (CPBNPs) that displayed multiple applications for cancer therapy [14]. These nanoparticles were biocompatible to different normal cell lines. Further, the CPBNPs exhibited fluorescence quenching of doxorubicin selectively. The DOX conjugated CPBNPs (CPBNPs-DOX) showed improved in vitro anticancer effects compared to pristine DOX in various cancer cell lines (SK-OV-3, MCF-7, and B16F10). G0/G1 phase arrest and activation of apoptotic proteins (Bax and cleaved caspase-3 observed by western blot analysis) in B16F10 melanoma cancer cell lines were observed upon treatment with CPBNPs-DOX that support the anti-cancer activity. Moreover, TEM and EDAX analysis showed the presence of nanoparticles inside cytoplasm and nucleus of the nanoparticles/nanoconjugates treated B16F10 cells. Considering these observations, PBNPs and other analogues of PBNPs can be promising drug delivery candidates for the treatment of different cancers in near future.

9.4.1.2 Gene delivery

Gene therapy refers to the administration of nucleic acids into animal cells to treat genetic disorders [64]. The application of nanotechnology also extended to gene therapy for the delivery of therapeutic nucleic acids into desired area. Direct introduction of nucleic acids into the body raises the several complications such as toxicity, host immune response and degradation in serum nucleases [65]. Researchers used viral vectors to transfect foreign nucleic acids into animal cells and later observed that viral vectors could generate the several complications when used in vivo. Hence, nonviral vector based gene delivery vehicles are required for the treatment of gene related diseases. In this regard, nanotechnology gained the great importance as nucleic acid delivery vehicles that overcome the above limitations. The present section describes about gene delivery applications of PBNPs briefly.

In recent years, PBNPs gained the tremendous importance as gene delivery vehicles. For instance, Li et al. prepared NIR active PBNPs functionalized with chitosan and showed the gene delivery applications in cultured HeLa cells under NIR laser irradiation [56]. The authors described that the chitosan-functionalized PBNPs have the ability to absorb the NIR light which can be converted into heat. As a result, enhanced uptake of the nanoparticles was seen in the HeLa cells, which strengthens higher gene transfection efficacy (at 808 nm laser light irradiation with power density of 2 W cm^{-2}) of the nanoparticles. This study, demonstrated that chitosan-functionalized PBNPs can be future gene delivery vectors with NIR light control. Similarly, in another study, Li et al. designed chitosan coated PBNPs system containing DNA packed microbubbles (MBs@CS/PB/DNA) and demonstrated the enhanced gene delivery by both photothermal and ultrasound mediated targeted microbubble disruption [66]. MBs@CS/PB/DNA system releases the DNA by applying the higher-intensity ultrasound irradiation which causes the absorption of NIR light leading to generation of heat that helps in cancer cells killing. All together these studies describe the gene delivery efficacy of PBNPs for treatment of cancer.

9.4.1.3 Photothermal therapy

Photothermal or thermal ablation is one of the modern cancer therapies which utilize photothermal agents and near infra-red laser to generate the heat inside the cancer cells. Photothermal therapy (PTT) is a noninvasive therapy aimed to burn the cancer cells by the generation of heat (>50°C) upon laser irradiation. A variety of nanomaterials including gold nanorods, gold nanoshells, tungsten oxide nanowires, copper sulfides, palladium nanosheets, bismuth selenide nanoplates and carbon nanotubes demonstrated PTT applications [17–20]. Very recently, scientists discovered PBNPs as successful new generation photothermal agents because of facile synthesis procedure, low cost, high biocompatibility and exceptional photothermal conversion competence in NIR window.

PBNPs are considered as potent photothermal agents compared to other metal nanoparticles due to charge transfer transition between Fe(II) and Fe(III) [13]. Fu et al. prepared the Prussian blue (PB) nanoshell-based nanotheranostic agents by growing them on the superparamagnetic Fe_3O_4 nanocores and demonstrated the cancer therapy using targeted photothermal effect of the PBNPs core shells under magnetic resonance imaging (MRI) guidance. The authors described the contrast enhancement of Fe_3O_4@PB core–shell nanoparticles in both in vitro and in vivo (Fig. 9.3). Administration of this nanoparticles system in tumor-bearing nude mice exhibited the tumor growth inhibition due to the thermal ablation effect of Fe_3O_4@PB nanoparticles under near-infrared exposure [67].

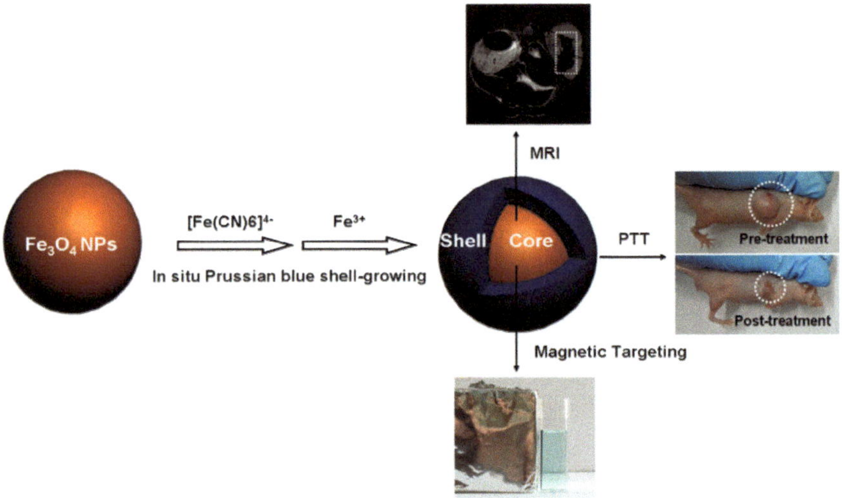

Figure 9.3 Illustration of the synthesis procedure and theranostic applications of the multifunctional Fe_3O_4@PB NPs. Reprinted with permission from [67]. Copyright © 2014, American Chemical Society.

In another study, Li et al. developed a targeted MRI-based photothermal therapy using glypican-3 antibody as targeting agent and PBNPs as MRI as well as photothermal agents in the treatment of hepatocellular carcinoma [33]. The authors demonstrated that glypican-3 conjugated PBNPs internalized into HepG2 cells through receptor mediated endocytosis pathway and showed efficient cytotoxicity and MRI contrast enhancement. In another interesting study, Jing et al. established PBNP-based thermo chemotherapy containing hyaluronic acid as targeting ligand and 10-hydroxycamptothecin as chemotherapeutic agent (Fig. 9.4) [68].

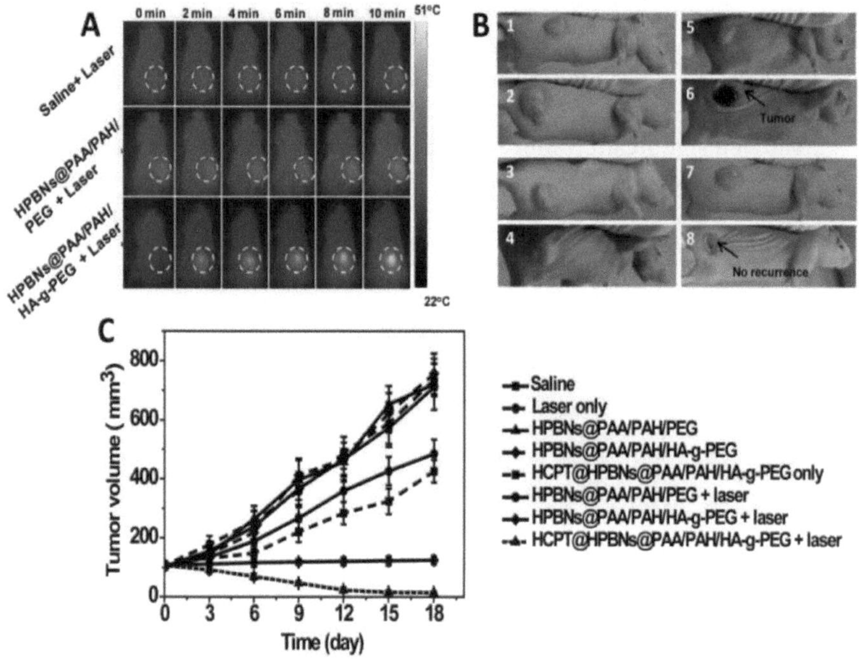

Figure 9.4 (A) IR thermal images of Hela tumor-bear mice at different time intervals under the 808-nm laser irradiation (0.8 W/cm^2). (B) The representative tumor photographs of each group were recorded before and after treatment: (1) saline group before therapy; (2) saline group at 18 day after therapy, (3) HCPT@ HPBNs@PAA/PAH/HA-g-PEG group before therapy; (4) HCPT@ HPBNs@PAA/PAH/HA-g-PEG group at 18 day after therapy, (5) HPBNs@PAA/PAH/HA-g-PEG + laser group before therapy; (6) HPBNs@PAA/PAH/HA-g-PEG + laser group at 18 day after therapy, (7) CPT@HPBNs@PAA/PAH/HA-g-PEG + laser group before therapy; (8) HCPT@HPBNs@PAA/PAH/HA-g-PEG + laser group at 18 day after therapy. (C) The growth of Hela tumors in nude mice after various treatments. Reprinted with permission from [68]. Copyright © 2016, IVY Spring (Open access article).

The authors claimed that modification of PBNPs surface with hyaluronic acid and poly ethylene glycol (PEG) significantly increased the stability and long circulation period in blood. Further, the PBNPs also exhibited the CD44 receptor targeting potential in Hela cells and efficient photothermal effect as well as light triggered release of 10-hydroxycamptothecin. Cheng et al. prepared theranostic pegylated Prussian blue nanocubes (PBNCs) that exhibited both cancer imaging and photothermal therapy applications [57]. Intravenous administration of this pegylated PBNCs showed higher accumulation in tumor tissue. Further, application of NIR laser irradiation in the PBNCs injected mice showed significant tumor reduction that suggests the potent clinical significance of PBNCs. This study holds the promising strategy with PBNCs in cancer therapy by working as image guided cancer therapy.

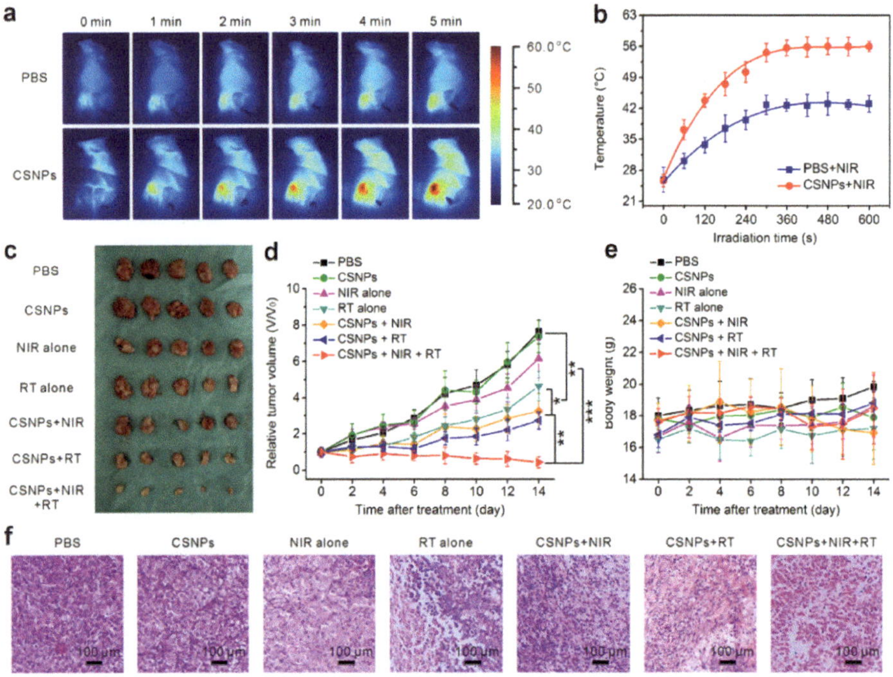

Figure 9.5 In vivo analysis of PTT–RT synergistic therapy. Thermographs (a) and corresponding temperature changes (b) of mice bearing 4T1 xenograft tumors during NIR laser exposure after intravenous injection of CSNPs (10 mg/mL, 100 μL) for 24 h. (c) Representative photos of harvested tumors from mice at day 14 after different treatments (n = 5). Tumor growth inhibition curves (d) and body weight changes (e) for various treatments (n = 5). (f) H&E staining results at day 14 after different treatments. All of the experiments were conducted under the same conditions (PTT: 808 nm, 5 min, 1.5 W/cm^2; RT: 4 Gy). Error bars: standard deviation of mean for at least three replicates. P values: *, $P < 0.05$; **, $P < 0.01$; ***, $P < 0.001$. Reprinted with permission from [36]. Copyright © 2017, American Chemical Society.

In another recent report, Dou et al. demonstrated the multifunctional nanotheranostic applications (magnetic resonance-computed tomography: MR-CT imaging and synergistic photothermal and radiosensitive therapy: PTT-RT therapy) of PB@Au core-satellite nanoparticles (CSNPs) [36]. PBNPs serve as efficient T1- and T2-weighted MR contrast imaging and PTT agent. On the other hand, AuNPs enables CT enhancement and radiosensitization. CSNPs were able to target tumor localization by passively targeted accumulation observed by MR-CT imaging. CSNPs exhibited good biocompatibility. Moreover, CSNPs (100 μL of 10 mg/mL) showed excellent tumor regression ability in mice bearing 4T1 breast xenograft tumors upon NIR laser exposure (PTT: 808 nm, 5 min, 1.5 W/cm^2; RT: 4 Gy) (Fig. 9.5). Together the above-mentioned literature supports the strong PTT potential of PBNPs and PBANPs for cancer treatment.

9.4.2 Imaging and Diagnostics

In recent times, biomedical imaging has gained massive interest due to its analytical and diagnostic capability at the cellular, sub-cellular and molecular level. Different imaging techniques (MRI, CT, SPECT, PET, ultrasound and fluorescence) were utilized to unveil several critical issues in the diagnosis of cancer by utilizing composites of PBNPs. These noninvasive imaging techniques allow real-time visualization of living systems and help in the diagnosis of various diseases including cancer. Further, these imaging techniques provide crucial information about the disease condition. PBNPs and PBANPs were extensively studied for various imaging purposes due to their unique physicostructural properties [39, 53, 57]. The present section describes the advances in the PBNP-based bio-imaging in the cancer diagnosis.

9.4.2.1 MRI imaging

MRI is a powerful diagnostic tool for molecular and cellular imaging of the brain, the central nerve systems (CNS) and other body parts. Further MRI helps in detecting tumors area by imaging the soft tissues with improved resolution. Due to lack of differences in relaxation time, MRI contrast agents help to increase the clarity of the images. The basic principle of MRI is the utilization of nuclear magnetic resonance (NMR), radio waves, and field gradients together with relaxation (T_1 and T_2) of proton spins in an external magnetic field [69]. Accordingly, the conventional contrast agents are classified as T_1 and T_2 contrast agents and used based on the function of paramagnetic (different gadolinium complexes: Magnevist, ProHance, Multihance, etc.) or superparamagnetic (super paramagnetic iron oxide nanoparticles: SPION) properties [39]. However, with the vast development of research in nanoscience and technologies scientists are trying to develop multimodal imaging agent that can avoid various limitation of conventional imaging agents such as sensitivity, spatial resolution, depth of penetration, and safety. In this context, PBNPs and PBANPs hold enormous potential as a T_1-weighted MRI contrast agent. In 1968, Ito and coworkers demonstrated that PB exhibited ferromagnetic nature (T_c = 5.6 K) due to its parallel alignment of the magnetic moments using polarized neutron diffraction studies [70]. PB is a face-centered cubic crystal complex contains two different iron centers Fe^{3+} (nitrogen bound and high spin; $S = 5/2$) and Fe^{2+} (carbon bound and low spin; $S = 0$), which are bridged by the CN^- groups [53]. Hence these nanocomplexes (contains five unpaired electrons in one Fe^{2+}-C≡N-Fe^{3+} unit) has the ability to shorten the transversal and longitudinal relaxation times (T_1 or T_2) of protons from bulk water. Also, these nanocomplexes are accessible to water co-ordination that is important for the development of T_1-weighted MRI contrast agent with high relaxivity [53].

Several published reports demonstrated the MRI contrast imaging properties of PBNPs and PBAs towards in vitro and in vivo systems [53, 57]. Recently, Cheng et al. demonstrated the development of a biocompatible pegylated theranostics PBNCs with intrinsic paramagnetic property and powerful near-infrared (NIR) absorbance that can be useful for in vivo bimodal imaging-guided photothermal therapy [57]. The high r_1 relaxivity and a low r_2/r_1 ratio (longitudinal relativities: r_1 = 6.4 mM^{-1} S^{-1}, and transverse relativities: r_2 = 8.1 mM^{-1} S^{-1}, with the r_2/r_1 = 1.26) of PBNCs is ideal for applying it as a useful T_1-weighted MR contrast agent. Importantly, PBNCs was found to be biocompatible up to 0.2 mg/mL, which increases the utility for theranostics applications. In vivo T_1-weighted magnetic resonance (MR: a 9.4-T MR imaging system) imaging of PBNCs (1 mg/mL, 200 mL) in mice bearing murine breast tumor show high tumor accumulation following 12 h of systemic Intravenous (i.v.) administration (Fig. 9.6a–d). Hence PBNPS can be successfully used for the MRI imaging property. Other notable published reports demonstrated the MRI-contrast imaging (T_1 or T_2) properties of PBNPs in various cancer cells (HepG2, HeLa) as either targeting or nontargeting manner [33, 53]. Another recent study by Dumont et al. showed MRI imaging capabilities of biofunctionalized antibody conjugated (with anti-human eotaxin-3) gadolinium-containing PBNPs (GdPBNPs) in an in vitro inflammatory disease model (eosinophilic esophagitis: EoE) [39]. Importantly, the relaxivities of GdPBNPs are nine times higher compared to clinically used agent Magnevist. Further, these nanoparticles (at a dose of 0.18 mg) can be utilized for the quantitative detection of eosinophils using MRI at 3 T from a mixture of squamous epithelial cells (OE-21) and eosinophils (EoL-1) cells. Hence, all these recent reports demonstrated the excellent MRI imaging properties of PBNPs and PBAs that can be successfully used as imaging and cancer diagnostics agents in near future.

In another recent report, Dumont et al. demonstrated the MRI imaging and targeted fluorescence imaging of manganese-containing Prussian blue nanoparticles (MnPBNPs) in pediatric brain tumor model (PBTs) [44]. Importantly, PBTs are a leading cause of death in children. The authors synthesized core–shell-based nanoparticles where core lattice structure contains paramagnetic Mn^{2+} ions that generate MRI contrast (both positive and negative). The bio-functionalized shell is made of fluorescent avidin that helps fluorescence imaging as well as attachment biotinylated ligands (biotinylated antibodies targeting neuron-glial antigen 2 or biotinylated transferrin) that target PBTs. These nanoparticles and nanobioconjugates were thoroughly characterized by several physicochemical techniques. Moreover, they demonstrated the MRI and fluorescence imaging of MnPBNPs in vitro as well as in an orthotopic mouse model of PBTs using ex vivo fluorescence imaging by measuring their organ and brain bio-distribution.

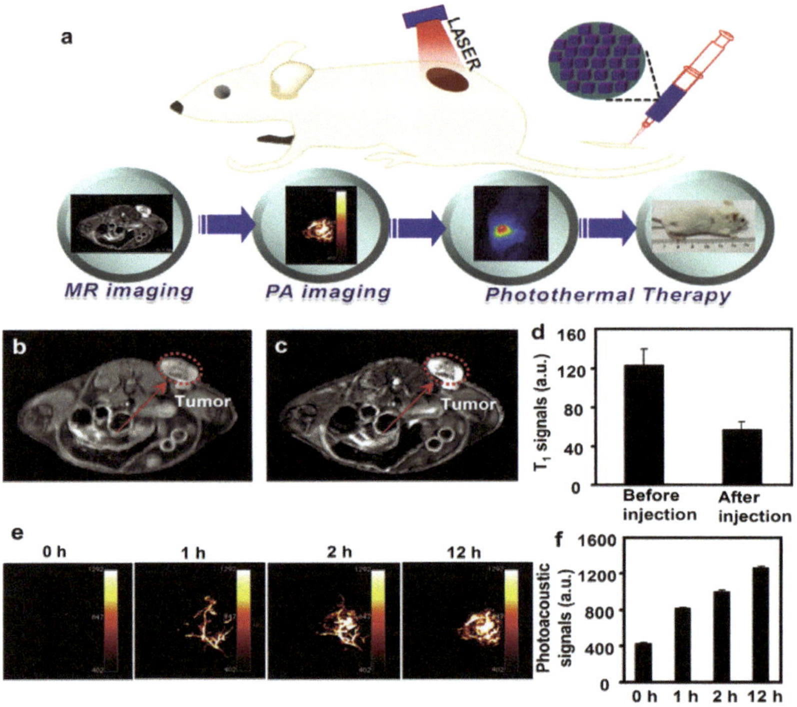

Figure 9.6 In vivo dual-model imaging in 4T1-tumor bearing mice. (a) A scheme of PBNCs used for bimodal imaging and PTT. (b, c) T1 MR images of tumors on mice taken pre-injection (b) and 12 h post i.v. injection of PB-PEG NCs (c). Obvious brightening effect was observed in the tumor region. (d) T1-MR signals in the tumors before and 12 h after i.v. injection with PB-PEG NCs. (e) PAT images of tumors on mice before and after. i.v. injection with PB-PEG NCs taken at different time points. (f) Photoacoustic signals in the tumors from mice before and after i.v. injections of PB-PEG NCs taken at different time points. Reprinted with permission from [57]. Copyright © 2014 Elsevier Ltd.

9.4.2.2 Photoacoustic imaging

Photoacoustic imaging and photoacoustic tomography (PAT) has attracted widespread interests as a noninvasive imaging modality with prominent optical contrast, outstanding spatial resolution and considerably improved imaging depth that can be utilized in brain structural and functional imaging, tumor angiogenesis monitoring and breast cancer imaging [71, 72]. Interestingly, PAT can be utilized for deep tissue imaging (using diffused and ballistic light) and photoacoustic signal-to-noise ratio (SNR) that can be further improved by smart selection of excitation laser wavelength within the near-infrared (NIR) region (in order to cancel out the light scattering effect) [73]. However, the photoacoustic signal using PAT remains weak for deep tissues imaging due to the low intrinsic absorption (in NIR region) of endogenous chromophores. In the past years, several nanomaterials (copper sulfide NPs, gold NPs, carbon nanotubes, etc.) were developed as potential NIR absorber for the photoacoustic imaging [74–76]. However, several problems such as high cost, nonbiodegradability, long-term toxicity and complicity restrict the use of these conventional nanomaterials for imaging. In this context, PBNPs were explored as a new generation contrast agent for NIR light mediated PAT owing to their strong NIR laser absorption properties [57, 77].

Recently, Cheng et al. demonstrated PAT imaging of the breast tumor after a single injection (i.v.) of pegylated PBNCs (10 mg/kg) into tumor bearing mice (Fig. 9.6e) [57]. The tumor vasculatures lightened up using the strong photoacoustic signals over time after administration of PBNCs owing to EPR effects (Fig. 9.6e,f). Moreover, the strong photoacoustic signal was distributed to overall tumor region with time. Thus PAT imaging can provide the detailed intra-tumoral distribution of the nanomaterials which is highly useful for cancer theranostics. In another recent study, Jing et al. demonstrated the in vitro and in vivo PA imaging of core/shell gold@Prussian blue nanoparticles [Au@PBNPs] (size = 17.8 nm) with high molar extinction coefficient and photostability [77]. Au@PBNPs could produce strong photoacoustic signal under the depth of chicken muscles (4.3 cm) upon exposure to NIR laser pulses (at a wavelength of 765 nm) in a concentration dependent manner (80–40 mg mL^{-1}). Additionally, Au@PBNPs (200 mL, 8 mg mL^{-1}) was used for the PA imaging of HT-29 (colon colorectal adenocarcinoma) bearing mice after a single intravenous injection (Fig. 9.7). Interestingly the PA images of tumor showed enhanced signal intensity after 22 h of post injection of Au@PBNPs compared to the intrinsic contrast image that indicates the accumulation of Au@PBNPs due to the EPR effect (Fig. 9.7).

Hence, Au@PB NPs could act as a prospective contrast agent for deep tissues noninvasive PA imaging. Furthermore, it can provide crucial information about tumor structure, location and best time-window for the photothermal therapy based on the accumulation of nanoparticles. Similarly, Liang et al. demonstrated the superior in vitro (from under chicken breast muscles at 60 mg mL^{-1} dose) and in vivo (mouse brain at 20 mg kg^{-1} dose i.v.) PAT imaging of PBNPs upon strong absorption of NIR laser pulses (at wavelength of 765 nm) [78]. It was observed that PBNPs could image the mouse brain vasculature with greater clarity after 5–30 min post injection. More importantly, no apparent acute toxicity was observed to the various organs in mice observed by H&E staining increasing the potential utility of PBNPs as a photoacoustic imaging agent. All these studies indicate the potential photoacoustic imaging properties of PBNPs and PBAs.

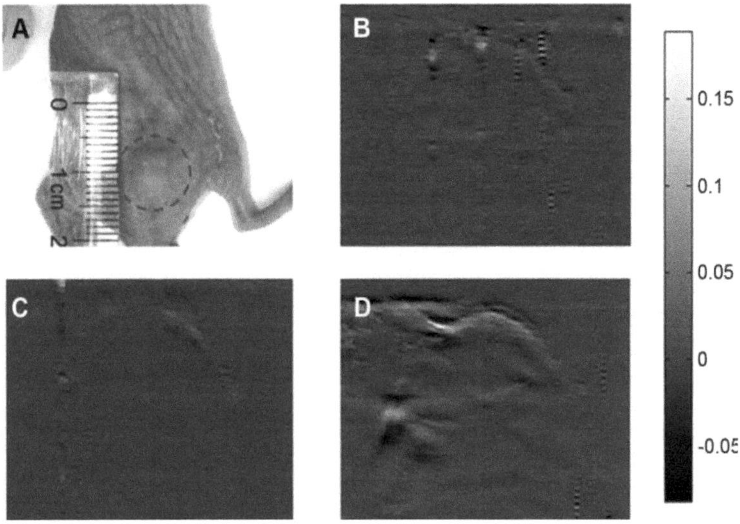

Figure 9.7 (A) Photograph of the nude mice tumor before the data acquisition. PAT images acquired (B) before, (C) 2 h after, and (D) 22 h after the tail intravenous injection of Au@PBNPs. Reprinted with permission from [77]. Copyright © 2014 Elsevier Ltd.

9.4.2.3 CT imaging

X-ray computed tomography (CT) is another very important and advantageous imaging technology due to high resolution and acquisition of 3D visual image for locating tumors [79].

Gold nanostructures were extensively explored as CT imaging agents due to morphology-dependent absorption of NIR light and high absorption coefficient of X-ray. [80] Recently, Jing et al. demonstrated the in vitro and in vivo CT imaging properties of core/shell gold@Prussian blue nanoparticles [Au@PB NPs] (size = 17.8 nm) at different concentrations due to high attenuation of gold [77]. CT values of Au@PB NPs increased linearly with the concentrations of Au and are much higher (at 40 mg mL^{-1}) compared to iodine-based Ultravist along with 10-fold higher X-ray attenuation coefficient (Fig. 9.8A,B).

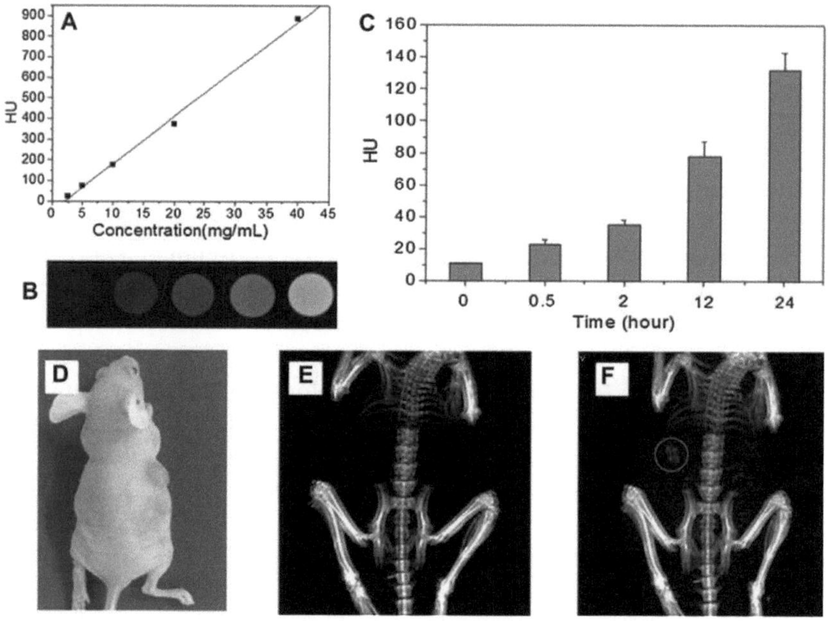

Figure 9.8 (A) X-ray attenuation intensity (HU) as a function of the concentration of Au@PB NPs. (B) CT images of Au@PB NPs at different concentrations. (C) X-ray attenuation intensity (HU) of the tumor region before and after intravenous of Au@PB NPs. (D) Photograph of HT-29 tumor-bearing nude mouse. 3D-reconstructed CT images of HT-29 tumor-bearing nude mouse before (E) and 24 h (F) after intravenous injection of Au@PB NPs. Reprinted with permission from [77]. Copyright © 2014 Elsevier Ltd.

The CT contrast enhancement of the tumor site (HT-29 tumor-bearing nude mouse) upon single i.v. injection of Au@PB

NPs (500 mg kg^{-1}) was quantified by measuring the CT values in a time-dependent manner. CT values of tumor site increased linearly with time and reached highest value at 24 h of post injection (Fig. 9.8C) possibly due to EPR effects. Beside this 3D-reconstructed technique was applied to locate the tumor site. The morphology and location of the colon tumor was clearly visualized at 24 h after administration of Au@PB NPs owing to significant enhancement of CT value (Fig. 9.8D–F). Together these results established Au@PB as an X-ray CT contrast agent for precise location of the tumor. More importantly, PBNPs can be utilized as bimodal imaging (PAT-CT or PAT-MRI) agents for cancer theranostics applications.

9.4.2.4 Fluorescence imaging

Confocal and fluorescence are the tools for the bio-imaging and diagnostics. PBNPS could be functionalized with any fluorescent tagging agents or imaging agents. Dumont et al. demonstrated the fluorescence imaging of eosinophilic cells using bio-functionalized antibody conjugated (with anti-human eotaxin-3) gadolinium-containing PBNPs (GdPBNPs) as evidenced by the green fluorescence [39, 53]. The nanoparticles showed specific targeting of eotaxin-3 receptors on the eosinophilic cells observed by the laser scanning confocal microscopy images. Other published reports also showed the fluorescence imaging of Texas Red cadaverine conjugated PBNPs inside HeLa cells [53]. In order to demonstrate the ability of PBNPs to act as delivery vehicle the authors conjugated the Texas Red cadaverine (Texas Red C5) dye using EDAC coupling reaction. The Texas Red conjugated PBNPs (TRPBNPs) were thoroughly dialyzed to totally remove unattached dye molecules. Finally, the authors showed the presence of red fluorescence of TRPBNPs inside HeLa cells observed by confocal microscopy as evidenced by the presence of red fluorescent signal. Together these results confirmed that functionalized PBNPs could be utilized as a bio-imaging agent using the fluorescence properties.

9.4.2.5 Contrast agent for SPECT scintigraphy

Single-photon emission computed tomography imaging (SPECT) is another powerful imaging and diagnostic tool for radio-imaging popularly used in nuclear medicine for myocardial imaging and cancer diagnostics. In this technique radioisotopes like $^{201}Tl^+$ is internalized and emits radiation which is further captured and imaged by an external detector. However, fast clearance from body, high costs and other complicity restrict the use of these isotopes to some extent. PB and PBNPs have unique chelating properties of heavy metals and radio-active metals. Utilizing this property, Perrier et al. demonstrated the fabrication of ultra-small ^{201}Tl-labeled PBNPs (<5 nm) stabilized by glucose-functionalized aminotriethyleneglycol (β-GluTEG-NH_2) or amino-polyethylene glycol and their application as potential radio-nanoprobes for in vivo SPECT imaging [81]. Importantly, pegylation can increase the blood circulation time of the nanomaterials and prohibit quick glomerular clearing. The uptake kinetics of $^{201}Tl^+$/PB@PEG-NH_2 and $^{201}Tl^+$/PB@β-GluTEG-NH_2 was monitored upon single i.v. injection of these nanomaterials in mice organs (lung, liver and kidney) in a time-dependent manner (from 20 min to 48 h) using SPECT imaging. The results exhibited the time-dependent distribution kinetics of these nanomaterials from transient pass in the pulmonary compartments to lung, liver and finally to glomerular excretion from kidneys. Hence, $^{201}Tl^+$ labeled PBNPs could be utilized for the SPECT/CT imaging radio nanoprobes for the imaging of various cancers and other diseases.

9.4.2.6 Ultrasound (US) imaging

Ultrasound (US) imaging has gained enormous attention and clinical acceptance due to several reasons including real-time imaging, noninvasiveness, high safety, excellent soft-tissue resolution and convenience [82]. However, due to large particle size and poor biological stability, conventional US imaging agents (hollow mesoporous silica nanoparticles, Au nanocages, organic microbubbles, liposomes) present various challenges. More recently, the perfluorocarbon has been often applied clinically due to their high tissue impedance mismatch, US imaging enhancement, FDA-approved safety and uncomplicated excretion [83]. In this context, Jia et al. demonstrated activated US imaging and PTT of n-perfluoropentane (PFP)-encapsulated hollow mesoporous Prussian blue nanocubes (HPB-PFP) for in vivo tumor diagnosis and regression [29]. The HPB shell absorbed NIR light and generated heat that can further raised the temperature of tumor tissue. Consequently, the heat helped in the gasification and bubbling of encapsulated liquid PFP. Moreover, these PFP bubbles caused tissue impedance mismatch, which ultimately exhibited ultrasound imaging (in vitro and in vivo) (Fig. 9.9).

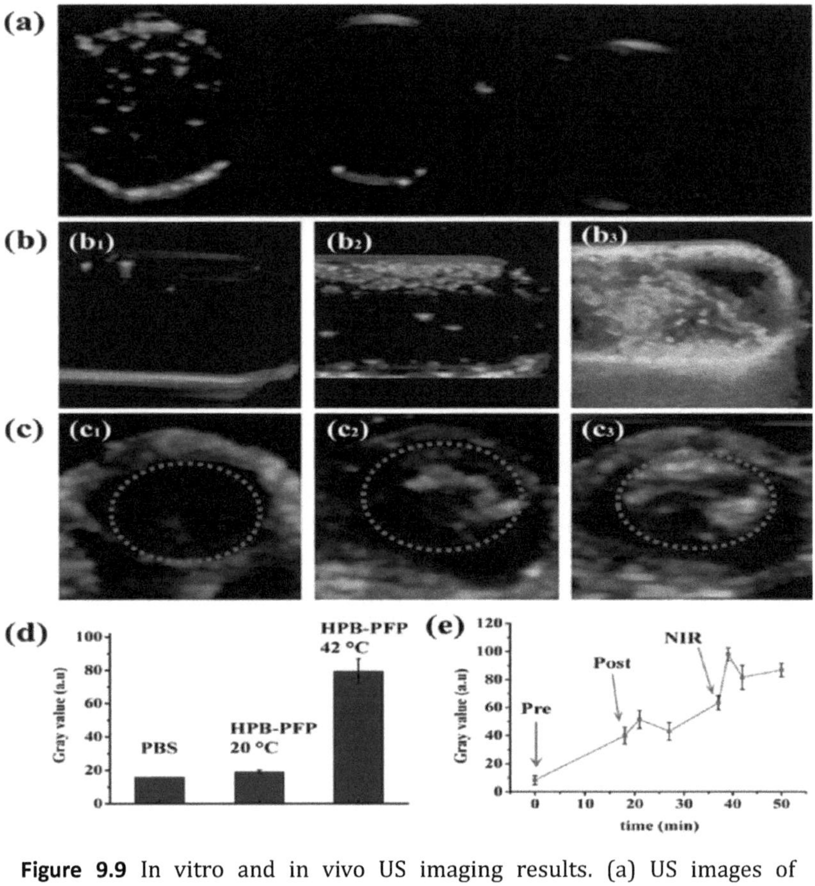

Figure 9.9 In vitro and in vivo US imaging results. (a) US images of HPB-PFP/HPB/PBS at 42°C (left, HPB-PFP (100 ppm); middle, HPB (100 ppm); right, PBS). (b) b1, PBS at 20°C; b2, HPB-PFP solution at 20°C; b3, HPB-PFP solution at 42°C. (c) In vivo US images of tumor site (red dotted circle) (c1) before and (c2) after injection of HPB-PFP solution (30 μL, 200 ppm) and (c3) after being irradiated by 808 nm laser light at 2 W/cm^2 for 5 min. (d) Gray values of in vitro US imaging corresponding to those in panel b. (e) Gray values of in vivo US imaging corresponding to those in panel c, for a duration of 50 min from preinjection. Reprinted with permission from [29]. Copyright © 2015, American Chemical Society.

9.5 Future Opportunities, Challenges and Conclusion

The future challenge is to synthesize and develop novel and nontoxic PBNPs and PBANPs for multifunctional cancer theranostics applications. Hence, extensive safety assessment of these nanoparticles must be evaluated in a preclinical model to satisfy regulations before administration in the human body or clinical trial. Moreover, it is urgently required to perform potential long-term toxicity, biodegradability, immunogenicity, excretion, metabolic fate (in vitro and in vivo), pharmacokinetics and pharmacodynamics studies in mouse model before use of these nanomaterials in human body. Beside these, necessary information including diffusion, uptake, retention, active targeting, clearance of PBNPs and PBANPs should also be carefully studied. Several published reports support the biocompatible nature (in vitro and in vivo) of PBNPs and PBANPs [14, 53, 54]. Due to this, PB, PBNPs and PBANPs are widely used in several applications including energy storage, catalysis, chelating agent, imaging and cancer theranostics applications. The cyanide functional group in the PB remains strongly bound with iron metal and does not show any toxicity on biological system [84]. Most importantly, PB capsule (Radiogardase) has been approved by the FDA as an oral antidote for the removal of radioactive and nonradioactive heavy metal (cesium and thallium) poisoning from human body. Recently, several groups, including ours, demonstrated the nontoxicity of different nanomaterials (EHNs, AuNPs, silica nanoparticles, etc.) in pre-clinical model after performing acute, chronic and genotoxicity studies [85–87]. However, other groups showed toxicity of various nanoparticles (cerium oxide nanoparticles, PtNPs, etc.) that limited the use of these nanoparticles in clinics [88, 89]. Hansen et al. recently summarized several published data (obtained from 400 studies) regarding the toxicity of nearly 965 nanomaterials [90]. Hence, thorough and careful toxicological analysis needs to be evaluated before the use of these nanomaterials in clinics. The nanomedicine approach allows scientists to develop efficient nano-engineered materials that could be the most prospective alternative technologies for the treatment of cancer related diseases. Since PBNPs and PBANPs are nontoxic and already used clinically as an antidote for thallium and cesium contamination. Finally, we strongly believe that these nanomaterials could be useful for potential multifunctional cancer theranostics applications in the near future.

Acknowledgments

This work was supported by 12th Five Year Plan (FYP) project (ADD: CSC0302), CSIR, New Delhi, to C.R.P. C.R.P is also thankful to DST, Nanomission, New Delhi (SR/NM/NS-1252/2013; GAP 0570) for partial financial support for this research. The authors want to thank CSIR-IICT for providing research facility (IICT manuscript number is IICT/pubs./ 2018/259). S.M. and B.R.R. are thankful to CSIR, New Delhi for senior research fellowship.

Abbreviations

AuNPs: Gold nanoparticles
Au@PB NPs: Gold@Prussian blue nanoparticles
β-GluTEG-NH_2: β-glucose-functionalized aminotriethyleneglycol
CD44: Cluster of differentiation 44
CPBNPs: Copper Prussian blue nanoparticles
CNS: Central nerve systems
CSNPs: Core-satellite nanoparticles
CT: Computer tomography
CTAB: Cetyl trimethyl ammonium bromide
DNA: Deoxyribonucleic acid
dODN: decoy oligodeoxy nucleotide
DOX: Doxorubicin
EDAC: 1-Ethyl-3-(3-dimethylaminopropyl)carbodiimide
EDAX: Energy dispersive of X rays
EoE: Eosinophilic esophagitis
EHNs: Europium hydroxide nanorods
EPR: Enhanced permeability and retention
GdPBNPs: Gadolinium-containing Prussian blue nanoparticles
HAS: Human serum albumin
H&E: Hematoxylin and eosin
HPBNCs-PFP: Hollow mesoporous Prussian blue nanocubes
MBs: Microbubbles
MCF-7: Michigan Cancer Foundation-7
MnPBNPs: Manganese containing Prussian blue nanoparticles

MR: Magnetic resonance
MRI: Magnetic resonance imaging
MUA: Mercaptoundecanoic acid
NIR: Near-infrared radiation
NMR: Nuclear magnetic resonance
PAT: Photoacoustic tomography
PB: Prussian blue
PBAs: Prussian blue analogues
PBANPs: Prussian blue analogue nanoparticles
PBNCs: Prussian blue nanocubes
PBNPs: Prussian blue nanoparticles
PBTs: Pediatric brain tumor model
PEG: Poly ethylene glycol
PET: Positron emission tomography
PFP: Perfluoro pentane
PtNPs: Platinum nanoparticles
PTT: Photothermal therapy
PVP: Polyvinyl pyrrolidone
SPECT: Single-photon emission computed tomography
SPION: Super paramagnetic iron oxide nanoparticles
SNR: Signal-to-noise ratio
STAT3: Signal transducer and activator of transcription 3
TEM: Transmission electron microscopy
TRPBNPs: Texas red conjugated PBNPs
US FDA: Food and Drug Administration
US: Ultrasound
XRD: X-Ray diffraction

References

1. Kong, B., Tang, J., Selomulya, C., Li, W., Wei, J., Fang, Y., Wang, Y. C., Zheng, G. F., Zhao, D. Y. (2014). Oriented mesoporous nanopyramids as versatile plasmon-enhanced interfaces, *J. Am. Chem. Soc.*, **136**, pp. 6822–6825.
2. Bartoll, J. (2008). The early use of Prussian blue in paintings. Art 2008 Proceedings of the 9th International Conference on Non-Destructive Testing of Art, Jerusalem, Israel.
3. Marte, F., Careaga, V. P., Mastrangelo, N., de Faria, D. L. A., Maier, M. S. (2014). The Sibyls from the church of San Pedro Telmo: A micro-Raman spectroscopic investigation, *J. Raman Spectrosc.*, **45**, pp. 1046-1051.

4. Ludi, A. (1981). Prussian blue, an inorganic evergreen, *J. Chem. Educ.*, **58**, p. 1013.
5. Monk, P. M. S., Mortimer, R. J., Rosseinsky, D. R. (2007). *Electrochromism and Electrochromic Devices* (Cambridge University Press, Cambridge).
6. Hu, C. W., Kawamoto, T., Tanaka, H., Takahashi, A., Lee, K. M., Kao, S. Y., Liao, Y. C., Ho, K. C. (2016). Water processable Prussian blue-polyaniline: Polystyrene sulfonate nanocomposite (PB-PANI: PSS) for multi-color electrochromic applications, *J. Mater. Chem. C.*, **4**, pp. 10293–10300.
7. Giese, W. W. (1988). Ammonium-ferric-cyano-ferrate(II) (Afcf) as an effective antidote against radiocesium burdens in domestic-animals and animal derived foods, *Brit. Vet. J.*, **144**, pp. 363–369.
8. Pai, V. (1987). Acute thallium poisoning. Prussian blue therapy in 9 cases. *West Indian Med. J.*, **36**, pp. 256–258.
9. Giese, W. W.(1989). Countermeasures for reducing the transfer of radiocesium to animal derived foods, *Sci. Total Environ.*, **85**, pp. 317–327.
10. Risica, S., Campos Venuti, G., Rogani, A., Baronciani, D., Petrone, M. (1992). Caesium contamination in human milk and transfer factor from diet, *Analyst*, **117**, 511–514.
11. FDA. Prussian Blue (ferric hexacyanoferrate (II)) for Treatment of Internal Contamination with Thallium or Radioactive Cesium. http://wwwfdagov/Drugs/EmergencyPreparedness/BioterrorismandDrugPreparedness/ucm130334htm. 2015.
12. FDA. Questions and Answers on Prussian Blue. http://wwwfdagov/Drugs/EmergencyPreparedness/BioterrorismandDrugPreparedness/ucm130337htm-2. 2011.
13. Fu, G., Liu, W., Feng, S., Yue, X. (2012). Prussian blue nanoparticles operate as a new generation of photothermal ablation agents for cancer therapy, *Chem. Commun (Camb)*, **48**, pp. 11567–11569.
14. Mukherjee, S., Rao, B. R., Sreedhar, B., Paik, P., Patra, C. R. (2015). Copper Prussian blue analogue: Investigation into multifunctional activities for biomedical applications, *Chem. Commun.*, **51**, pp. 7325–7328.
15. Zhang, W., Hu, S. L., Yin, J. J., He, W. W., Lu, W., Ma, M., Gu, N., Zhang, Y. (2016). Prussian blue nanoparticles as multienzyme mimetics and reactive oxygen species scavengers, *J. Am. Chem. Soc.*, **138**, pp. 5860–5865.
16. Perera, V. S., Yang, L. D., Hao, J. H., Chen, G. J., Erokwu, B. O., Flask, C. A., Zavalij, P. Y. Basilion, J. P., Huang, S. P. D. (2014). Biocompatible

Nanoparticles of $KGd(H_2O)_2[Fe(CN)_6]$center dot H_2O with extremely high T-1-weighted relaxivity owing to two water molecules directly bound to the Gd(III) center, *Langmuir*, **30**, pp. 12018–12026.

17. Lee, C., Hwang, H. S., Lee, S., Kim, B., Kim, J. O., Oh, K. T., Lee, E. S., Choi, H. G., Youn, Y. S. (2017). Rabies virus-inspired silica-coated gold nanorods as a photothermal therapeutic platform for treating brain tumors, *Adv. Mater.*, DOI: 10.1002/adma.201605563.

18. Yang, T., Tang, Y., Liu, L., Lv, X., Wang, Q., Ke, H., Deng, Y., Yang, H., Yang, X., Liu, G., Zhao, Y., Chen, H. (2017). Size-dependent Ag_2S nanodots for second near-infrared fluorescence/photoacoustics imaging and simultaneous photothermal therapy, *ACS Nano*, **11**, pp. 1848–1857.

19. Wang, W., Tang, Q., Yu, T., Li, X., Gao, Y., Li, J., Liu, Y., Rong, L., Wang, Z., Sun, H., Zhang, H., Yang, B. (2017). Surfactant-free preparation of Au@resveratrol hollow nanoparticles with photothermal performance and antioxidant activity, *ACS Appl. Mater. Interfaces*, **9**, pp. 3376–3387.

20. Peng, H., Tang, J., Zheng, R., Guo, G., Dong, A., Wang, Y., Yang, W. (2017). Nuclear-targeted multifunctional magnetic nanoparticles for photothermal therapy, *Adv. Healthc. Mater.*, doi: 10.1002/adhm.201601289.

21. Keggin, J. F., Miles, F. D. (1936). Structures and formulæ of the Prussian blues and related compounds, *Nature*, **137**, pp. 577–578.

22. Ware, M. (2008). Prussian blue: Artists' pigment and chemists' sponge, *J. Chem. Educ.*, **85**, pp. 612–621.

23. Itoi, M., Jike, T., Nishio-Hamane, D., Udagawa, S., Tsuda, T., Kuwabata, S., Boukheddaden, K., Andrus, M. J., Talham, D. R. (2015). Direct observation of short-range structural coherence during a charge transfer-induced spin transition in a CoFe Prussian blue analogue by transmission electron microscopy, *J. Am. Chem. Soc.*, **137**, pp. 14686–14693.

24. Zhang, L., Wu, H. B., Madhavi, S., Hng, H. H., Lou, X. W. (2012). Formation of Fe_2O_3 microboxes with hierarchical shell structures from metal-organic frameworks and their lithium storage properties, *J. Am. Chem. Soc.*, **134**, pp. 17388–17391.

25. Murray, L. J., Dinca, M., Long, J. R. (2009). Hydrogen storage in metal-organic frameworks, *Chem. Soc. Rev.*, **38**, pp. 1294–1314.

26. Durham, B., Anderson, T. J., Switzer, J. A., Endicott, J. F., Glick, M. D. (1977). Cobalt(II)-mediated oxygenation of a macrocyclic ligand.

X-ray structures of the cobalt(II) and cobalt(III) products, *Inorg. Chem.*, **16**, pp. 271–278.

27. Farah, A. M., Billing, C., Dikio, C. W., Dibofori-Orji, A. N., Oyedeji, O. O., Wankasi, D., Mtunzi, F. M., Dikio, E. D. (2013). Synthesis of Prussian blue and its electrochemical detection of hydrogenperoxide based on cetyltrimethylammonium bromide (CTAB) modified glassy Carbon Electrode, *Int. J. Electrochem. Soc.*, **8**, pp. 12132–12146.

28. Giner-Casares, J. J., Henriksen-Lacey, M., Coronado-Puchau, M., Liz-Marzan, L. M. (2016). Inorganic nanoparticles for biomedicine: Where materials scientists meet medical research, *Mater. Today*, **19**, pp. 19–28.

29. Jia, X., Cai, X., Chen, Y., Wang, S., Xu, H., Zhang, K., Ma, M., Wu, H., Shi, J., Chen, H. (2015). Perfluoropentane-encapsulated hollow mesoporous Prussian blue nanocubes for activated ultrasound imaging and photothermal therapy of cancer, *ACS Appl. Mater. Interfaces,* **7**, pp. 4579–4588.

30. Sahu, A., Lee, J. H., Lee, H. G., Jeong, Y. Y., Tae, G. (2016). Prussian blue/serum albumin/indocyanine green as a multifunctional nanotheranostic agent for bimodal imaging guided laser mediated combinatorial phototherapy, *J. Control. Release,* **236**, pp. 90–99.

31. Feng, S. S., Li, X. D., Ma, F., Liu, R. F., Fu, G. L., Xing, S., Yue, X. L. (2016). Prussian blue functionalized microcapsules for effective removal of cesium in a water environment, *RSC Adv.*, **6**, pp. 34399–34410.

32. Song, X. X., Liu, Z. J., Tang, Q. (2013). Folate grafted Prussian blue entrapped with Gadolinium(III) as a new contrast agent for tumor-targeted magnetic resonant imaging, *J. Nanosci. Nanotechnol.*, **13**, pp. 5233–5239.

33. Li, Z. L., Zeng, Y. Y., Zhang, D., Wu, M., Wu, L. J., Huang, A. M., Yang, H. H., Liu, X. L., Liu, J. F. (2014). Glypican-3 antibody functionalized Prussian blue nanoparticles for targeted MR imaging and photothermal therapy of hepatocellular carcinoma, *J. Mater. Chem. B*, 2, pp. 3686–3696.

34. Paul, G., Prado, Y., Dia, N., Riviere, E., Laurent, S., Roch, M., Vander Elst, L., Muller, R. N., Sancey, L., Perriat, P., Tillement, O., Mallah, T., Catala, L. (2014). Mn-II-containing coordination nanoparticles as highly efficient T-1 contrast agents for magnetic resonance imaging, *Chem. Commun.,* **50**, pp. 6740–6743.

35. Chen, F. Y., Gu, Z. J., Yi, J. W., Xu, X. Z., Kulkarni, P., Tang, Q. (2016). The evaluation of lanthanum trapped Prussian blue as a phosphate

binding agent with reduced bone uptake, *New J. Chem.*, **40**, pp. 2644–2648.

36. Dou, Y., Li, X., Yang, W., Guo, Y., Wu, M., Liu, Y., Li, X., Zhang, X., Chang, J. (2017). PB@Au core-satellite multifunctional nanotheranostics for magnetic resonance and computed tomography imaging in vivo and synergetic photothermal and radiosensitive therapy, *ACS Appl. Mater. Interfaces*, **9**, pp. 1263–1272.
37. Cai, X. J., Gao, W., Ma, M., Wu, M. Y., Zhang, L. L., Zheng, Y. Y., Chen, H. R., Shi, J. L. (2015). A Prussian blue-based core-shell hollow-structured mesoporous nanoparticle as a smart theranostic agent with ultrahigh pH-responsive longitudinal relaxivity, *Adv. Mater.*, **27**, pp. 6382–6389.
38. Wang, L., Zhang, Q., Chen, S., Xu, F., Chen, S., Jia, J., Tan, H., Hou, H., Song, Y. (2014). Electrochemical sensing and biosensing platform based on biomass-derived macroporous carbon materials, *Anal. Chem.*, **86**, pp. 1414–1421.
39. Dumont, M. F., Hoffman, H. A., Yoon, P. R., Conklin, L. S., Saha, S. R., Paglione, J., Sze, R. W., Fernandes, R. (2014). Biofunctionalized gadolinium-containing Prussian blue nanoparticles as multimodal molecular imaging agents, *Bioconjug. Chem.*, **25**, pp. 129–137.
40. Jang, J. T., Nah, H., Lee, J. H., Moon, S. H., Kim, M. G., Cheon, J. (2009). Critical enhancements of MRI contrast and hyperthermic effects by dopant-controlled magnetic nanoparticles, *Angew. Chem. Int. Edit.*, **48**, pp. 1234–1238.
41. Perera, V. S., Chen, G. J., Cai, Q., Huang, S. P. D. (2016). Nanoparticles of gadolinium-incorporated Prussian blue with PEG coating as an effective oral MRI contrast agent for gastrointestinal tract imaging, *Analyst*, **141**, pp. 2016–2022.
42. Kandanapitiye, M. S., Gott, M. D., Sharits, A., Jurisson, S. S., Woodward, P. M., Huang, S. P. D. (2016). Incorporation of gallium-68 into the crystal structure of Prussian blue to form (KGa_xFe_{1-x})-Ga-68$[Fe(CN)_6]$ nanoparticles: Toward a novel bimodal PET/MRI imaging agent, *Dalton T.*, **45**, pp. 9174–9181.
43. Perera, V. S., Hao, J. H., Gao, M., Gough, M., Zayalij, P. Y., Flask, C., Basilion, J. P., Huang, S. D. (2011). Nanoparticles of the novel coordination polymer $KBi(H_2O)_2[Fe(CN)_6]$center dot H_2O as a potential contrast agent for computed tomography, *Inorg. Chem.*, **50**, pp. 7910–7912.
44. Dumont, M. F., Yadavilli, S., Sze, R. W., Nazarian, J., Fernandes, R. (2014). Manganese-containing Prussian blue nanoparticles for imaging of pediatric brain tumors, *Int. J. Nanomed.*, **9**, pp. 2581–2595.

45. Razzaque, S., Hussain, S. Z., Hussain, I., Tan, B. (2016). Design and utility of metal/metal oxide nanoparticles mediated by thioether end-functionalized polymeric ligands, *Polymers*, **8**, p. 156.

46. Singh, D., McMillan, J. M., Kabanov, A. V., Sokolsky-Papkov, M., Gendelman, H. E. (2014). Bench-to-bedside translation of magnetic nanoparticles, *Nanomed. UK*, **9**, pp. 501–516.

47. Jang, S. C., Haldorai, Y., Lee, G. W., Hwang, S. K., Han, Y. K., Roh, C., Huh, Y. S. (2015). Porous three-dimensional graphene foam/Prussian blue composite for efficient removal of radioactive (137) Cs, *Sci. Rep.*, **5**, article number: 17510.

48. Vipin, A. K., Fugetsu, B., Sakata, I., Isogai, A., Endo, M., Li, M. D., Dresselhaus, M. S. (2016). Cellulose nanofiber backboned Prussian blue nanoparticles as powerful adsorbents for the selective elimination of radioactive cesium, *Sci. Rep.*, **6**, article number: 37009.

49. Siegel, R. L., Miller, K. D., Jemal, A. (2017). Cancer statistics, *CA Cancer J. Clin.*, **67**, pp. 7–30.

50. Urruticoechea, A., Alemany, R., Balart, J., Villanueva, A., Vinals, F., Capella, G. (2010). Recent advances in cancer therapy: An overview, *Curr. Pharm. Des.*, **16**, pp. 3–10.

51. Patra, C. R., Bhattacharya, R., Wang, E. F., Katarya, A., Lau, J. S., Dutta, S., Muders, M., Wang, S. F., Buhrow, S. A., Safgren, S. L., Yaszemski, M. J., Reid, J. M., Ames, M. M., Mukherjee, P., Mukbopadhyay, D. (2008). Targeted delivery of gemcitabine to pancreatic adenocarcinoma using cetuximab as a targeting agent, *Cancer Res.*, **68**, pp. 1970–1978.

52. Siddiqui, I. A., Adhami, V. M., Chamcheu, J. C., Mukhtar, H. (2012). Impact of nanotechnology in cancer: Emphasis on nanochemoprevention, *Int. J. Nanomed.*, **7**, pp. 591–605.

53. Shokouhimehr, M., Soehnlen, E. S., Hao, J. H., Griswold, M., Flask, C., Fan, X. D., Basilion, J. P., Basu, S., Huang, S. P. D. (2010). Dual purpose Prussian blue nanoparticles for cellular imaging and drug delivery: A new generation of T-1-weighted MRI contrast and small molecule delivery agents, *J. Mater. Chem.*, **20**, pp. 5251–5259.

54. Patra, C. R. (2016). Prussian blue nanoparticles and their analogues for application to cancer theranostics, *Nanomedicine (Lond.)*, **11**, pp. 569–572.

55. Li, Z. L., Hu, Y., Jiang, T. T., Howard, K. A., Li, Y. G., Fan, X. L., Sun, Y., Besenbacher, F., Yu, M. (2016). Human-serum-albumin-coated Prussian blue nanoparticles as pH-/thermotriggered drug-delivery vehicles for cancer thermochemotherapy, *Part Part Syst. Char.*, **33**, pp. 53–62.

56. Li, X. D., Liang, X. L., Ma, F., Jing, L. J., Lin, L., Yang, Y. B., Feng, S. S., Fu, G. L., Yue, X. L., Dai, Z. F. (2014). Chitosan stabilized Prussian blue nanoparticles for photothermally enhanced gene delivery, *Colloids Surf. B*, **123**, pp. 629–638.

57. Cheng, L., Gong, H., Zhu, W. W., Liu, J. J., Wang, X. Y., Liu, G., Liu, Z. (2014). PEGylated Prussian blue nanocubes as a theranostic agent for simultaneous cancer imaging and photothermal therapy, *Biomaterials*, **35**, pp. 9844–9852.

58. Ahmad, M. Z., Akhter, S., Jain, G. K., Rahman, M., Pathan, S. A., Ahmad, F. J., Khar, R. K.(2010). Metallic nanoparticles: Technology overview & drug delivery applications in oncology, *Expert Opin. Drug Deliv.*, **7**, pp. 927–942.

59. Cho, K., Wang, X., Nie, S., Chen, Z. G., Shin, D. M. (2008). Therapeutic nanoparticles for drug delivery in cancer, *Clin. Cancer Res.*, **14**, pp. 1310–1316.

60. Bae, Y. H., Park, K. (2011). Targeted drug delivery to tumors: Myths, reality and possibility, *J. Control. Release*, **153**, pp. 198–205.

61. Danhier, F., Feron, O., Preat, V. (2010). To exploit the tumor microenvironment: Passive and active tumor targeting of nanocarriers for anti-cancer drug delivery, *J. Control. Release*, **148**, pp. 135–146.

62. Lian, H. Y., Hu, M, Liu, C. H., Yamauchi, Y., Wu, K. C. W. (2012). Highly biocompatible, hollow coordination polymer nanoparticles as cisplatin carriers for efficient intracellular drug delivery, *Chem. Commun.*, **48**, pp. 5151–5153.

63. Wang, S. J., Chen, C. S., Chen, L. C. (2013). Prussian blue nanoparticles as nanocargoes for delivering DNA drugs to cancer cells, *Sci. Technol. Adv. Mat.*, **14**.

64. Kumar, S. R., Markusic, D. M., Biswas, M., High, K. A., Herzog, R. W. (2016). Clinical development of gene therapy: Results and lessons from recent successes, *Mol. Ther. Methods Clin. dev.* **3**, 16034.

65. Nayerossadat, N., Maedeh, T., Ali, P. A. (2012). Viral and nonviral delivery systems for gene delivery, *Adv. Biomed. Res.*, **1**, doi: 10.4103/2277-9175.98152.

66. Li, X. D., Yue, X. L., Wang, J. R., Liang, X. L., Jing, L. J., Lin, L., Yang, Y. B., Feng, S. S., Qian, Y. J., Dai, Z. F. (2016). Prussian blue nanoparticle-loaded microbubbles for photothermally enhanced gene delivery through ultrasound-targeted microbubble destruction, *Sci. Bull.*, **61**, pp. 148–156.

67. Fu, G., Liu, W., Li, Y., Jin, Y., Jiang, L., Liang, X., Feng, S., Dai, Z. (2014). Magnetic Prussian blue nanoparticles for targeted photothermal

therapy under magnetic resonance imaging guidance, *Bioconj. Chem.*, **25**, pp. 1655–1663.

68. Jing, L. J., Shao, S. M., Wang, Y., Yang, Y. B., Yue, X. L., Dai, Z. F. (2015). Hyaluronic acid modified hollow Prussian blue nanoparticles loading 10-hydroxycamptothecin for targeting thermochemotherapy of cancer, *Theranostics*, **6**, pp. 40–53.

69. Kim, J., Piao, Y., Hyeon, T. (2009). Multifunctional nanostructured materials for multimodal imaging, and simultaneous imaging and therapy, *Chem. Soc. Rev.*, **38**, pp. 372–390.

70. Clemente-Leon, M., Coronado, E., Soriano-Portillo, A., Colacio, E., Dominguez-Vera, J. M., Galvez, N., Madueno, R., Martin-Romero, M. T. (2006). Magnetic Langmuir-Blodgett films of ferritin with different iron contents, *Langmuir*, **22**, pp. 6993–7000.

71. Wang, L. V., Hu, S. (2012). Photoacoustic tomography: In vivo imaging from organelles to organs, *Science*, **335**, pp. 1458–1462.

72. Pu, K. Y., Shuhendler, A. J., Jokerst, J. V., Mei, J. G., Gambhir, S. S., Bao, Z. N., Rao, J. H. (2014). Semiconducting polymer nanoparticles as photoacoustic molecular imaging probes in living mice, *Nat. Nanotechnol.*, **9**, pp. 233–239.

73. Ku, G., Wang, L. V. (2005). Deeply penetrating photoacoustic tomography in biological tissues enhanced with an optical contrast agent, *Opt. Lett.*, **30**, pp. 507–509.

74. Wang, Y. W., Xie, X. Y., Wang, X. D., Ku, G., Gill, K. L., O'Neal, D. P., Stoica, G., Wang, L. V. (2004). Photoacoustic tomography of a nanoshell contrast agent in the in vivo rat brain, *Nano Lett.*, **4**, pp. 1689–1692.

75. De La Zerda, A., Zavaleta, C., Keren, S., Vaithilingam, S., Bodapati, S., Liu, Z., Levi, J., Smith, B. R., Ma, T. J., Oralkan, O., Cheng, Z., Chen, X. Y., Dai, H. J., Khuri-Yakub, B. T., Gambhir, S. S. (2008). Carbon nanotubes as photoacoustic molecular imaging agents in living mice, *Nat. Nanotechnol.*, **3**, pp. 557–562.

76. Ku, G., Zhou, M., Song, S. L., Huang, Q., Hazle, J., Li, C. (2012). Copper Sulfide nanoparticles as a new class of photoacoustic contrast agent for deep tissue imaging at 1064 nm, *ACS Nano*, **6**, pp. 7489–7496.

77. Jing, L., Liang, X., Deng, Z., Feng, S., Li, X., Huang, M., Li, C., Dai, Z. (2014). Prussian blue coated gold nanoparticles for simultaneous photoacoustic/CT bimodal imaging and photothermal ablation of cancer, *Biomaterials*, **35**, pp. 5814–5821.

78. Liang, X. L., Deng, Z. J., Jing, L. J., Li, X. D., Dai, Z. F., Li, C. H., Huang, M. M. (2013). Prussian blue nanoparticles operate as a contrast

agent for enhanced photoacoustic imaging, *Chem. Commun.,* **49**, pp. 11029–11031.

79. Lee, N., Cho, H. R., Oh, M. H., Lee, S. H., Kim, K., Kim, B. H., Shin, K., Ahn, T. Y., Choi, J. W., Kim, Y. W., Choi, S. H., Hyeon, T. (2012). Multifunctional Fe_3O_4/TaOx core/shell nanoparticles for simultaneous magnetic resonance imaging and X-ray computed tomography, *J. Am. Chem. Soc.,* **134**, pp. 10309–10312.

80. Kim, D., Park, S., Lee, J. H., Jeong, Y. Y., Jon, S. (2007). Antibiofouling polymer-coated gold nanoparticles as a contrast agent for in vivo x-ray computed tomography imaging, *J. Am. Chem. Soc.,* **129**, pp. 7661–7665.

81. Perrier, M., Busson, M., Massasso, G., Long, J., Boudousq, V., Pouget, J. P., Peyrottes, S., Perigaud, C., Porredon-Guarch, C., de Lapuente, J., Borras, M., Larionova, J., Guari, Y. (2014). Tl-201(+)-labelled Prussian blue nanoparticles as contrast agents for SPECT scintigraphy, *Nanoscale,* **6**, pp. 13425–13429.

82. Tranquart, F., Grenier, N., Eder, V., Pourcelot, L. (1999). Clinical use of ultrasound tissue harmonic imaging, *Ultrasound Med. Biol.,* **25**, pp. 889–894.

83. Liberman, A., Wu, Z., Barback, C. V., Viveros, R., Blair, S. L., Ellies, L. G., Vera, D. R., Mattrey, R. F., Kummel, A. C., Trogler, W. C. (2013). Color Doppler ultrasound and gamma imaging of intratumorally injected 500 nm iron-silica nanoshells, *ACS Nano,* **7**, pp. 6367–6377.

84. Yang, Y., Brownell, C., Sadrieh, N., May, J., Del Grosso, A., Place, D., Leutzinger, E., Duffy, E., He, R., Houn, F., Lyon, R., Faustino, P. (2007). Quantitative measurement of cyanide released from Prussian Blue, *Clin. Toxicol.,* **45**, pp. 776–781.

85. Bollu, V. S., Nethi, S. K., Dasari, R. K., Rao, S. S., Misra, S., Patra, C. R. (2016). Evaluation of in vivo cytogenetic toxicity of europium hydroxide nanorods (EHNs) in male and female Swiss albino mice, *Nanotoxicology,* **10**, pp. 413–425.

86. Liu, T., Li, L., Teng, X., Huang, X., Liu, H., Chen, D., Ren, J., He, J., Tang, F. (2011). Single and repeated dose toxicity of mesoporous hollow silica nanoparticles in intravenously exposed mice, *Biomaterials,* **32**, pp. 1657–1668.

87. Patra, C. R., Abdel Moneim, S. S., Wang, E., Dutta, S., Patra, S., Eshed, M., Mukherjee, P., Gedanken, A., Shah, V. H., Mukhopadhyay, D. (2009). In vivo toxicity studies of europium hydroxide nanorods in mice. *Toxicol. Appl. Pharmacol.,* **240**, pp. 88–98.

88. Aalapati, S., Ganapathy, S., Manapuram, S., Anumolu, G., Prakya, B. M. (2014). Toxicity and bio-accumulation of inhaled cerium oxide nanoparticles in CD1 mice, *Nanotoxicology*, **8**, pp. 786–798.

89. Elder, A., Yang, H., Gwiazda, R., Teng, X., Thurston, S., He, H., Oberdorster, G. (2007). Testing nanomaterials of unknown toxicity: An example based on platinum nanoparticles of different shapes, *Adv. Mater.*, **19**, pp. 3124–3129.

90. Hansen, S. F., Larsen, B. H., Olsen, S. I., Baun, A. (2007). Categorization framework to aid hazard identification of nanomaterials, *Nanotoxicology*, **1**, pp. 1–8.

Chapter 10

Prussian Blue Type Nanoparticles for Biomedical Applications

Joulia Larionova, Yannick Guari, and Jérôme Long

Institut Charles Gerhardt Montpellier (ICGM), University of Montpellier, Montpellier, France

Joulia.larionova@umontpellier.fr

10.1 Introduction

Prussian blue (PB) type nanoparticles belong to a wide class of inorganic nanoparticles, such as metal, metal oxide or quantum dots nano-objects, which present unique size-and shape-dependent physical and chemical properties arising from the high surface-to-volume ratio, quantum size effects and electrodynamic interactions. On the other hand, they benefit from all the advantages of molecule-based materials, consisting of well-determined and flexible molecule-based structures, adjustable composition, tunable physical and chemical properties, low density, porosity and relatively soft chemical ways to their synthesis

Prussian Blue-Type Nanoparticles and Nanocomposites: Synthesis, Devices, and Applications
Edited by Yannick Guari and Joulia Larionova

ISBN 978-981-4800-05-1 (Hardcover), 978-0-429-02474-0 (eBook)
www.panstanford.com

through the self-assembly reactions of molecular precursors at room temperature (Fig. 10.1). These nano-objects are made up of transition metal ions or lanthanides connected through bridging-cyano ligands to give three-dimensional coordination networks, the propagation of which is usually stopped by various stabilizing agents giving nano-objects of various size and shape [1–4]. The authentic PB nanoparticles present the general formula $A_xFe_y^{III}[Fe^{II}(CN)_6]_z$ (where A is a monovalent cation) and numerous Prussian blue Analogous (PBA) can be written according to the formula $A_xM_y[M'(CN)_6]_z$ (where M and M′ are transition metal ions or lanthanides). In their bulk form, these compounds present the NaCl-type face centered cubic (*fcc*) structure when transition metal ions are used, while more sophisticated structural organizations are formed with lanthanides.

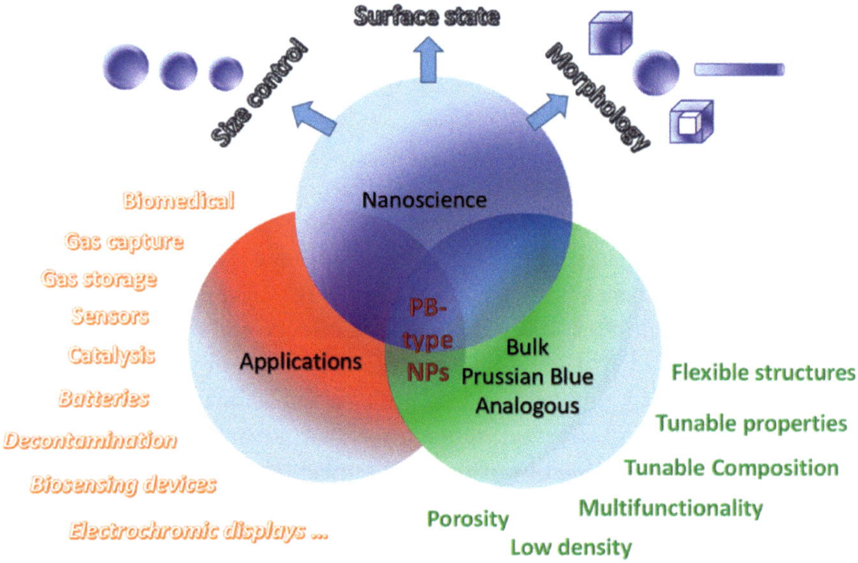

Figure 10.1 Schematic representation of PB type nanoparticles, which benefits from the advantages of the cross-section between the bulk molecule-based materials (PB and PBA) and the nanoscience giving rise to the appearance of nano-objects with unique physical and chemical properties useful for various applications.

The first PB or PBA nano-objects have been reported by S. Mann and colleagues and S. P. Moulik and colleagues in 1999 and

2000, respectively, by using reverse micellar systems [5, 6]. Since then, the synthesis and investigation of these nano-sized systems have attracted a great deal of attention of numerous research teams worldwide resulting in a multitude of PB type nano-objects with different chemical compositions. They are available in a wide size range from a few to several tens of nanometers, with different morphologies and controlled surface state. Two main approaches have been used to this purpose: (i) the self-assembly reaction between appropriated precursors in the presence of stabilizing agents, and (ii) the controlled kinetic growth of the cyano-bridged network without stabilizers and the post-synthetic anchoring of the latter on the nanoparticles surface. The first approach usually provides small nanoparticles with sizes less than 10 nm. The size control may be achieved by using various stabilizing agents, such as polyethylene glycols with different molecular weights and functionalities, carbohydrates, polymers and others, as well as by the modulation of the precursor-to-stabilizer molar ratio. The second approach gives nano-objects having sizes ranging from several tens to several hundreds of nanometers. In this case, the size is usually controlled at the growing step by the precursor's addition rate, concentration, and the presence of ionic salts.

As other inorganic nano-sized systems, they exhibit unique, size and morphology dependent magnetic, photo-magnetic, optic, host–guest, sorption properties making them very interesting for potential technological applications, such as gas capture, gas storage, catalysis, sensors, electrochromic devices, decontamination and others (Fig. 10.1). Among them, the PB nanoparticles were proposed for applications in biology and medicine. Due to the fact that bulk PB presents very high stability due to an extremely low dissociation constant ($K_d = 3 \times 10^{-41}$) in aqueous media and at low pH, it has been approved by the Food and Drug Administration (FDA) for human use as an antidote for a radioactive or non-radioactive Cs^+ and Tl^+ poisoning. As an example, the daily dose for an adult in the case of a radioactive Cs^+ contamination constitutes *ca.* 10 g, illustrating the low toxicity of PB. This exceptional stability, low toxicity, various useful properties, as well as their tunability permitting the adjustment of the physical properties trough the modulation of the nanoparticles composition explain the growing interest for these nano-objects in relation with biomedical applications [7]. During the past decade, they have been proposed as contrast agents for several types of imaging, as nanocarriers for drug delivery, as therapeutic agents for the cancer treatment, as multifunctional nanoprobes for different types of imaging, as agents for theranostic (combination of diagnosis with therapy) or as efficient antidotes for Cs^+ or Tl^+ treatments (Fig. 10.2). An important effort has been focused on cell uptake investigations of these nanoparticles and their behavior in living organisms. Particular attention has also been devoted to the synthesis of core–shell or hybrid nanoparticles for which one of the components is PB in combination with gold, iron oxide, or quantum dots nanoparticles in order to design highly challenging multifunctional systems. In the next sections, we will focus on the recent advances of these nano-objects for imaging, including Magnetic Resonance Imaging (MRI), optic, photoacoustic imaging and scintigraphy as well as photothermal therapy (PTT) and drug delivery.

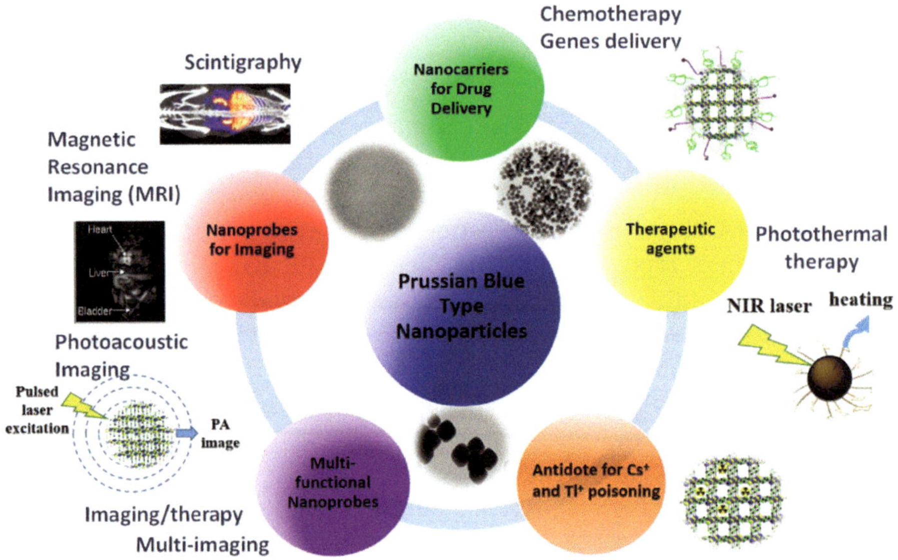

Figure 10.2 Schematic representation of various biomedical applications proposed for PB type nanosystems.

10.2 Prussian Blue Type Nanoparticles as Contrast Agents for Imaging

Among investigations on the new contrast agents for different imaging methodologies where PB-type nanoprobes can be used due to their exciting magnetic, optical, or host–guest properties, probably the first and the most studied ones, belong to the various attempts to design efficient nanoprobes for MRI.

10.2.1 MRI

MRI is a powerful noninvasive imaging technique of modern medicine used to visualize internal body structures [8]. The basic physical principle of nuclear magnetic resonance (NMR) imaging consists of the application of magnetic field gradients to encode the signal in all three spatial directions. The NMR signal can be weighted by the water density and the longitudinal (T_1) and/or transversal (T_2) relaxation times of the tissues [9]. Contrast agents (CAs) are currently used in order to improve an image quality by locally increasing the nuclear relaxation rates [10]. Their efficiency are usually evaluated on the basis of their ability to influence the longitudinal, r_1, and transverse, r_2, relaxivities, and then the r_{2p}/r_{1p} ratio that determine the action of the CA on the image. For $r_{2p}/r_{1p} < 2$, the CA is positive and called T_1-relaxing and provides a brilliant spot on the MR image in the organ or tissues where it is delivered. In this line of thought, paramagnetic Gd^{3+} ($S = 7/2$)-based macrocycles are the currently used commercial positive CAs in clinical practice. For instance, commercial positive CAs based on gadolinium chelate complexes called ProHance® (or Gadoteridol) presents r_1 values of about 3.5 $mM^{-1}s^{-1}$ at 1.5 T and at 298 K [11–13]. The relaxivity mechanism for this family of CAs is relatively well described by the Solomon–Bloembergen–Morgan (SBM) theory. For r_{2p}/r_{1p} values higher than 2, the CA is called negative T_2-relaxing and it provides a dark signal on the image. Such case is usually obtained when using superparamagnetic iron-based nanoparticles, such as Endorem®. In the recent years, extensive fundamental and industrial research focused on the development of new efficient CAs with an increased effect on the protons' relaxivity allowing administration at a lower dose or imaging of low-concentration targets. Since the positive CAs are the mostly used in clinical practice, new families of T_1—relaxing CAs with improved relaxivity, lower administration dose and local contrast effects—were developed by the covalent anchoring of Gd^{3+} complexes to various nanostructures, polymers [14, 15], or by design of various nanoparticles including gadolinium oxide nanoparticles [16–20] and metal organic framework nano-objects [21–29]. Regarding PB type nanoparticles, first reports on the relaxivity measurements have been made for citrate coated PB nanoparticles of 13 nm, and have demonstrated their relatively modest values of the longitudinal relaxivity (for instance 0.079 $mM^{-1}s^{-1}$ at 11.7 T [30] and 0.2 $mM^{-1}s^{-1}$ at 1.5 T [31]) in comparison with Gd^{3+}-based CAs. However, other studies have reported abnormally high r_1 values for PB nano-objects of 4.7 $mM^{-1}s^{-1}$ [32] and 6.4 $mM^{-1}s^{-1}$ (at 3 T) with a size of 60 nm in the latter case [33], which are probably due to the different way in which the molecular weight of the nanoparticles has been estimated with respect to the PB formula ($Fe_4[Fe(CN)_6]_3$ vs. $Fe[Fe(CN)_6]_{3/4}$). Due to the fact that PB nanoparticles present a high stability in a wide range of pH (from 2 to 7) and a relatively low toxicity, they have been further investigated in their doped form with high spin paramagnetic Mn^{2+} ($S = 5/2$) or Gd^{3+} ($S = 7/2$) ions in order to increase the longitudinal relaxivity and then to provide the possibility to administrate the CA with lower concentrations. Thus, several Mn^{2+} doped PB nano-objects have been reported: PB/Mn^{2+} nanoparticles of 33 nm conjugated with anti-tumoral glial antigen [34], 60–160 nm cubic PB nanoparticles with different amounts of Mn^{2+} varying from 0 to 25% and coated with polyethylene glycol (PEG) [35] and 4–20 nm Mn^{2+}/In^{3+}-based PB analogue nanoparticles also with a varying amount of Mn^{2+} and stabilized with dextran [36]. These doped nano-sized systems exhibit improved r_1 values in comparison with the benchmark values observed for commercial CAs, such as Prohance® for instance (Table 10.1), but they strongly depend on the size, coating and amount of high spin paramagnetic ions (Table 10.1). Thus, it is very difficult to establish the clear influence of these parameters on the relaxivity. It is noteworthy that controversial results have been found in the case of Mn^{2+} doping: The higher r_1 value has been obtained for the smaller Mn^{2+} amount in the case of Mn^{2+}/In^{3+} doped PB nanoparticles [36], while the opposite situation (the higher r_1 value for the higher amount of Mn^{2+}) has been reported in the case of Mn^{2+} doped PB of 60–160 nm [35]. Moreover, the question of Mn^{2+} leaching in living organisms remains open and no clear discussion about the mechanisms related to the observed relaxivities has been provided so far.

Gd^{3+}-doped Prussian blue nanoparticles with Gd^{3+} located in the interstitial sites of the PB *fcc* structure with a homogeneous or a gradient distribution of the lanthanide ions within the nanoparticles leads to improved r_1 values, which are several orders of magnitude higher than those of commercial CAs (Table 10.1) [31, 37, 38]. For Gd^{3+}-doped nanoparticles of ca. 270 nm, it has been demonstrated that the longitudinal relaxivity increases up to 37.9 $mM^{-1}s^{-1}$ with the amount of lanthanide ions. In particular, it has been shown that the Gd^{3+} ions located in the lattice sites of PB are coordinated by two water molecules, which can be efficiently exchanged [38]. Concerning the mechanism, it has been proposed that the positive contrast in MR images might be explained by the contribution of the protons of the coordinated or interstitial water molecules, which interact with the paramagnetic ions through an inner-sphere mechanism [32]. However, as for Mn^{2+}-doped systems, the mechanism responsible for the observed relaxivities has not clearly been understood.

Several nano-sized systems containing Gd^{3+} ion with related structures have also been investigated. In this respect, the citrate coated $KGd(H_2O)_4[Fe^{II}(CN)_6]$ nanoparticles of 25 nm exhibit a very high r_1 value of 35.2 $mM^{-1}s^{-1}$. A series of ultra-small $Gd(H_2O)_4[Fe^{III}(CN)_6]$ nanoparticles coated with various biocompatible stabilizing agents, such as PEG, D-Mannitol, glucosamine, show r_1 values in the range 10–13 $mM^{-1}s^{-1}$ (Table 10.1). In contrast with the Gd^{3+}-doped PB systems, the Gd^{3+} ion in such structures acts as a node in the coordination network and is therefore linked to cyano-ligand via coordination bonds. The high longitudinal relaxivity in these systems has been explained by the presence of: (i) two coordinated water molecules per each Gd^{3+} site in the lanthanide's inner sphere, (ii) about one hundred Gd^{3+} ions per nanoparticle (S = 7/2) linked through cyano-bridges with Fe^{3+} ions (S = 1/2) giving rise to the appearance of a paramagnetic behavior at room temperature with a relatively high magnetic moment per nanoparticle, (iii) crystallized water molecules in the pores, (iv) efficient exchange between coordinated, crystallized and solvent water molecules, (v) an optimal size of the nanoparticles (2–4 nm) permitting high water accessibility and favoring slow molecular rotation (Fig. 10.3).

Table 10.1 Comparison of longitudinal relaxivity for PB type nanoparticles

Formula	Size (nm)	Coating	r_1 ($mM^{-1}s^{-1}$)	Ref.
ProHance®	—	—	3.5 (1.5 T)	—
$Fe_4[Fe(CN)_6]_3$	13	Citrate	0.79 (11.7 T)	[30]
$Fe_4[Fe(CN)_6]_3$	13	Citrate	0.2 (1.5 T)	[31]
$Fe_4[Fe(CN)_6]_3$	22	Citrate, antibody	0.14 (9.4 T)	[39]
$Fe_4[Fe(CN)_6]_3$	3.5	PEG-NH_2	0.10 (7 T)	[40]
$K_{0.8}Fe_4[Fe(CN)_6]_{3.2}$	—	Avidin antibody	4.7 (3 T)	[32]
$K_{2.8}Fe_4[Fe(CN)_6]_{3.7}$	33	Avidin	7.9 (3 T)	[34]
PB	60	PEG/PAH/PAA	6.4 (3T)	[33]
PB	60	Citrate/PAH/PAA	5.73 (3T)	[35]
$K_{0.53}Gd_{0.89}Fe_4[Fe(CN)_6]_{3.8}$	33	Avidin, antibody	38.5 (3T)	[32]
$K_{0.3}Gd_{0.2}Fe[Fe(CN)_6]$	58	Citrate	12.3 (1.4 T)	[37]
$Gd_{0.1}Fe_{3.9}[Fe(CN)_6]_3$	13	Citrate	7.6 (7 T)	[31]
Gd^{3+}-doped PB	270	PVP	Up to 37.9 depending on the amount of Gd^{3+}	[38]
Mn^{2+}-doped PB with [Mn]/([Mn]+[Fe]) = 4.1, 15.3, 26.8%	80 to 160	Citrate/PAH/PAA	5.85–7.64 (3T)	[35]
$K_xMn_{0.05}In_{0.05}[Fe(CN)_6]_y$	4	Dextran	15 (1.5 T)	[36]
$K_{0.6}Mn_{0.7}Fe_4[Fe(CN)_6]_{3.5}$	33	Avidin	15.8 (3 T)	[41]
$Gd(H_2O)_4[Fe^{III}(CN)_6]$	6.4	Chitosan	6–7 (4.7 T)	[42]
$Gd(H_2O)_4[Fe^{III}(CN)_6]$	2.5	D-mannitol, PEG	9.9–13.3 (4.7T)	[40, 43, 44]
$KGd(H_2O)_4[Fe^{II}(CN)_6]$	25	Citrate	35.2 (1.4 T)	[45]

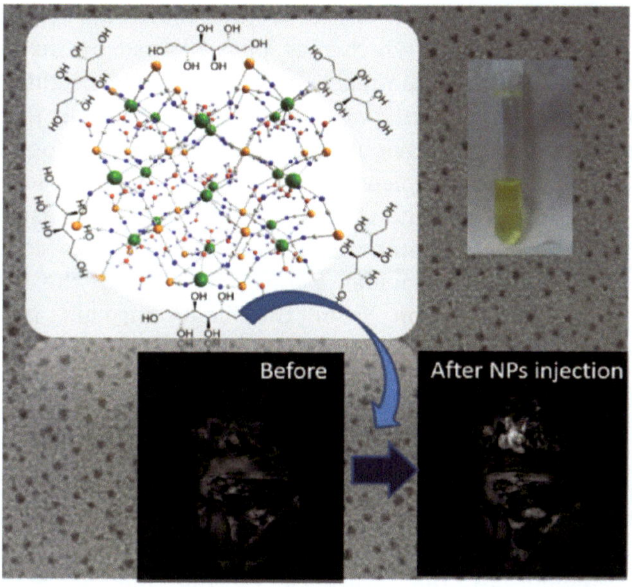

Figure 10.3 (Top) TEM and schematic representation of $Gd^{3+}/[Fe(CN)_6]^{3-}/$ D-mannitol nanoparticles. (Bottom) T_1-weighted images of coronal slices of mice acquired before and after nanoparticles intravenous injection [43].

All these nanoparticles provide a clear positive effect (brightness) on T_1-weighted MR images of phantoms or in vitro, which depends on the nanoparticles concentration.

In vivo MRI has been reported only on four investigated systems, including PEGylated PB and Mn^{2+}-doped nanoparticles [33, 35], Gd^{3+}-doped PB nanoparticles and hollow nano-objects [38] and $Gd(H_2O)_4[Fe^{III}(CN)_6]$ stabilized by D-mannitol [43]. In the case of PEGylated systems reported by Liu and colleagues, an increase (about twice) of the MR signal intensity has been observed in the tumor after 12 h following intravenous injection of the nanoparticles of about 60 nm. However, no information about the nanoparticles biodistribution and elimination has been given. In the case of small $Gd(H_2O)_4[Fe^{III}(CN)_6]$ nanoparticles of 3.4 nm stabilized by D-mannitol, a significant T_1-signal enhancement has been observed 20 min. after intravenous injection of the nanoparticles in the heart, the aorta, the liver, the vessels, the vena cava, the kidneys and the veins leading to the bladder. After 60 min, the signal starts to decrease indicating their entry in the tissues. The maximal concentration of the nanoparticles is found in the vena cava indicating its particular affinity with the liver, which may be explained by the presence of D-mannitol at the nanoparticles' surface. The bright signal of the blood pool indicates that these nanoparticles can flow in the vessels without an obvious uptake by the reticuloendothelial system (i.e. liver and spleen) for a relatively long period of time (60 min) and can therefore be used as intravascular MR contrast agent. In comparison, the use of the commercial contrast ProHance® provides a lower increase of the MR signal and a faster excretion from the body through glomerular filtration clearance (about 20 min).

10.2.2 Others Types of Imaging

PB type nanoparticles have also been proposed as new contrast agents for Near Infrared (NIR) light mediated photoacoustic imaging due to their strong absorption in the 600–900 nm region owing to the presence of the metal-to-metal ions charge transfer band (from Fe^{2+} to Fe^{3+}) and their high photostability [46]. Photoacoustic tomography is an emerging imaging technique based on the acoustic detection of the optical absorption from endogenous chromophores present in the living organisms (such as hemoglobin) or exogenous contrast agents (such as organic dyes or nanoparticles) permitting to generate high-resolution anatomical/vascular images. The photoacoustic effect converts absorbed optical energy into acoustic energy [47]. Photoacoustic imaging presents numerous advantages over other imaging techniques, such as a relatively high deep penetration (several centimeters), high spatial resolution (100 mm), rich optical contrast, absence of speckle artifacts, employment of nonionizing laser, rapidity and a relatively low price. Among different contrast agents, the most often used in vivo are various fluorescent organic dyes (for instance indocyanine green, Alexa Fluor 750, Evans blue, etc.), Au and Ag nano-objects with a variety of shapes and sizes (nanospheres, nanoshells, nanorods, nanocages, nanoclusters, nanostars, etc.), quantum dots and carbon nanotubes. It has been demonstrated that gold nanoparticles present an absorption cross-section five orders of magnitude higher than dyes and a longer body circulation. All these contrast agents present a high absorption in the visible or in the sought-after NIR region permitting them to increase the contrast of the photoacoustic signal. In addition, they may improve imaging depth or accuracy and provide a specific information. In this line of thought, PB nanoparticles exhibit a large absorption band between 600 and 900 nm, which can be modulated by doping with transition metal ions, such as Mn^{2+}. It has been shown that citrate coated PB nanoparticles of 40 nm are able to enhance the photoacoustic signal under 4.3 cm thickness of chicken breast muscle in vitro. The in vivo experiences confirmed approximately threefold signal enhancement after nanoparticles injection and their relatively prolonged blood circulation time (45 min) (Fig. 10.4 shows the photoacoustic image of a mice brain) [46, 48]. The accumulation in the tumor and then a local photo-acoustic signal intensity increase have also been observed after intravenous injection of 18 nm Au/PB nanoparticles after 22 h [48]. The advantages of the PB nanoparticles over other contrast agents include their very high photostability, low cost, non-toxicity, easy surface modification, and a high photoacoustic conversion rate, which is of the same order of magnitude as Au nanoparticles.

Another type of imaging for which PB type nanoparticles have been proposed as contrast agents is single-photon emission computed tomography (SPECT). Scintigraphy constitutes a powerful diagnostic technique widely used in nuclear medicine, wherein the radioisotope is internalized and the emitted radiation is captured by an external detector. More than two decades ago, $^{201}Tl^{+}$ was the only radioactive element available to clinicians for myocardial imaging and, despite the appearance of ^{99}Tc, remains extensively used for different types of myocardial diagnostics by intravenous administration of an aqueous TlCl solution [49, 50]. In this case, about 90% of radioactive thallium ions are cleared from the blood after the first cycle of circulation and only 10% of the injected activity is used for imaging. Radioactive ^{201}Tl-labeled PB ultra-small nanoparticles of 2–3 nm coated with biocompatible ligands made from glycol chains and/or carbohydrates covalently anchored via functional amino groups at the nanoparticles surface have been used as radiophores. They present a prolonged blood circulation and a glomerular clearing combined with an excellent colloidal stability in physiological solutions. The presence of the radioactive thallium isotope ($^{201}Tl^{+}$) in the tetrahedral sites of the PB structure makes these nanoparticles efficient as radiotracers for the scintigraphy of liver in comparison with commercial $^{201}TlCl$ (Fig. 10.5), which is very rapidly (few minutes) concentrated in the kidneys.

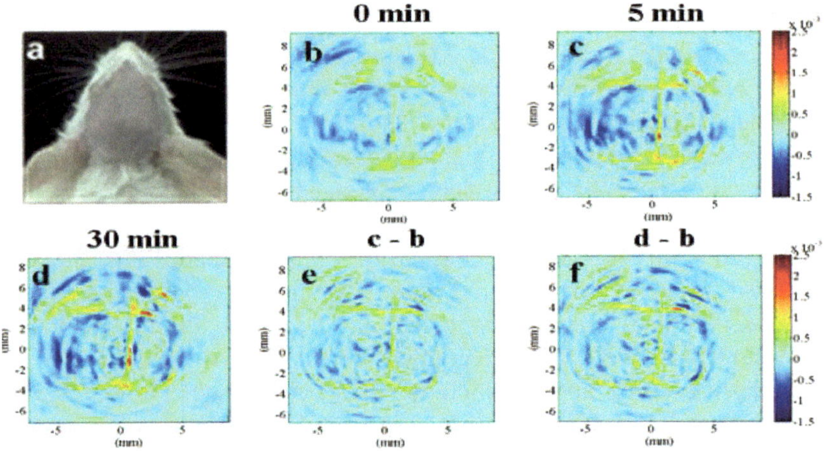

Figure 10.4 Photoacoustic maging of a mouse brain employing PB nanoparticles and NIR light at a wavelength of 765 nm. (a) Photograph of the mouse brain obtained before the data acquisition; Photoacoustic image acquired (b) before, (c) 5 min after, and (d) 30 min after the tail intravenous injection of PB nanoparticles; (e) and (f) Differential images obtained by subtracting the pre-injection image from the post-injection images (e = c – b; f = d – b). Reproduced from Ref. [46] with permission from the Royal Society of Chemistry.

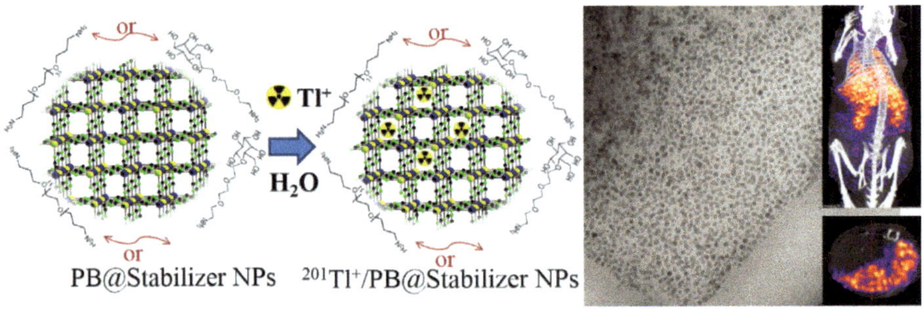

Figure 10.5 (Left) Schematic representation of the $^{201}Tl^{+}$/PB@stabilizer nanoparticles synthesis; (right) TEM image of the $Na_{0.45}Fe[Fe(CN)_6]_{0.86}$@$(PEG\text{-}NH_2)_{0.22}$ nanoparticles and SPECT/CT image of mice 20 min. after *iv* injection of the $^{201}Tl^{+}$/PB@PEG-NH_2 (longitudinal (top) and transversal (down) planes).

10.3 Prussian Blue Type Nanoparticles as Therapeutic or Theranostic Agents

PB type nanoparticles have extensively been explored as nanocarriers for drug and genes delivery, as new therapeutic agents for photothermal therapy (PTT) of cancer or other diseases, for Cs^+ and Tl^+ poisoning and as a multifunctional platform for combining different types of therapy.

In this line of thought, the delivery of drugs, DNA and genes by nano-sized PB and its analogues has been investigated in vitro, in particular for cancer therapy, due to the possibility of capturing small anticancer molecules in the microporosity of the cyano-bridged network, or by attaching the latter to the nanoparticles' surface. Thus, an inhibition of the cancer cells' proliferation has been demonstrated in vitro using PB or its copper analogue nanocarriers via the efficient delivery of different drugs. For instance, in the case of $Cu_{2.72}K_{0.55}[Fe(CN)_6]_2$ nano-objects of 55 nm loaded with the FDA approved anti-cancer drug doxorubicin (DOX), in vitro investigations have demonstrated an improved inhibition of the cancer cells' proliferation compared to DOX, due to a continuous drug release up to 14 days, which is not trivial for conventional nanocarriers [52]. PB nanoparticles of 75 nm conjugated with a DNA drug (decoy oligodeoxynucleotide, which is able to inhibit the signal transducer and activator of transcription), have been successfully internalized by human prostate carcinoma cells by endocytosis. They have presented an increased cell-killing ability depending on the amount of the drug conjugated [53]. Similarly, hollow PB nanoparticles of 110 nm have been used for the intracellular *cis*-platin delivery in bladder cancer cells [54].

Moreover, several PB analogue nanoparticles themselves exhibit valuable therapeutic properties related to their anti-cancer or anti-microbial activity or their ability to selectively capture or exchange alkali and transition metal ions. Thus, PVP coated silver-based PB analogue nanoparticles, $Ag_3[Fe(CN)_6]$/PVP, exhibit a significant inhibition of different cancer cells' proliferation in vitro and a inhibition of a tumor growth in mice model, as well as excellent antibacterial activity towards several bacteria [55]. Moreover, a selective Cu^{2+} to Zn^{2+} ion exchange ability of PVP-coated $K_2Zn_3[Fe(CN)_6]_2$ nanoparticles of 44 nm related to the Irving–Williams series' stability has been proposed to be used for treatment of Wilson's disease [56].

In the case of radioactive Cs^+ contamination or Tl^+ poisoning, recent developments in pharmacokinetic and nanomedicine have demonstrated that the use of nanoscaled PB and PBA improves the kinetic and efficacy of the decontamination in comparison with the corresponding bulk materials. In this connection, two types of nanocomposites involving PB nanoparticles included into silica or carbon sponge matrixes have been proposed as an effective absorbent for the decontamination of Cs^+ and Tl^+ ions in the gastrointestinal tract of rats [7, 57–59]. Another system based on the ultra-small PB nanoparticles stabilized by the Aonys® oil acts as the in vivo Cs^+ uptake agent and presents the higher efficiency with a significant dose effect and different elimination ways than commercially available PB (Radiogardase®) (Fig. 10.6) [60, 61].

The most extensive development of PB type nanoparticles for therapeutic features is linked with their exceptional potential for laser-assisted photothermal ablation making them very interesting as photothermal agents for the cancer therapy. Photothermal therapy (PTT) is a newly developed hyperthermia strategy for the therapeutic treatment of tumor cells, which employs Near Infrared (NIR) photoabsorbers to generate heat upon NIR laser irradiation. It presents among others, interesting advantages over conventional therapeutic methods, such as high specificity, precise spatial-temporal selectivity, and minimal invasiveness [62–64]. Therefore, the absorption of NIR laser irradiation in biological tissues is minimal and the penetration depth is optimal. Thus, the principle involves "burning" cancer cells by laser irradiation at low energy wavelengths in the presence of a NIR photo-adsorbing agent. The therapeutic efficacy of PTT significantly depends on the conversion of light to heat with photothermal agents, especially of a nano-size. For these reasons, to be efficient, PTT agents should present: (i) strong absorption in the NIR region (with high molar extinction coefficient); (ii) high PTT conversion efficiency; (iii) high PTT stability; (iv) low toxicity, biocompatibility. Up to date, various Au nano-objects including Au nanorods, nanoshells, spherical nanoparticles, etc., carbon nanotubes, graphene oxide nanoparticles or upconversion nanocrystals have been found very promising for this purpose [65].

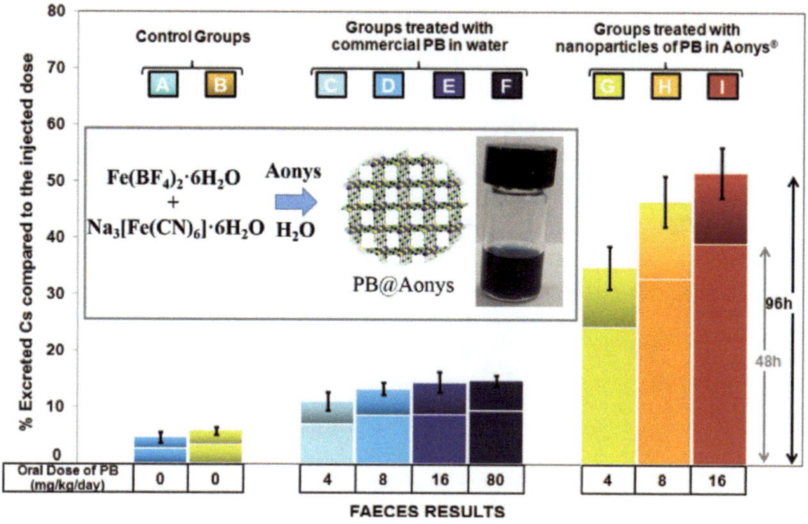

Figure 10.6 Percentage of Cs^+ recovery in feces after decorporation of non-treated mice (left histograms), mice treated with the commercial Radioardase® (middle histograms), mice treated with PB@Aonys (right histograms) (the initial injected Cs^+ dose is 100%) showing: (i) an important increase of the Cs^+ excretion by using PB@Aonys® treatment; (ii) an increase of the Cs^+ excretion as the concentration of the PB increased (the dose effect) for PB@Aonys® and a tendency to saturation for Radiogardase® treatment, and (iii) an inversion in the elimination ways in comparison with non-treated or PB treated rats: after the PB@Aonys® treatment, a more important amount of Cs^+ is eliminated by the feces (34.9–51.7%) and a less significant part (21.4–26.1%) is excreted by urine. Insert: schematic synthesis and photograph of the PB@Aonys® system.

In this line of thought, PB nanoparticles are well known for their excellent absorption properties in the NIR spectral domain because they show metal-to-metal ions charge transfer band (from Fe^{2+} to Fe^{3+}) situated between 650 and 900 nm. For this reason, they have extensively been explored as new potential PTT agents in vitro and in vivo. It has been shown that PB nanoparticles of 42 and 65 nm stabilized by citrate upon irradiation (at 808 or 705 nm) rapidly and efficiently convert the laser irradiation into thermal energy that conducts to an important temperature increase [41, 66]. The efficiency depends on the nanoparticles' concentration, their chemical composition, the laser power, and the irradiation time. It has been demonstrated that nano-sized PB presents an higher efficiency in comparison with Au nanorods and a better photothermal stability than organic dyes used as photothermal agents [66]. For instance, the PTT conversion efficiency calculated for the hollow PB nanocubes is equal to 41.4%, which is much higher than that obtained for Au nanorods (21%) [67]. Dumont *et al.* showed in vivo that 24 h after intratumoral injection of PB and NIR irradiation, the tumors in mice appeared necrotic in comparison with non-treated mice (Fig. 10.7) [41].

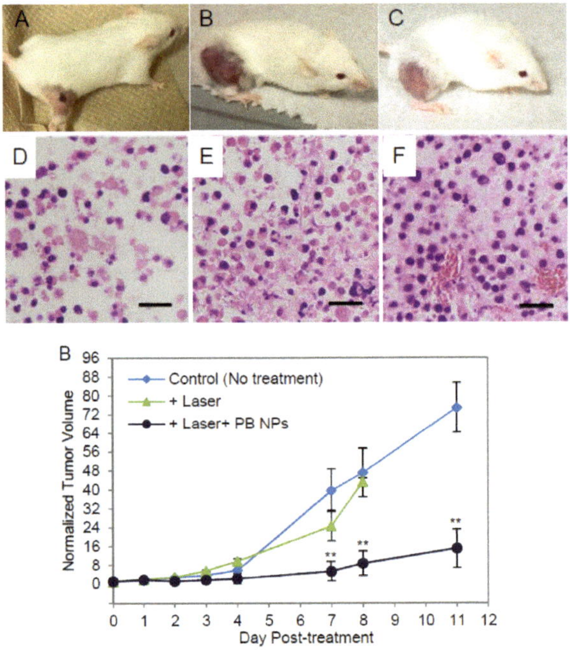

Figure 10.7 (Left) Photograph (top panel) and histological analyses of the effect of PB-based PTT agents on neuroblastoma tumor-bearing mice for: (A) NIR irradiation after intra-tumor injection of PB nanoparticles, (B) NIR irradiation (without nanoparticles), (C) Control mice and the respective H&E staining demonstrated marked tumor debulking (D–F). (Right) Normalized tumor volume as a function of time after treatment showed a marked reduction in tumor growth rate after NIR irradiation treatment with nanoparticles. Reproduced from Ref. [41] with permission from the Royal Society of Chemistry.

A complete eradication of tumor cells with only the PTT treatment is, however, rather difficult due to the suboptimal laser energy in deep tissues related to the light scattering and absorption effects, a limited light penetration and the insufficient tumor target specificity of PTT agents. For this reason, PTT treatment may be combined with drug or gene delivery to enhance the therapeutic efficiency through synergic effects between the PTT and chemotherapy. In this spirit, several works report on the design of PB nanoparticles stabilized by gelatin containing doxorubicin (DOX) [68] or conjugated with pDNA [69] or antigen specific cytotoxic lymphocytes [70]; or hollow nanocubes loaded with anti-cancer drugs such as DOX [67, 71] or 10-hydroxycamptothecin [72]. In all cases, an important improvement in the killing effect or gene transfection efficiency with NIR irradiation has been observed. As an example, hollow PB nanoparticles of 236 nm loaded with DOX exhibit a pH and temperature dependent slow drug release during 48 h. The therapeutic efficacy increases from 13.6% to 51.2% when the chemotherapy is combined with PTT. Moreover, in vivo investigations have shown not only a significant decrease of the tumor volume after the combined chemo-PTT therapy treatment, but also a complete disappearance of the tumor several weeks after treatment [67]. Another interesting approach has been developed on the design of PB-based nanocarriers for NIR PTT-assisted CO gas release for the synergic treatment in vivo of malignant tumors [73]. This combined treatment actually allows a synergic effect to suppress the tumor growth. A particular attention in the recent years has been devoted to the design of multifunctional PB-type nanoparticles possessing simultaneously diagnostic and therapeutic abilities (i.e. theranostic agents). In this line of thought, PB nanoparticles or hollow nano-objects able to act as efficient PTT agents and/or loaded with drugs were doped with paramagnetic Mn^{2+} or Gd^{3+} ions to render them responsive for MRI [31, 35, 38, 67, 74]. MRI may also be associated with a photoacoustic imaging that conducts to design multimodal nanoprobes (Fig. 10.8) [38]. This strategy should enable the monitoring of the accumulation of therapeutic agents in the tumor and its eradication.

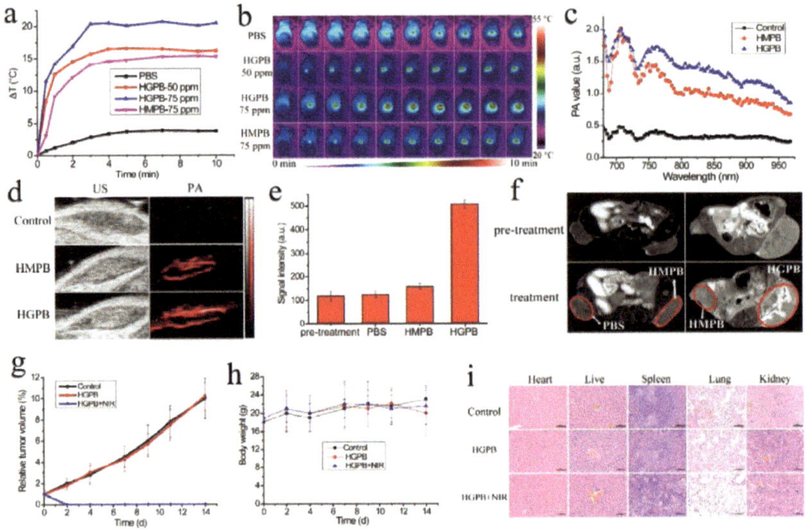

Figure 10.8 Simultaneously enhanced dual-mode imaging and tumor PTT in vivo performed with Gd^{3+}-doped PB nanoparticles. (a) Temperature curves of tumor before and after treatment (intratumoral injection) with PB nanoparticles (PBS), hollow mesoporous PB nanoparticles (HMPB), and hollow Gd^{3+}-doped PB nanoparticles (HGPB); (b) Corresponding thermal images of tumor before and after injection of these nanoparticles; (c) Photoacoustic values at different wavelengths (680–970 nm) before and after injection of hollow mesoporous PB (HMPB) and Gd^{3+}-doped hollow PB (HGPB) nanoparticles; (d) Photoacoustic values at 808 nm before and after injection; (e) MRI signal intensity and (f) MRI images of tumor before and after injection of nanoparticles; (g) Relative tumor volumes of mice in different groups; (h) Body weight of mice in different groups. (i) H&E staining of heart, liver, spleen, lung, and kidney of mice after treatment for 14 days. Reprinted with permission from Cai X, Gao W, Zhang L, Ma M, Liu T, Du W, et al. Enabling Prussian blue with tunable localized surface plasmon resonances: Simultaneously enhanced dual-mode imaging and tumor photothermal therapy. *ACS Nano.* 2016;10:11115-26. Copyright 2016 American Chemical Society.

10.4 Hybrid Nanoparticles Containing Nano-Sized Prussian Blue Combined with Metal or Metal Oxides as Theranostic Agents

A new recently emerging challenge in the field of PB nanoparticles consists of designing multifunctional hybrid nano-objects combining metal or metal oxides with PB in various forms, such as core–shell or decorated nano-objects. Such intricate materials combine the properties of each components within a single nanosystem and then exhibit diverse physical responses when subjected to various external stimuli, permitting to achieve multifunctional platforms for nanomedicine. The mostly often-used approach consists of attempts to design Fe_3O_4 core@PB shell nano-objects by the in situ growth of the PB shell on the surface of the magnetite nanoparticles. Such systems can combine the size and the shape-dependent magnetic properties of the nano-sized magnetite (superparamagnetism, high r_2 relaxivity making them to be interesting as T_2-weighted MRI contrast agents, possibility of a magnetic guidance) with the PTT and photoacoustic features of PB in order to obtain magneto-optical theranostic agents. This approach offers the possibility of multimodal imaging and the combination of imaging with a therapeutic treatment (Fig. 10.9). Several systems consisting of Fe_3O_4 nanoparticles of 15–25 nm surrounded by a PB shell of a few nanometers without stabilizing agents [75] or covered with gelatin filled with DOX [76] or enwrapped by chitosan conjugated with DNA [77] have been reported as multifunctional platforms for the cancer treatment. As an example, surfactant free Fe_3O_4@PB nano-objects of around 15–20 nm have been synthesized and investigated in the aim to design MRI-guided targeted PTT agents. They present a transverse r_2 relaxivity value of 58.9 $M^{-1}s^{-1}$ appropriate for their use as T_2-weighted MRI contrast agents due to the presence of the magnetite core. Besides, they exhibit a high PTT effect for tumor ablation and tumor growth inhibition (85.9%) under NIR irradiation thanks to the strong absorption of the PB shell [75]. However, although all these nano-objects present the characteristics and properties of both components, it should be noted that this approach gave rather aggregates of nanoparticles without clear distinction between the magnetite core and the PB shell.

Another system which has recently been developed by several teams consists of the combination of gold with PB (Fig. 10.10). Given that nano-sized gold exhibits size and shape dependent optical properties (Surface Plasmon Resonance band) and may be used as contrast agent in Computed Tomography (CT), its combination with PB permits also realizing multifunctional platforms able to combine imaging (CT, photoacoustic) and therapy (PTT) in a single nano-object. In this way, Au core@PB shell nanoparticles of 18 nm, with, however, not well-distinguished core and shell, have been designed and their ability to act as contrast agents for CT and photoacoustic imaging, as well as PTT therapeutic agents for HT-29 tumor has been demonstrated [48]. Note that the synthetic approach used for these nanoparticles is closed to what has been employed for the above mentioned magnetite-based/PB systems and consists of the in situ growth of the PB shell on the citrate stabilized Au nanoparticles, that can probably explain the problem of aggregation. An interesting approach to the synthesis of non-aggregated and well-defined Au core@PB shell nanoheterostructures has been developed through the time-controlled growth of the PB on the cyanide stabilized Au nanoparticles [78]. In that case, nanoheterostructures with a spherical Au core of 20 nm and a PB shell of 27 nm have been obtained (Fig. 10.10). These nanoparticles coated with a phospholipidic bilayer have been labeled with the radioactive isotope $^{201}Tl^{+}$ to obtain nanoprobes for SPEC-CT scintigraphy. Finally, theranostic materials for MRI/CT imaging combined with synergic PTT and radiosensitive therapy have been proposed on the base of a PB core of 87 nm surrounded by Au satellite nanoparticles of 13 nm [79].

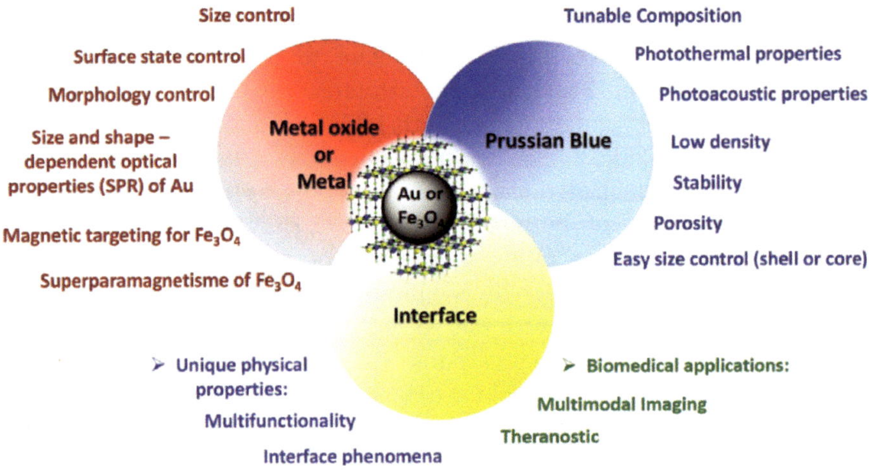

Figure 10.9 Advantages of hybrid metal or metal oxide core@PB shell nanoparticles design.

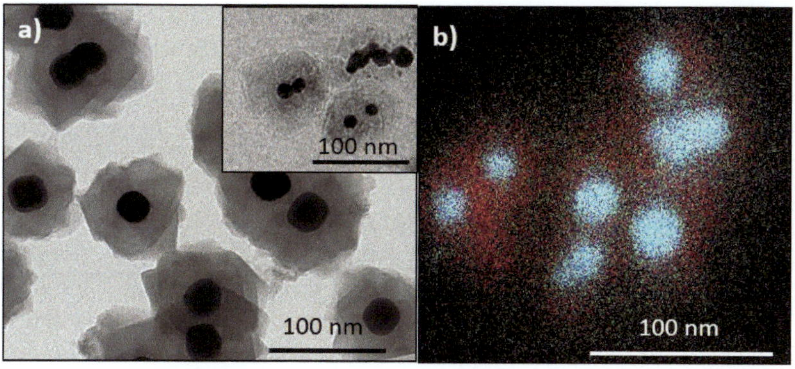

Figure 10.10 (a) TEM image of Au core@PB shell nanoparticles surrounded by a phospholipidic bilayer with a well-visible Au core of 20 nm and a PB shell of 27 nm. Inset: a CryoTEM image of these nanoparticles. (b) STEM-HAADF image of Au core@PB shell nanoparticles with EDS mapping showing the Fe (red), P (green) and Au (blue) atoms.

10.5 Conclusion

PB type nanoparticles possess unique and exciting physical and chemical properties making them highly promising for biomedical applications. In this connection, numerous PB type nano-objects have been designed and investigated in order to provide a proof of concept for their efficient employment as nanoprobes for various type of imaging, such as MRI, photoacoustic imaging or scintigraphy or as therapeutic agents for photothermal therapy and Cs^+ and Tl^+ ions capture. They indeed can compete with commercial contrast agents or radiotracers for certain type of imaging and may be more efficient as photo adsorbing agents for PTT than other inorganic nano-objects, such as for instance gold nanorods. Moreover, a fine-tuning of the nanoparticles composition, their size, their shape and their surface state allows the design of multifunctional nano-objects combining different types of imaging (for instance MRI and photoacoustic tomography), or combining imaging with therapy (for instance PTT treatment with photoacoustic or MR imaging) in order to visualize the therapeutic agent accumulation in the targeted organs or tissues. Systematic toxicological investigations in vitro show a low toxicity of these nanoparticles and their high stability in biological media. All these investigations constitute the first step towards the use of PB type nanoparticles as a new emerging family of nanomedicines, but important developments are highly needed to prove their clinical relevance.

The first point to outline is that nano-sized PB and its analogous have extensively been investigated in terms of their synthesis and a large range of nanoparticles of different composition, size (starting from few to several hundred nanometers) surrounded by various stabilizing agents have been reported up to day [80]. However, in the most mentioned works devoted to biomedical applications, the synthesis is often limited by the citrate stabilized PB nanoparticles of 20–60 nm and there is a general lack of systematic investigations on the size, composition and shape dependent photothermal, photoacoustic, MRI related properties, their toxicity, and their behavior in the body. Moreover, a limited number of stabilizing ligands have been used in combination with PB. In a general way, the ligand choice depends on the future application and in this way, highly biocompatible, water soluble, and covalently anchored to the nanoparticles surface ligands providing the possibility of prolonged blood circulation and glomerular clearing are good candidates for the coating of imaging's nanoprobes. On the other hand, a therapeutic treatment requires targeting functions. Consequently, an important development of nanoparticles conjugated with ligands able to specifically target specific tissues or organs should also be provided.

The second point concerns a general lack of in vivo investigations on PB nano-objects in order to establish their therapeutic and imaging efficiency in the living organisms as well as to provide an insight on their biological behavior. The latter is the consequence of a complex combination related to the nanoparticles' composition, their size, their morphology, and their surface state. So far, only a few articles address in vivo investigations after intravenous injection of the nanoparticles, which show a relatively long blood circulation of around 45–60 min [43] for small nano-objects and a rapid tumor accumulation through the EPR effect in the case of large (236 nm) nanoparticles [81]. Furthermore, the nanoparticles excretion is very poorly investigated. We can cite only two works on PB nanoparticles elimination by urine and feces after per os administration [82, 83], and one work showing the glomerular clearing of ultra-small nanoparticles after intravenous injection [51]. Thus, the becoming in the organism of such PB nanoparticles remains an opened question that needs to be addressed in order to envisage their use as nanomedicine.

A rapid development of PB type nano-objects for biomedical applications will permit an important breakthrough in the nanoparticles' design allowing the synthesis of smart PB based multifunctional nanoparticles, which can be pharmacologically and biologically active and which possess several physical properties allowing the association of diagnosis with therapeutic treatment.

Acknowledgments

The authors thank the University of Montpellier and CNRS for financial support.

References

1. Larionova J., Guari Y., Sangregorio C., Guerin C. Cyano-bridged coordination polymer nanoparticles. *New J. Chem.*, 2009, 33, 1177–1190.
2. Catala L., Volatron F., Brinzei D., Mallah T. Functional coordination nanoparticles. *Inorg. Chem.*, 2009, 48, 3360–3370.
3. Dujardin E., Mann S. Morphosynthesis of molecular magnetic materials. *Adv. Mater.*, 2004, 16, 1125–1129.
4. Catala L., Mallah T. Nanoparticles of Prussian blue analogs and related coordination polymers: From information storage to biomedical applications. *Coord. Chem. Rev.*, 2017, 346, 32–61.
5. Moulik S. P., De G. C., Panda A. K., Bhowmik B. B., Das A. R. Dispersed molecular aggregates. 1. Synthesis and characterization of nanoparticles of $Cu_2[Fe(CN)_6]$ in H_2O/AOT/n-heptane water-in-oil microemulsion media. *Langmuir.*, 1999, 15, 8361–8367.
6. Vaucher S., Li M., Mann S. Synthesis of Prussian blue nanoparticles and nanocrystal superlattices in reverse microemulsions. *Angew. Chem. Int. Ed.*, 2000, 39, 1793–1796.
7. Long J., Guari Y., Guerin C., Larionova J. Prussian blue type nanoparticles for biomedical applications. *Dalton Trans.*, 2016, 45, 17581–17587.

8. Rinck P. *Magnetic Resonance in Medicine.* The textbook of the European magnetic resonance forum. Oxford: Black Wave Sci. Publ, 1993.
9. Callaghan P. T., Clark C. J., Forde L. C. Use of static and dynamic NMR microscopy to investigate the origins of contrast in images of biological tissues. *Biophys. Chem.*, 1994, 50, 225–235.
10. Carretta P., Lascialfari A. *NMR-MRI, μSR and Mössbauer Spectroscopies in Molecular Magnets*: Springer Milan, 2007.
11. Merbach A. S., Helm L., Toth E. *The Chemistry of Contrast Agents in Medical Magnetic Resonance Imaging*, 2nd ed., Wiley, 2013.
12. Caravan P., Ellison J. J., McMurry T. J., Lauffer R. B. Gadolinium(III) chelates as MRI contrast agents: Structure, dynamics, and applications. *Chem. Rev.*, 1999, 99, 2293–2352.
13. Aime S., Botta M., Fasano M., Geninatti Crich S., Terreno E. 1H and 17O-NMR relaxometric investigations of paramagnetic contrast agents for MRI. Clues for higher relaxivities. *Coord. Chem. Rev.*, 1999, 185, 321–333.
14. Werner E. J., Datta A., Jocher C. J., Raymond K. N. High-relaxivity MRI contrast agents: Where coordination chemistry meets medical imaging. *Angew. Chem. Int. Ed.*, 2008, 47, 8568–8580.
15. Aime S., Castelli D. D., Crich S. G., Gianolio E., Terreno E. Pushing the sensitivity envelope of lanthanide-based magnetic resonance imaging (mri) contrast agents for molecular imaging applications. *Acc. Chem. Res.*, 2009, 42, 822–831.
16. Johnson N. J. J., Oakden W., Stanisz G. J., Scott Prosser R., van Veggel F. C. J. M. Size-tunable, ultrasmall $NaGdF_4$ nanoparticles: Insights into their T1 MRI contrast enhancement. *Chem. Mater.*, 2011, 23, 3714–3722.
17. Park J. Y., Baek M. J., Choi E. S., Woo S., Kim J. H., Kim T. J., et al. Paramagnetic ultrasmall gadolinium oxide nanoparticles as advanced T1 MRI contrast agent: Account for large longitudinal relaxivity, optimal particle diameter, and in vivo T1 MR images. *ACS Nano*, 2009, 3, 3663–3669.
18. Hifumi H., Yamaoka S., Tanimoto A., Akatsu T., Shindo Y., Honda A., et al. Dextran coated gadolinium phosphate nanoparticles for magnetic resonance tumor imaging. *J. Mater. Chem.*, 2009, 19, 6393–6399.
19. Bridot J.-L., Faure A.-C., Laurent S., Rivière C., Billotey C., Hiba B., et al. Hybrid gadolinium oxide nanoparticles: Multimodal contrast agents for in vivo imaging. *J. Am. Chem. Soc.*, 2007, 129, 5076–5084.

20. Evanics F., Diamente P. R., van Veggel F. C. J. M., Stanisz G. J., Prosser R. S. Water-soluble GdF_3 and GdF_3/LaF_3 nanoparticles physical characterization and NMR relaxation properties. *Chem. Mater.*, 2006, 18, 2499–2505.

21. Pereira G. A., Peters J. A., Almeida Paz F. A., Rocha J., Geraldes C. F. G. C. Evaluation of [Ln(H_2cmp)(H_2O)] metal organic framework materials for potential application as magnetic resonance imaging contrast agents. *Inorg. Chem.*, 2010, 49, 2969–2974.

22. McKinlay A. C., Morris R. E., Horcajada P., Férey G., Gref R., Couvreur P., et al. BioMOFs: Metal–organic frameworks for biological and medical applications. *Angew. Chem. Int. Ed.*, 2010, 49, 6260–6266.

23. Horcajada P., Chalati T., Serre C., Gillet B., Sebrie C., Baati T., et al. Porous metal-organic-framework nanoscale carriers as a potential platform for drug delivery and imaging. *Nat Mater.*, 2010, 9, 172–178.

24. Rowe M. D., Chang C.-C., Thamm D. H., Kraft S. L., Harmon J. F., Vogt A. P., et al. Tuning the magnetic resonance imaging properties of positive contrast agent nanoparticles by surface modification with RAFT polymers. *Langmuir*, 2009, 25, 9487–9499.

25. Nishiyabu R., Hashimoto N., Cho T., Watanabe K., Yasunaga T., Endo A., et al. Nanoparticles of adaptive supramolecular networks self-assembled from nucleotides and lanthanide ions. *J. Am. Chem. Soc.*, 2009, 131, 2151–2158.

26. Nishiyabu R., Aimé C., Gondo R., Noguchi T., Kimizuka N. Confining molecules within aqueous coordination nanoparticles by adaptive molecular self-assembly. *Angew. Chem. Int. Ed.*, 2009, 48, 9465–9468.

27. Taylor K. M. L., Rieter W. J., Lin W. Manganese-based nanoscale metal–organic frameworks for magnetic resonance imaging. *J. Am. Chem. Soc.*, 2008, 130, 14358–14359.

28. Rieter W. J., Pott K. M., Taylor K. M. L., Lin W. Nanoscale coordination polymers for platinum-based anticancer drug delivery. *J. Am. Chem. Soc.*, 2008, 130, 11584–11585.

29. Horcajada P., Serre C., Maurin G., Ramsahye N. A., Balas F., Vallet-Regí M., et al. Flexible porous metal-organic frameworks for a controlled drug delivery. *J. Am. Chem. Soc.*, 2008, 130, 6774–6780.

30. Shokouhimehr M., Soehnlen E. S., Khitrin A., Basu S., Huang S. D. Biocompatible Prussian blue nanoparticles: Preparation, stability, cytotoxicity, and potential use as an MRI contrast agent. *Inorg. Chem. Commun.*, 2010, 13, 58–61.

31. Shokouhimehr M., Soehnlen E. S., Hao J., Griswold M., Flask C., Fan X., et al. Dual purpose Prussian blue nanoparticles for cellular

imaging and drug delivery: A new generation of T1-weighted MRI contrast and small molecule delivery agents. *J. Mater. Chem.*, 2010, 20, 5251–5259.

32. Dumont M. F., Hoffman H. A., Yoon P. R. S., Conklin L. S., Saha S. R., Paglione J., et al. Biofunctionalized gadolinium-containing Prussian blue nanoparticles as multimodal molecular imaging agents. *Bioconjug. Chem.*, 2014, 25, 129–137.

33. Cheng L., Gong H., Zhu W., Liu J., Wang X., Liu G., et al. PEGylated Prussian blue nanocubes as a theranostic agent for simultaneous cancer imaging and photothermal therapy. *Biomaterials*, 2014, 35, 9844–9852.

34. Dumont M. F., Yadavilli S., Sze R. W., Nazarian J., Fernandes R. Manganese-containing Prussian blue nanoparticles for imaging of pediatric brain tumors. *Int. J. Nanomed.*, 2014, 9, 2581–2595.

35. Zhu W., Liu K., Sun X., Wang X., Li Y., Cheng L., et al. Mn^{2+}-Doped Prussian blue nanocubes for bimodal imaging and photothermal therapy with enhanced performance. *ACS Appl. Mater. Interfaces*, 2015, 7, 11575–11582.

36. Paul G., Prado Y., Dia N., Riviere E., Laurent S., Roch M., et al. MnII-containing coordination nanoparticles as highly efficient T1 contrast agents for magnetic resonance imaging. *Chem. Commun.*, 2014, 50, 6740–6743.

37. Li Y., Li C. H., Talham D. R. One-step synthesis of gradient gadolinium ironhexacyanoferrate nanoparticles: A new particle design easily combining MRI contrast and photothermal therapy. *Nanoscale*, 2015, 7, 5209–5216.

38. Cai X., Gao W., Zhang L., Ma M., Liu T., Du W., et al. Enabling Prussian blue with tunable localized surface plasmon resonances: Simultaneously enhanced dual-mode imaging and tumor photothermal therapy. *ACS Nano*, 2016, 10, 11115–11126.

39. Li Z., Zeng Y., Zhang D., Wu M., Wu L., Huang A., et al. Glypican-3 antibody functionalized Prussian blue nanoparticles for targeted MR imaging and photothermal therapy of hepatocellular carcinoma. *J. Mater. Chem. B*, 2014, 2, 3686–3696.

40. Perrier M., Kenouche S., Long J., Thangavel K., Larionova J., Goze-Bac C., et al. Investigation on NMR relaxivity of nano-sized cyano-bridged coordination polymers. *Inorg. Chem.*, 2013, 52, 13402–13414.

41. Hoffman H. A., Chakrabarti L., Dumont M. F., Sandler A. D., Fernandes R. Prussian blue nanoparticles for laser-induced photothermal therapy of tumors. *RSC Adv.*, 2014, 4, 29729–29734.

42. Guari Y., Larionova J., Corti M., Lascialfari A., Marinone M., Poletti G., et al. Cyano-bridged coordination polymer nanoparticles with high nuclear relaxivity: Toward new contrast agents for MRI. *Dalton Trans.*, 2008, 28, 3658–3660.

43. Perrier M., Gallud A., Ayadi A., Kennouche S., Porredon C., Gary-Bobo M., et al. Investigation of cyano-bridged coordination nanoparticles Gd^{3+}/[Fe(CN)6]3-/d-mannitol as T1-weighted MRI contrast agents. *Nanoscale*, 2015, 7, 11899–11903.

44. Kenouche S., Perrier M., Bertin N., Larionova J., Ayadi A., Zanca M., et al. In vivo quantitative NMR imaging of fruit tissues during growth using Spoiled Gradient Echo sequence. *Magn. Reson. Imaging,* 2014, 32, 1418–1427.

45. Perera V. S., Yang L. D., Hao J., Chen G., Erokwu B. O., Flask C. A., et al. Biocompatible nanoparticles of $KGd(H_2O)_2[Fe(CN)_6]\cdot H_2O$ with extremely high T1-weighted relaxivity owing to two water molecules directly bound to the Gd(III) center. *Langmuir*, 2014, 30, 12018–12026.

46. Liang X., Deng Z., Jing L., Li X., Dai Z., Li C., et al. Prussian blue nanoparticles operate as a contrast agent for enhanced photoacoustic imaging. *Chem. Commun.*, 2013, 49, 11029–11031.

47. Xia J., Yao J., Wang L. V. Photoacoustic tomography: Principles and advances. *Electromagn. Waves (Cambridge, Mass)*, 2014, 147, 1–22.

48. Jing L., Liang X., Deng Z., Feng S., Li X., Huang M., et al. Prussian blue coated gold nanoparticles for simultaneous photoacoustic/CT bimodal imaging and photothermal ablation of cancer. *Biomaterials,* 2014, 35, 5814–5821.

49. Weich H. F., Strauss H. W., Pitt B. The extraction of thallium-201 by the myocardium. *Circulation*, 1977, 56, 188–191.

50. Lebowitz E., Greene M. W., Fairchild R., Bradley-Moore P. R., Atkins H. L., Ansari A. N., et al. Thallium-201 for medical use. I. *J. Nucl. Med.* 1975, 16, 151–155.

51. Perrier M., Busson M., Massasso G., Long J., Boudousq V., Pouget J. P., et al. $201Tl^+$-labelled Prussian blue nanoparticles as contrast agents for SPECT scintigraphy. *Nanoscale*, 2014, 6, 13425–13429.

52. Mukherjee S., Rao B. R., Sreedhar B., Paik P., Patra C. R. Copper Prussian blue analogue: Investigation into multifunctional activities for biomedical applications. *Chem. Commun.*, 2015, 51, 7325–7328.

53. Shao-Jen W., Chun-Sheng C., Lin-Chi C. Prussian blue nanoparticles as nanocargoes for delivering DNA drugs to cancer cells. *Sci. Technol. Adv. Mater.*, 2013, 14, 044405.

54. Lian H.-Y., Hu M., Liu C.-H., Yamauchi Y., Wu K. C. W. Highly biocompatible, hollow coordination polymer nanoparticles as cisplatin carriers for efficient intracellular drug delivery. *Chem Commun.*, 2012, 48, 5151–5153.

55. Sudip M. C., Ranjan Patra. Biocompatible polymer coated silver prussian blue nanoparticles (SPB-NPS: AG3[Fe (CN)6]). 2016.

56. Kandanapitiye M. S., Wang F. J., Valley B., Gunathilake C., Jaroniec M., Huang S. D. Selective ion exchange Governed by the Irving–Williams series in $K_2Zn_3[Fe(CN)_6]_2$ nanoparticles: Toward a designer prodrug for Wilson's disease. *Inorg. Chem.*, 2015, 54, 1212–1214.

57. Torad N. L., Hu M., Imura M., Naito M., Yamauchi Y. Large Cs adsorption capability of nanostructured Prussian blue particles with high accessible surface areas. *J. Mater. Chem.*, 2012, 22, 18261–18267.

58. Hong J.-Y., Oh W.-K., Shin K.-Y., Kwon O. S., Son S., Jang J. Spatially controlled carbon sponge for targeting internalized radioactive materials in human body. *Biomaterials*, 2012, 33, 5056–5066.

59. Timchalk C., Creim J. A., Sukwarotwat V., Wiacek R., Addleman R. S., Fryxell G. E., et al. In vitro and in vivo evaluation of a novel ferrocyanide functionalized nanoporous silica decorporation agent for cesium in rats. *Health Phys.*, 2010, 99, 420–429.

60. Maurel J. C. E. C., Lavaud C., Guari Y., Guari J., Long J. In situ preparation of cyano-bridged metal nanoparticles within a biocompatible reverse micellar system. WO2017005899 A1. 2017.

61. Lavaud C., Kajdan M., Compte E., Maurel J.-C., Lai Kee Him J., Bron P., et al. In situ synthesis of Prussian blue nanoparticles within a biocompatible reverse micellar system for in vivo Cs^+ uptake. *New J. Chem.*, 2017, 41, 2887–2890.

62. Zou L. W., H., He B., Zeng L., Tan T., Cao H., He X., Zhang Z., Guo S., Li Y. Current approaches of photothermal therapy in treating cancer metastasis with nanotherapeutics. *Theranostics*, 2016, 6, 762–772.

63. Shanmugam V., Selvakumar S., Yeh C.-S. Near-infrared light-responsive nanomaterials in cancer therapeutics. *Chem. Soc. Rev.*, 2014, 43, 6254–6287.

64. Alkilany A. M., Thompson L. B., Boulos S. P., Sisco P. N., Murphy C. J. Gold nanorods: Their potential for photothermal therapeutics and drug delivery, tempered by the complexity of their biological interactions. *Adv. Drug Del. Rev.*, 2012, 64, 190–199.

65. Kim H., Chung K., Lee S., Kim D. H., Lee H. Near-infrared light-responsive nanomaterials for cancer theranostics. *Wiley Interdisciplinary Rev. Nanomed. Nanobiotechnol.*, 2016, 8, 23–45.

66. Fu G., Liu W., Feng S., Yue X. Prussian blue nanoparticles operate as a new generation of photothermal ablation agents for cancer therapy. *Chem. Commun.*, 2012, 48, 11567–11569.

67. Cai X., Jia X., Gao W., Zhang K., Ma M., Wang S., et al. A versatile nanotheranostic agent for efficient dual-mode imaging guided synergistic chemo-thermal tumor therapy. *Adv. Funct. Mater.*, 2015, 25, 2520–2529.

68. Xue P., Cheong K. K. Y., Wu Y., Kang Y. An in-vitro study of enzyme-responsive Prussian blue nanoparticles for combined tumor chemotherapy and photothermal therapy. *Colloids Surf. B Biointerfaces*, 2015, 125, 277–283.

69. Li X.-D., Liang X.-L., Ma F., Jing L.-J., Lin L., Yang Y.-B., et al. Chitosan stabilized Prussian blue nanoparticles for photothermally enhanced gene delivery. *Colloids Surf. B Biointerfaces*, 2014, 123, 629–638.

70. Burga R. A., Patel S., Bollard C. M., Y Cruz C. R., Fernandes R. Conjugating Prussian blue nanoparticles onto antigen-specific T cells as a combined nanoimmunotherapy. *Nanomedicine*, 2016, 11, 1759–1767.

71. Wu M., Wang Q., Liu X., Liu J. Highly efficient loading of doxorubicin in Prussian blue nanocages for combined photothermal/chemotherapy against hepatocellular carcinoma. *RSC Adv.*, 2015, 5, 30970–30980.

72. Jing L., Shao S., Wang Y., Yang Y., Yue X., Dai Z. Hyaluronic acid modified hollow Prussian blue nanoparticles loading 10-hydroxycamptothecin for targeting thermochemotherapy of cancer. *Theranostics*, 2016, 6, 40–53.

73. Li W.-P., Su C.-H., Tsao L.-C., Chang C.-T., Hsu Y.-P., Yeh C-S. Controllable CO release following near-infrared light-induced cleavage of iron carbonyl derivatized Prussian blue nanoparticles for CO-assisted synergistic treatment. *ACS Nano*, 2016, 10, 11027–11036.

74. Cai X., Gao W., Ma M., Wu M., Zhang L., Zheng Y., et al. A Prussian blue-based core–shell hollow-structured mesoporous nanoparticle as a smart theranostic agent with ultrahigh pH-responsive longitudinal relaxivity. *Adv. Mater.*, 2015, 27, 6382–6389.

75. Fu G., Liu W., Li Y., Jin Y., Jiang L., Liang X., et al. Magnetic Prussian blue nanoparticles for targeted photothermal therapy under magnetic resonance imaging guidance. *Bioconjug. Chem.*, 2014, 25, 1655–1663.

76. Xue P., Bao J., Wu Y., Zhang Y., Kang Y. Magnetic Prussian blue nanoparticles for combined enzyme-responsive drug release and photothermal therapy. *RSC Adv.*, 2015, 5, 28401–28409.

77. Xue P., Bao J., Zhang L., Xu Z., Xu C., Zhang Y., et al. Functional magnetic Prussian blue nanoparticles for enhanced gene transfection and photothermal ablation of tumor cells. *J. Mater. Chem. B,* 2016, 4, 4717–4725.

78. Maurin-Pasturel G., Long J., Guari Y., Godiard F., Willinger M.-G., Guerin C., et al. Nanosized heterostructures of Au@Prussian blue analogues: Towards multifunctionality at the nanoscale. *Angew. Chem. Int. Ed.*, 2014, 53, 3872–3876.

79. Dou Y., Li X., Yang W., Guo Y., Wu M., Liu Y., et al. PB@Au core–satellite multifunctional nanotheranostics for magnetic resonance and computed tomography imaging in vivo and synergetic photothermal and radiosensitive therapy. *ACS Appl. Mater. Interfaces,* 2017, 9, 1263–1272.

80. Félix G., Nicolazzi W., Salmon L., Molnár G., Perrier M., Maurin G., et al. Enhanced cooperative interactions at the nanoscale in spin-crossover materials with a first-order phase transition. *Phys. Rev. Lett.,* 2013, 110, 235701.

81. Jia X., Cai X., Chen Y., Wang S., Xu H., Zhang K., et al. Perfluoropentane-encapsulated hollow mesoporous Prussian blue nanocubes for activated ultrasound imaging and photothermal therapy of cancer. *ACS Appl. Mater. Interfaces,* 2015, 7, 4579–4588.

82. Guari Y., Larionova J., Long J., Maurel J.-C., Lavaud C., Compte E. In situ preparation of cyano-bridged metal nanoparticles within a biocompatible reverse micellar system. EP15306135. 2015.

83. Timchalk C., Creim J. A., Sukwarotwat V., Wiacek R., Addleman R. S., Fryxell G. E., et al. In vitro and in vivo evaluation of a novel ferrocyanide functionalized nanoporous silica decorporation agent for cesium in rats. *Health Phys.*, 2010, 99, 420–429.

Index